A winter squash.

A group of three variegated box-elder trees used as specimens.

Cucumbers grown in a pot will produce fruit.

Tuberous begonias are reliable annual shade plants.

Dumbcane (*Dieffenbachia*) is produced in a green-house for sale as houseplants.

Okra, a warm-season vegetable popular in the South.

Practical Horticulture

THIRD EDITION

Laura Williams Rice
Robert P. Rice, Jr.

Prentice Hall
Upper Saddle River, New Jersey 07458

Library of Congress Cataloging-in-Publication Data

Rice, Laura Williams.
  Practical horticulture / Laura Williams Rice, Robert P. Rice, Jr.
  —3rd ed.
      p.   cm.
  Includes bibliographical references and index.
  ISBN 0-13-260688-7
  1. Horticulture.   2. Gardening.   3. Indoor gardening.   I. Rice,
Robert P.   II. Title.
SB318.R48   1996
  635—dc20   96-18250
                                                            CIP

Chapter Opening Illustration Credits: 1—courtesy of American Airlines; 2—courtesy of the New York Public Library Picture Collection; 3—National Garden Bureau; 4—Laura Rice; 5—Soil Conservation Service; 6—National Garden Bureau; 7—USDA; 8—USDA; 9—courtesy of All-American Selections; 10—Laura Rice; 11—courtesy of Toro Company, Irrigation Division, Riverside California; 12—courtesy of Toro Company, Irrigation Division; Riverside California; 13—USDA; 14—Laura Rice; 15—courtesy of British Indoor Gardens; 16—courtesy of Pan American Seed Company; 17—Laura Rice; 18—Laura Rice; 19—Elinor S. Beckwith; 20—USDA.

Acquisition Editor: Charles Stewart
Editorial Production Services and Interior Design: WordCrafters
    Editorial Services, Inc.
Cover Designer: Miguel Ortiz
Prepress Manufacturing Buyer: Ilene Sanford
Managing Editor: Mary Carnis
Director of Production and Manufacturing: Bruce Johnson

© 1997, 1993, 1986 by Prentice Hall, Inc.
A Simon & Schuster Company
Upper Saddle River, NJ 07458

Printed in the United States America
10   9   8   7   6   5   4   3

ISBN 0-13-260688-7

Prentice-Hall International (UK) Limited, *London*
Prentice-Hall of Australia Pty. Limited, *Sydney*
Prentice-Hall Canada Inc., *Toronto*
Prentice-Hall Hispanoamericana, S.A., *Mexico*
Prentice-Hall of India Private Limited, *New Delhi*
Prentice-Hall of Japan, Inc., *Tokyo*
Simon & Schuster Asia Pte. Ltd., *Singapore*
Editora Prentice-Hall do Brasil, Ltda., *Rio de Janeiro*

Contents

Preface

Since the initial publication of *Practical Horticulture* in 1980, horticulture has changed substantially in some of its branches, but very little in others. This third edition has been carefully prepared to provide up-to-date information for the 1990s. New photos have been added and others updated. The lists of Selected References at the end of each chapter have been updated. However, many books from the 1970s remain classic texts in their fields and have not been superceded by any superior works. For that reason, they have been retained as references.

One major new subject that has been included in this edition is a section on floral arrangement which follows the indoor landscaping section in Chapter 19. Chapter 19 is now titled "Decorating with Growing Plants and Fresh Flowers" to reflect this change of content. All basic techniques for a homeowner to produce floral arrangements are described and illustrated.

Examples are given of American floral design forms such as the circular and triangular arrangement types. However, more avant garde designs such as vegetative, pavé, waterfall, and the like are also described.

Second, the pesticide section of Chapter 13, on "Diagnosing and Treating Outdoor Plant Disorders," has been expanded to include more pesticide theory and the new concept of "biorational" control. The list of pesticides allowed for home use provided in Chapter 13 has been updated to conform to current regulations.

The authors sincerely hope that the effort that went into revision will result in a more up-to-date and useful text for both teachers and students of horticulture.

Laura Williams Rice
San Luis Obispo, California

Part I

Fundamentals of Horticulture

Introduction

The word "horticulture" originates from Latin. The Latin word "hortus," meaning garden, is the root of the first part; "culturea," which composes the second part, means cultivation. Together they give a fairly accurate translation: garden cultivation.

Today horticulture is only one branch of the broad field of agricultural plant sciences. But it is among the most important, dealing with the growing of fruits, vegetables, and both indoor and outdoor ornamental plants. Two other branches of agricultural plant sciences, forestry and crop science, encompass the remaining plants in cultivation. Trees grown for other than ornamental use are usually covered under forestry. Grains, pasture grasses, and fiber crops such as cotton are discussed under agronomy or crop science.

Plant pathology (the study of plant diseases), entomology (the study of insects), and plant breeding are allied to all branches of agriculture. Their contribution to the successful cultivation of plants is immeasurably important.

Contrasted to these applied plant sciences is botany, the pure science of plants. Botanists frequently deal with noncultivated plants, researching plant classification and internal plant processes, among other topics. However, they also sometimes work with food or ornamental plants as do horticulturists.

The Divisions of Professional Horticulture

Because plants are grown for a variety of uses, the art and science of horticulture has many divisions. The most familiar are listed here, along with types of employment opportunities found in each area.

Fruit Production

Fruit production, called *pomology,* covers the growing of tree fruits such as apples and citrus. It also deals with the "small fruits," including blueberries, grapes, and strawberries.

In the pomology industry, orchard supervisors establish and maintain fruit orchards throughout the country. Other pomologists produce young fruit trees for sale to commercial orchards. Vineyard supervisors specialize in the cultivation of grapes (*viticulture*), and their crops can be destined for either fresh eating or winemaking. In universities, research pomologists dedicate their efforts to improving fruit quality or yield.

Vegetable Production

Vegetable production is called *olericulture.* Olericulture covers the cultivation of all of the vegetables and also such crops as melons and rhubarb which are customarily grown in vegetable gardens.

Because most vegetables are cultivated outdoors, most vegetable growers work in field vegetable production. They are responsible for the crop from planting through harvest, and their tasks may include weed control, insect and disease control, and irrigation. However, in some areas of the country vegetables are grown in greenhouses. Vegetable growers in this specialization must understand both greenhouse operation and vegetable cultivation to perform their jobs.

Landscape Horticulture

The outdoor ornamental-plant facet of horticulture is called *environmental horticulture* or *landscape horticulture.*

The landscape horticulture industry employs thousands of people in hundreds of job specialties. Workers in nursery production propagate and grow millions of ornamental trees, shrubs, and groundcovers annually and ship them throughout the country. Horticulturally trained designers then develop landscape plans for residences and commercial buildings which are implemented by landscape installers or crews. Later, landscape maintenance personnel maintain the plants in healthy condition by watering, pruning, fertilizing, and controlling pests.

In garden centers, nursery salespeople furnish plants and gardening supplies to the public. They in turn are supplied with plants by salespeople representing the wholesale plant growers.

The field of landscape horticulture also includes nonindustry occupations. Botanic gardens and arboretums employ landscape horticulturists for teaching and maintaining the gardens. City park and street tree programs may require people having training in landscape horticulture.

Turf

This is the branch of horticulture dealing with turfgrass culture. People who work in this field can be involved in breeding new turfgrass varieties, controlling insects and diseases, or maintaining turf for golf courses, parks, and lawns.

Floriculture

The production and sale of field-grown flowers for cutting or greenhouse-grown flowers and plants is a branch of horticulture known as *floriculture*.

Floriculturists can be involved in either the wholesale or the retail parts of the industry. Many wholesale growers are in charge of greenhouses where they raise flowers and indoor plants throughout the winter and bedding plants (seedlings of flowers and vegetables) in the spring. But in warm areas like Florida, California, and Texas, outdoor production of cut flowers and foliage plants is common.

Much of the wholesale growers' output reaches the public through retail florists, where floral designers create arrangements, bouquets, and corsages.

Indoor plant shops and open-air flower kiosks are dynamic small businesses in the retail portion of floriculture. They are one way for a floriculturist to start his own business. Some indoor plant shops expand into plant rental businesses and employ people to care for leased plants in offices and hotels.

Recreational Horticulture

Two aspects of horticulture emphasize the emotional and recreational value of working with plants: horticulture therapy and home horticulture.

Horticulture Therapy

The value of horticulture as therapy for those with physical, intellectual, or emotional disabilities has been recognized for many years. The blind, deaf, wheelchair-confined, or otherwise physically handicapped can all participate in some horticultural activity because of the many senses that are involved when growing a plant. A blooming flower can be appreciated by sight by many people, but through touch by others, and through smell by nearly all.

For the emotionally disturbed, intellectually handicapped, or imprisoned, horticulture has been proven to be a constructive and rewarding activity which teaches patience and responsibility.

Home Horticulture

As a recreational activity, horticulture is the nation's most popular pastime. The satisfaction of working with soil and plants to produce home-grown vegetables, lush houseplants, or a beautiful landscape is the hobby of millions of people. Balcony gardens made up of dozens of small pots flourish in the cities, dispelling the notion that gardening is a suburban or farm activity.

There are only two prerequisites one must have to assure success in gardening: a sustained interest in growing plants and information on how to grow them. Contrary to popular belief, green thumbs are not born. They are made by experience and observation.

A local library or bookstore is one source of horticultural information. Because of the popularity of home horticulture, many books are currently available. They include information from basic gardening to such specialized topics as orchid and bonsai growing. Local nurseries are a second source of gardening information. Their employees are usually very willing and qualified to answer gardening questions.

Perhaps the most overlooked source of gardening help is the Cooperative Extension Service. This government agency is a federal, state, and county cooperative whose purpose is to offer free agricultural and home economics information. Almost every county in the United States has a Cooperative Extension Office, listed under county government in the phone book. The office title will vary among states but will usually be County

Extension Service, Cooperative Extension Service, Agriculture Extension Service, or Farm Advisor.

Some of the many services which may be provided by the Extension Service are free gardening publications, phone advice regarding gardening questions, and diagnosis of disease and insect problems. If the county office is large enough, it may provide additional services such as soil testing, public lectures, and recorded dial-a-garden-tip messages.

The County Extension Service satisfies the information requests of most gardeners. However, each state also has at least one Agricultural Experiment Station. The Experiment Station conducts research on agricultural topics sometimes including those of interest to home gardeners. The results are then published and made available to the public free of charge or for a small fee. Lists of these publications are issued regularly and can be ordered by writing directly to the Agriculture Extension Service of the state (see Table 1-1)

The United States Department of Agriculture issues many publications for gardeners which are either free or inexpensive. A list may be obtained by writing to the Superintendent of Documents, U.S. Government Printing Office, Washington, D.C. 20402.

Those whose interest has narrowed to a specific group of plants should note that there are over 60 plant societies ranging from The African Violet Society of America to The Terrarium Association. These societies usually issue newsletters dealing with their area of specialization, and members frequently cooperate in exchanging plants. See the appendix at the end of the book for a list of these plant societies.

Selected References for Additional Information

American Horticultural Society Staff. *North American Horticulture.* 2nd ed. New York: Macmillan, 1992.

American Horticultural Therapy Association, *AHTA Newsletter,* Gaithersburg, MD: The American Horticultural Therapy Association.

Daubert, J. R. *Horticultural Therapy for the Mentally Handicapped.* Glencoe, IL: Chicago Horticultural Society, 1981.

Daubert, J. R. and E. A. Rothert, Jr. *Horticultural Therapy at a Psychiatric Hospital.* Glencoe, Il: Chicago Horticultural Society, 1981.

Edlin H. L. *Man and Plants.* London: Aldus Books, 1967.

Janick, J. *Horticultural Science.* 4th ed. San Francisco: W. H. Freeman, 1986.

Table 1-1 *State Extension Services*

Alabama
Cooperative Extension Service
Auburn University
Auburn, AL 36830

Alaska
Cooperative Extension Service
University of Alaska
Fairbanks, AK 99775-5200

Arizona
Cooperative Extension Service
Forbes Building
University of Arizona
Tucson, AZ 85721

Arkansas
Cooperative Extension Service
University of Arkansas
P.O. Box 391
Fayetteville, AR 72703

California
Cooperative Extension Service
University of California
Davis, CA 95616

Colorado
Cooperative Extension Service
Colorado State University
Fort Collins, CO 80523

Connecticut
Cooperative Extension Service
University of Connecticut
1376 Storrs Road U-36 Room 205
Storrs, CT 06269-4036

Delaware
Delaware Cooperative Extension Service
University of Delaware
Townsend Hall
Newark, DE 19717-1303

Florida
Cooperative Extension Service
University of Florida
Institute of Food and Agricultural
 Sciences
McCarty Hall
Gainesville, FL 32611

Georgia
Cooperative Extension Service
University of Georgia
College of Agriculture
Athens, GA 30602

Hawaii
Cooperative Extension Service
University of Hawaii at Manoa
3050 Maile Way
Honolulu, HI 96822

Idaho
Cooperative Extension Service
University of Idaho
Moscow, ID 83843

Illinois
Cooperative Extension Service
University of Illinois
College of Agriculture
122 Mumford Hall
Urbana, IL 61801

(continued)

Table 1-1 *(continued)*

Indiana
Cooperative Extension Service
Purdue University
Agriculture Administration
 Building
West Lafayette, IN 47907

Iowa
Cooperative Extension Service
Iowa State University
110 Curtis Hall
Ames, IA 50011

Kansas
Cooperative Extension Service
Kansas State University
Umberger Hall
Manhattan, KS 66506

Kentucky
Cooperative Extension Service
University of Kentucky
S-107 Agriculture Science Building
 North
Lexington, KY 40546-0091

Louisiana
Cooperative Extension Service
Louisiana State University
Knapp Hall
Baton Rouge, LA 70803-1900

Maine
Cooperative Extension Service
University of Maine
Orono, ME 04469

Maryland
Cooperative Extension Service
University of Maryland
Symons Hall
College Park, MD 20742

Massachusetts
Cooperative Extension Service
University of Massachusetts
Stockbridge Hall
Amhurst, MA 01003

Michigan
Cooperative Extension Service
Michigan State University
106 Agriculture Hall
East Lansing, MI 48824-1039

Minnesota
Minnesota Extension Service
University of Minnesota
St. Paul, MN 55108

Mississippi
Cooperative Extension Service
Mississippi State University
P.O. Box 5446
Mississippi State, MS 39762

Missouri
Cooperative Extension Service
University of Missouri-Columbia
2-70 Agriculture Building
Columbia, MO 65211

Montana
Cooperative Extension Service
Montana State University
Bozeman, MT 59717

Nebraska
Cooperative Extension Service
University of Nebraska
Institute of Agriculture and Natural
 Resources
Lincoln, NE 68583

Nevada
Cooperative Extension
University of Nevada-Reno
Reno, NV 89557-0004

New Hampshire
University of New Hampshire
 Cooperative Extension
University of New Hampshire
Durham, NH 03824

New Jersey
Rutgers Cooperative Extension
Cook College
P.O. Box 231
New Brunswick, NJ 08903

New Mexico
Cooperative Extension Service
New Mexico State University
Box 3AE
Las Cruces, NM 88003

New York
Cornell Cooperative Extension
 System
Cornell University
Ithaca, NY 14853

North Carolina
Agriculture Experiment Station
North Carolina State University
Box 7602
Raleigh, NC 27695-7602

North Dakota
North Dakota State University
Extension Service
Fargo, ND 58105

Ohio
Cooperative Extension Service
2120 Fyffe Road
The Ohio State University
Columbus, OH 43210-1099

Oklahoma
Cooperative Extension Service
Oklahoma State University
139 Agriculture Hall
Stillwater, OK 74078

Oregon
Extension Service
Oregon State University
Ballard Extension Hall, 101
Corvallis, OR 97331-3606

Pennsylvania
Cooperative Extension Service
Pennsylvania State University
323 Agriculture Administration
 Building
University Park, PA 16802

Rhode Island
Cooperative Extension Service
University of Rhode Island
Woodward Hall
Kingston, RI 02881

South Carolina
Cooperative Extension Service
Clemson University
107 Barre Hall
Clemson, SC 29634

(continued)

Table 1-1 *(continued)*

South Dakota

Cooperative Extension Service
South Dakota State University
Box 2207
Brookings, SD 57007

Tennessee

Agriculture Extension Service
University of Tennessee
P.O. Box 1071
Knoxville, TN 37901**Texas**

Agricultural Extension Service
Texas A & M University
College Station, TX 77843

Utah

Cooperative Extension Service
Utah State University
Logan, UT 84322-4900

Vermont

Cooperative Extension Service
University of Vermont
Burlington, VT 05405

Virginia

Cooperative Extension Service
Virginia Polytechnic Institute and State
  University
Blacksburg, VA 24061

Washington

Cooperative Extension
Washington State University
Pullman, WA 99164

West Virginia

Cooperative Extension Service
West Virginia University
Morgantown, WV 26506

Wisconsin

Cooperative Extension Service
University of Wisconsin-Madison
Madison, WI 53706

Wyoming

Cooperative Extension Service
College of Agriculture
The University of Wyoming
P.O. Box 3354
Laramie, WY 82071

I

Climate and Plant Growth

All plants outdoors (and even those indoors to some extent) are affected by the climate associated with the geographic area in which they grow. Humans can alter growing conditions in a limited way (such as by irrigating to compensate for scanty rainfall), but for the most part plant growth is constrained by natural climatic restrictions. Therefore a knowledge of how basic climate factors affect plant growth is of prime concern for the gardener.

Climate

The elements combining to make up a climate include primarily (1) temperature, (2) precipitation, (3) humidity, (4) light, and (5) wind. Each factor has a wide range of variation and can have a dramatic effect on plant growth.

Temperature

Temperature, and particularly minimum winter temperature, largely determines the geographic range over which a plant will grow. The lowest temperature that a plant can withstand is called its *cold hardiness* or *cold tolerance* and is usually expressed in degrees Fahrenheit or centigrade.

For many plants a temperature of about 28°F (−2°C) is the minimum they can survive because at this temperature the liquid contents of the plant cells freeze and the plant dies. These plants are designated as *frost-tender* and include many vegetables and flowers.

Plants that are able to survive temperatures lower than freezing are called *frost-hardy*. They vary widely in their tolerance to subfreezing temperatures. Some will live to 10°F (−12°C), others to −5°F (−21°C) or even −30°F (−35°C). In some cases the woody portions or root system may be frost-hardy but not the flowers and leaves. For example, a late spring frost may kill the blossoms and young leaves of a fruit tree, but the branches will leaf out again and continue growing. Some garden flowers die to the ground in fall but grow back from the roots the following spring.

Although frost-hardy and frost-tender are the main classifications of cold tolerance, some houseplants and many tropical plants suffer from "chilling injury" at temperatures less than 50°F (10°C). This is why bananas turn black in the refrigerator.

Minimum winter temperature limits the areas where many plants can live, but for some plants lack of cold prevents survival. Fruit trees such as apple and cherry and flowers such as peony and tulip require a period of cold temperatures to grow normally. They cannot be readily grown in tropical areas because of the comparatively warm winters. (The requirement of plants for cold is discussed in Chapter 3 under "Vernalization.")

Frost

An understanding of frost types and the weather conditions that favor each are important because of the damage done to frost-tender plants.

A frost can be one of two types. The first type is *radiation frost* that occurs when the air is cool and calm and skies are clear. With this frost, warmth from the sun accumulated by the soil and plants during the day is lost at night as heat radiating upward. If the day was warm and the night only slightly below freezing, the reservoir of heat in the soil and plants may last through the night. If not, frost damage will result.

When plants are covered by a barrier that blocks the flow of heat to the sky, radiation frost can often be prevented. In nature, clouds form this barrier, and radiation frosts seldom occur on cloudy nights. On clear nights when a radiation frost is likely, protection of the plants by layers of newspaper, cloth, smoke, or plastic film can serve the same function.

Sprinkler irrigation to leave a protective film of water on plants is also used. Sprinklers are turned on as soon as the temperature drops to freezing and left on until the temperature rises above freezing in the morning. The warmth contained in the water is given off as the water freezes and transferred by conduction to the plants, preventing frost injury.

The second type of frost is caused by a cold air mass moving into an area. The air flows over the earth, carrying off heat from the plants, and the plants are frosted. This type of frost occurs irrespective of cloud cover.

Frost often leaves ice crystals on plants and the ground in the early morning. These crystal-depositing frosts are called *hoar frosts*. They occur when moisture from humid air condenses onto the plants because of the drop in temperature.

If the moisture content of the air is low, the air temperature can reach freezing without moisture condensation, but plants will still be injured. These frosts are called *black frosts,* because the first sign of their occurrence is the blackening of injured plants.

The appearance of frost in some areas but not in others close by is usually due to elevation differences. Cold air is heavier than warm air and will flow downward and settle in the lowest area. Accordingly, a low-lying frost pocket may experience a frost two or three weeks earlier than the surrounding area.

Precipitation

Precipitation can take many forms including rain, snow, hail, and sleet. The precipitation that falls as rain is of greatest value to outdoor plants.

Rain

Lack of rainfall all year or during the period when other weather conditions (mainly temperature) favor growth is the limiting factor to plant growth, particularly in the western United States. In these areas irrigation is essential to grow plants other than those naturally adapted to a drought period.

Excessive rain or rain occurring out of season can be as detrimental to plants as lack of rainfall. Excessive rain can kill plants adapted to dry areas, particularly if the water cannot drain away from the roots. Unsea-

sonable rain near fruit harvest can make strawberries watery-tasting and can split unpicked apples and grapes. Frequent rains also spread plant diseases by splashing microorganisms from one leaf to the next and trigger the growth of dormant disease organisms present on plants.

Snow

During northern winters, snow insulates plants against continual freezing and thawing, conditions more harmful than continuously subfreezing temperatures. By covering the surface of the soil, snow also reduces evaporation of water from the soil surrounding the roots of plants.

Sleet, Hail, and Freezing Rain

Sleet, hail, and freezing rain can be very injurious to plant health. Sleet and hail injure plants by tearing leaves and bruising or knocking off developing fruits. These forms of precipitation can fall with such force that tender plants will be completely beaten to the ground. Freezing rain frequently causes branch breakage on trees. As the rain hits cold plants, it quickly freezes. Layers of ice increase the weight of branches, causing them to bend and eventually to break. Heavy snowfall can also cause breakage due to its weight.

Dew

Dew is most likely to occur when the air is warm and humid. After sunset, the air temperature drops, and the atmosphere is unable to contain the amount of water it held during the daytime. The water then precipitates out over the area in the small droplets called dew.

Although its effect on the growth of most plants is not great, it can be a factor in the spread of diseases in turfgrass. It is also an important water source for plants that absorb water readily through their leaves.

Humidity

Humidity may also determine how well a plant will grow. Often, higher humidity improves plant growth by reducing the rate at which plants lose water from their leaves. But low humidity will not seriously hinder the growth of most outdoor plants, provided there is adequate water around the roots.

Figure 1-1 Fog in a redwood forest of coastal California. The moisture-laden air is the only source of water during the dry summer months. (*Photo by Rick Smith*)

Plants growing indoors in cold-winter climates are commonly damaged by insufficient humidity. Heated indoor air has an average relative humidity of only 10 to 25 percent. Since many plants used indoors are native to the humid tropics, the drop from the accustomed 50 to 90 percent to 25 percent can cause problems (see Chapter 17).

Fog and Mist

Two other forms of humidity, fog and mist, can greatly influence plant growth. In coastal California climates the moisture from fog and mist is absorbed by plants and supplements the meager rainfall (Figure 1-1). At the same time this vapor retards loss of water from leaves.

Light

Light duration, intensity, and quality affect growth of outdoor and indoor plants, and may control flowering and growth rate. Light duration is dependent on distance from the equator and season. Directly at the equator, day and night are equal in length all year.

North or south of the equator (the distance being marked by latitude lines), days become longer in the summer and shorter in winter. Also, the further from the equator, the greater the difference between the longest summer day and the shortest winter day. In the northernmost areas of the United States 16-hour days in the summer are normal.

The duration of daily light a plant receives is extremely important because it determines for what period of time the plant can manufacture the carbohydrate necessary for growth. This process, called *photosynthesis,* requires light. The more daylight hours, the more carbohydrate will be produced. The duration of the night also affects the start of some plant processes such as flowering (Chapter 3).

Light intensity or brightness also affects photosynthesis because, in general, the brighter the light intensity, the more photosynthesis occurs. Moderate to bright intensities are most beneficial to the majority of plants. Latitude will alter intensity. At the equator the sun passes directly overhead all year. But in the northern United States it is nearly overhead in summer but low in the southern sky in winter. Thus the winter intensity is lower because of the distance through the atmosphere the rays must travel. (See Chapter 16 for further discussion.)

The lesser amount of light in winter does not seriously affect outdoor plants. Some enter a resting phase at the onset of cooler temperatures. For others the light intensity is sufficient to continue photosynthesis and growth continues. However, indoor plants can be seriously retarded by the decrease in light because many are growing at subsistence light levels already. The dual handicaps of short daylight duration and low light intensity combine to make the climatic factor of light critical for indoor plants.

Wind

Wind is the fifth climatic element governing plant growth. Ocean wind limits the number of plants that can be grown near the shoreline, both by the salt spray, which browns plant leaves, and by the intensity, which whips and breaks leaves and branches. In dry areas, hot desert winds increase the rate of moisture loss from the soil and plant leaves, intensifying existing drought conditions.

Even in cold northern areas wind is an inhibiting factor to plant health. Subzero winds damage evergreens by removing moisture from the foliage, essentially freeze-drying it. In frozen ground the plant cannot replace the water, and injury similar to drought results.

Factors Modifying Climate

Natural factors modifying climate include distance from the equator, elevation, terrain, and the nearby presence of large bodies of water.

Elevation

Changes in elevation can give areas only a small distance apart completely different climates for plant growth. The higher the elevation of an area, the colder the average year-round temperature, with every rise of 300 feet (91.4 meters) causing an average temperature decrease of 1°F (0.5°C).

Terrain (Topography)

Changes in terrain also alter climate, particularly rainfall. Most rain storms move from west to east in North America. In a narrow range of hills, the west-facing slopes often receive considerably more rainfall than the east-facing slopes. The heaviness of the water in rain clouds prevents them from passing over the mountains until most of the precipitation has been released.

Sloping terrains can also vary several degrees in temperature from the top portion of the slope to the lower part or valley. Cool air, which is heavier than warm air, will flow down a slope and collect at the bottom, a phenomenon called *air drainage* (Figure 1-2). Consequently, plants growing higher on the slope are less likely to be damaged by cold than those at the bottom.

Bodies of Water

Large bodies of water such as the Great Lakes and oceans exert strong effects on the climate of nearby land. The enormous volume of water absorbs heat in the summer, making adjacent land cooler, and gives it off in the winter, raising the temperature. Thus they moderate temperatures all year, making the summer cooler and the winter warmer than in inland areas.

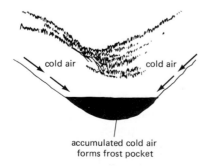

accumulated cold air
forms frost pocket

Figure 1-2 Typical cold-air drainage on sloping land.

On the North American continent, the influence of large bodies of water on agriculture is considerable. Warmer winter temperatures found along the Great Lakes make these regions large commercial growing areas for fruits such as grapes, cherries, and peaches, which do not grow as well further from the shores. Similarly, most of the Pacific coast is warm enough to grow certain vegetables all winter, but inland these crops would freeze.

Climate Modification by Humans

People can modify climate either intentionally or unintentionally. Among the most widespread unintentional causes of climate change are air pollution and consequent smog.

Air pollution can be composed of both gases (carbon monoxide, for example) and small airborne pieces of matter called *particulates*. Combined together they become smog, a persistent brown or yellow layer that hangs over some metropolitan areas.

One way smog affects plant growth is by temperature modification. A smog cloud acts like an insulation layer over an area and prevents heat from escaping to the atmosphere. Thus areas with smog cover have warmer summers than nearby areas without smog, and plants may be damaged by excess heat.

Smog also decreases the intensity of sunlight that reaches the ground, giving plants less light from which to manufacture carbohydrate. Particulates settle onto leaves and further decrease the light intensity reaching the plants.

The main way smog affects plant growth is through foliage injury by toxic gases. The gases enter the leaves, causing leaves to turn partially brown and sometimes to die. Plants vary in their susceptibility to smog injury, ranging from smog-tolerant to highly smog-intolerant. Plants classified as intolerant (many pines, for example) cannot be grown in smoggy areas.

Microclimates

Microclimates are small areas that have slightly different climate characteristics than the surrounding land. They may be less windy, shadier, moister, warmer, or in any other way different from the typical climate. These differences affect plants growing in the microclimate, sometimes helping and sometimes hindering growth.

Microclimates may be either artificially made or naturally occurring. They can be formed by the natural terrain and vegetation or unintentionally created by the erection of structures such as buildings, fences, and roadways. Although they often go unnoticed, they can be used by the observant gardener to provide

Figure 1-3 A microclimate for growing ferns. The north side of this house provides shelter from excess sun and wind. (*Photo by Rick Smith*)

the specialized growing conditions favored by different plants (Figure 1-3). In some cases they make it possible to grow plants not normally considered cold-hardy in that area.

Outdoor Microclimates

One example of a natural microclimate is a frost pocket. In addition to being colder, this microclimate might also be moister due to water runoff and would be suitable for moisture-loving plants not easily damaged by cold.

Another natural microclimate can be found under trees. The area is shadier and cooler in summer and would be good for shade-loving plants. It would also provide ideal conditions for growing indoor foliage plants outdoors in summer.

The area under the eaves of a house is a manmade microclimate found around most homes; plants there live under different environmental conditions than plants living away from the house. First, they are shielded from rainfall by an eave, so the soil will be drier. Moreover, the wall may afford protection from cold or drying winds, and it radiates warmth from the heated dwelling. During cool periods this additional warmth speeds up or prolongs growth, but if the wall faces south it may be excessively hot in summer and plant damage can occur (Figure 1-4). This is frequently the case with parking lot trees, which become overheated due to heat radiating from the asphalt.

At night when temperatures drop, objects that absorbed heat during the day begin losing it to the cooler night air. Heat radiating from the ground and buildings can protect plants from frost and extend the growing season.

Wind can be intensified or lessened in a microclimate. A plant by a wall will generally receive wind protection unless it is directly in the path of prevailing winds. However, when two walls are parallel, a wind tunnel can be formed. Wind conditions will be worse there than in a completely unprotected area. Such wind tunnels are frequently formed between closely spaced buildings.

Indoor Microclimates

Microclimates are usually thought of as being outdoors. However, indoor microclimates with varying temperature, light, and humidity can be found throughout a

Figure 1-4 Scorch on maple due to excess heat. (*Photo courtesy of Dr. R. E. Partyka, Columbus, OH*)

house or apartment. Because heat rises, upstairs rooms may be several degrees warmer than downstairs ones. A basement may be up to 20°F (11°C) cooler than the remainder of the home on a hot day, and the north side of the house may stay considerably cooler than the south side. Even within a single room, cool microclimates will be found adjacent to windows in cold winter areas because of cold entering through the glass.

Microclimates with varying humidity levels are also common. In the kitchen and bath, where water is used, evaporation raises the humidity. A basement is also usually humid.

Light microclimates are easy to detect, being governed by the size and location of windows. Obviously the further from a window a plant is, the less light it will receive. However, each foot or meter of distance can be thought of as a different microclimate. In addition, south-facing windows are maximum-light microclimates, east and west moderate, and north-facing low-light microclimates because of the angle of the sun in the northern hemisphere.

Plant-Growing Zones of North America

The question of where in the country a specific plant can be grown is important to every gardener. There is little to worry about when plants are bought at local nurseries because nurserymen seldom sell plants that do not grow in their area. But when a plant is ordered from a mail-order nursery, the consumer must check to make sure the plant he has selected will grow in his area.

Since minimum winter temperature is often what determines where a plant can be grown, knowing its minimum survival temperature is essential. This and other valuable information is usually given in the written descriptions of the plant. It can also be found in a gardening book from a library.

United States Department of Agriculture Plant Hardiness Zones

After determining the minimal survival temperature of a plant, it is then necessary to determine the minimum winter temperature of one's area. To simplify this procedure, the United States Department of Agriculture has prepared maps showing zones of minimum winter temperatures throughout the United States and Canada (see Figure 1-5). Eleven zones are outlined, ranging from completely frost-free areas in Florida, Texas, and California to a −50°F (−46°C) or colder zone in Canada. The map takes into account cold mountain climates in warmer zones, as shown in Colorado. But for the most part it is made up of bands which correlate minimum winter temperature with distance from the equator.

The USDA zone map is not the only plant hardiness map that exists, but it is the most widely used. Some states whose climate is influenced by many factor have designed more detailed maps. California is one such state, and the University of California at Davis lists 21 separate climate zones in the state due to the influences of the Pacific Ocean and mountain ranges.

Zone Maps Based on Overall Climate Characteristics

Several zone maps divide the country into general climate types. Such maps are helpful because they include such climatic factors as precipitation and light intensity as well as temperature. They give a more general view of the basic climate in large sections of the country. Such information is particularly valuable today when people move so frequently.

One such general climate classification developed by the late Dr. J. B. Edmond, Professor Emeritus, Mississippi State University, divides the continental United State into five regions (Figure 1-6).

1. *Northeast and Great Lakes Region.* This area is characterized by comparatively cool summers, long cold winters, and moderate rainfall. Light intensity is moderate in the summer and dim in the winter owing to a persistent cloud cover. Summer daytime temperatures may reach 90–95°F (32–35°C), but they are for

USDA PLANT
HARDINESS
ZONE MAP

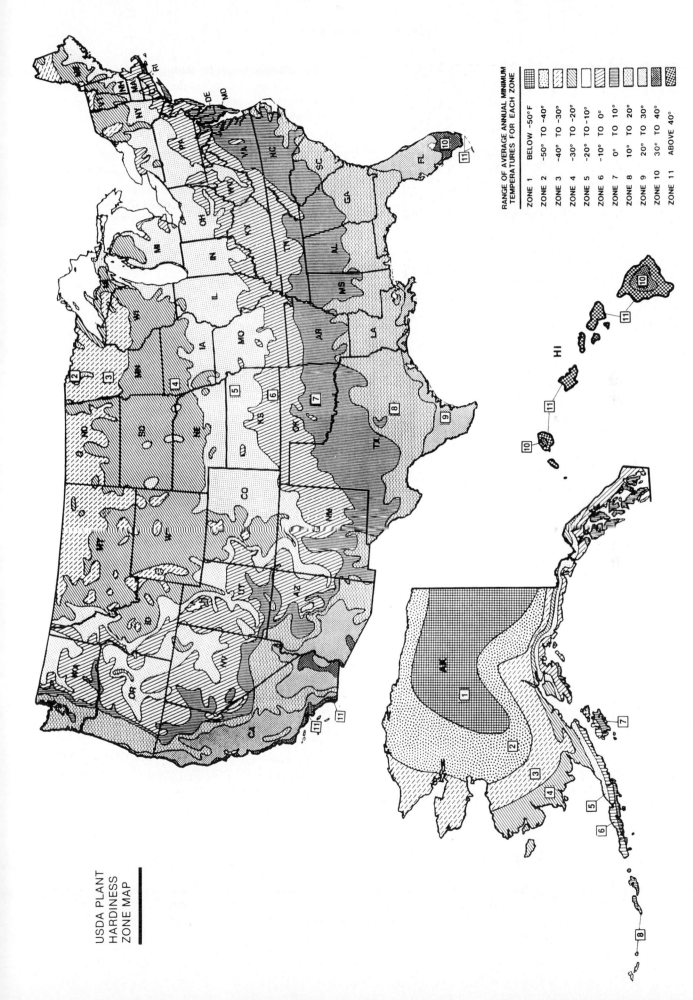

RANGE OF AVERAGE ANNUAL MINIMUM
TEMPERATURES FOR EACH ZONE

ZONE 1   BELOW -50°F
ZONE 2   -50° TO -40°
ZONE 3   -40° TO -30°
ZONE 4   -30° TO -20°
ZONE 5   -20° TO -10°
ZONE 6   -10° TO 0°
ZONE 7   0° TO 10°
ZONE 8   10° TO 20°
ZONE 9   20° TO 30°
ZONE 10   30° TO 40°
ZONE 11   ABOVE 40°

Figure 1-5

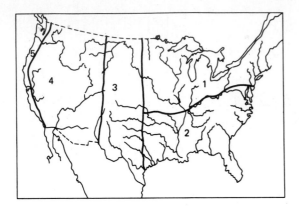

Figure 1-6 Climates of the continental United States: (1) cool-humid of the Northeast and Great Lakes, (2) warm-humid of the Southeast and Gulf Coast, (3) subhumid of the Great Plains, (4) arid of the intermountain region, (5) summer dry of the Pacific Coast. (*Map from* Fundamentals of Horticulture *by J. B. Edmond. Philadelphia: Blakiston, 1951. Reprinted by permission of McGraw-Hill*)

the most part cool enough to grow deciduous fruits (from trees that drop their leaves) such as apples, pears, and cherries. At the same time, temperatures in the summer are warm enough to grow most warm-season vegetables (such as tomatoes and beans). But summers are not long enough to grow many varieties of watermelons and sweet potatoes which require a long maturation. The rainfall throughout this region is usually sufficient and no irrigation is needed.

2. *Southeast and Gulf Coast Region.* This region has long warm summers, short mild winters, bright light intensity, and moderate to heavy rainfall. The temperature is quite variable: in the northern half yearly temperatures are cool enough for deciduous fruits; in the southern part they are warm enough for semitropical fruits like citrus and avocado. Winter temperatures in southern Florida are mild enough to grow vegetables all year, and much of the winter produce for the East is from there.

3. *Great Plains Region.* Wide fluctuation in temperature, scarce rainfall, and bright light intensity are typical of this region. Summer temperatures in the northern part of this large region may equal those in the southern section, but winter temperatures differ greatly. In the north they can drop as low as −30°F

(−34°C), whereas near the Gulf they are quite mild. The short growing season in summer limits the commercial production of crops in the north, but parts of Oklahoma and Texas produce vegetable crops throughout the year with the aid of irrigation.

4. *Intermountain Region.* This region has variable temperatures, scarce rainfall, and bright light intensity. The area is characterized by a wide range of elevations which, along with latitude, gives this region variable temperatures. In the southern part, summer temperatures are hot and the growing season is long. In the remainder, temperatures are cooler and the growing season is of moderate length.

5. *Pacific Coast Region.* The Pacific Coast divides into two sections lengthwise: the shoreline and the intermountain valleys directly over the coastal mountains. Summer temperatures are cool along the coast due to the Pacific Ocean but hot in the valleys east of the coastal mountains. Winter temperatures are mild throughout the region but particularly so along the ocean where frost is infrequent. Winter is the rainy season in this area, with seasonal rainfall ranging from very plentiful in the northern half to very scarce toward the Mexican border. The amount of rainfall also influences light intensities, making them moderate in the north but bright through the south. This region is very productive agriculturally, producing fruits and vegetables all year.

Selected References for Additional Information

Daubenmire, R. F. *Plants and Environment.* 3rd ed. New York: Wiley, 1974.

Moore, David M. *Plant Life* (*The Illustrated Encyclopedia of World Geography Series,* Vol. 4). New York: Oxford University Press, 1991.

Ray, Peter, Taylor A. Steeves, and Sarah A. Fultz. *Botany.* New York: Saunders, 1983.

Trewartha, G. I. *An Introduction to Climate.* 5th ed. New York: McGraw-Hill, 1980.

Unsworth, M. H. *Effects of Gaseous Air Pollution in Agriculture and Horticulture.* Woburn, MA: Butterworth Publications, 1982.

2

Botanical Nomenclature, Anatomy, and Physiology

Plant Nomenclature and Classification

Botanical nomenclature is the orderly classification and naming of plants. It is a subject that unnecessarily intimidates people unfamiliar with it because they believe one must know Latin in order to use the system. But the botanical naming system is not overly complex, and it does not require any background in Latin. In fact, a surprising number of common names are the same as botanical names, such as iris, fuchsia, and citrus.

There are several reasons for using botanical names in place of common names. The first reason is the universality of botanical names. Latin names are the standard worldwide system for communicating the identity of plants. The Latin name *Pyrus communis* would be recognized as the common pear by people in Asia, Africa, the Americas, or Europe, provided they were familiar with botanical nomenclature.

Precision is the second reason. Common names such as "daisy" or "ivy" can refer to any number of similar plants. If only common names were used to identify these plants, confusion would result.

Knowing the botanical name of a plant can also give clues to its growing requirements. For example, the common names bunny ears, beaver tail, and prickly pear do not indicate that these plants have anything in common, but their botanical names are similar and so is their culture.

The Origin and Construction of Botanical Names

The branch of botany which deals with the naming of plants is called taxonomy, and people doing the work are taxonomists. The naming system used dates back 250 years to the Swedish botanist Carolus Linnaeus, the "father of botany." He named and published the first references to many plants, using a naming method called the "binomial system." This system specifies that a plant name must have at least two parts.

For example, in the botanical name *Tagetes patula* (French marigold), *Tagetes* is called the *genus* (genera, plural). The second word, *patula*, is called the *specific epithet*. When combined, these two words form the plant *species*.

For ease of understanding, the genus can be thought of as the last name or surname of a plant. It indicates that the plant is related to other plants in that same genus. The specific epithet can be thought of as the first name of the plant, the name that distinguishes it from its relatives in the same genus.

In the two plants *Tagetes patula* and *Tagetes erecta*, for example, *Tagetes patula* is a small marigold with thumbnail-sized blossoms. *Tagetes erecta*, on the other hand, is a tall marigold with fist-sized blossoms. It is apparent that both plants are marigolds. But since they grow so unlike each other, they are given different specific epithets and hence become separate species.

For many wild plants, a two-part name is sufficient to distinguish between similar related plants. However, new plants which are mutations or bred from existing species are constantly being developed. To distinguish them from their parent plants, a third part called the *variety* or *cultivar* is added to the name. Common marigold cultivars, for example, would be *Tagetes patula* 'Lemondrop' and *Tagetes patula* 'Petite Harmony.'

A plant *variety* is a naturally occurring mutation or offspring that is significantly different from the parent and that becomes established in nature. For example, a species that normally only has white flowers might spontaneously mutate and a new variety with pink flowers would appear. A *cultivar* is a manmade and/or man-maintained variety, hence the name which is short for "cultivated variety." As an example, an orange tree might develop one branch that produces seedless fruit. But in nature this new kind of orange would not be able to reproduce because it has no seeds, and so the new type of orange would therefore not become established and so would not be called a variety. But through the actions of man who is able to reproduce the branch without seed, the cultivar would become established, and in fact this was the actual origin of seedless navel oranges.

The botanical names given to plants are not assigned haphazardly. A set of rules called the International Code of Botanic Nomenclature is adhered to strictly when new species are named. Although the genera and specific epithets must be in Latin-like form, the words are not always derived from that language. The names of prominent botanists, statesmen, and even explorers were often Latinized and incorporated in plant names. Frequently though, a Latin name conveys information about the plant it represents. The specific epithets *rubra, alba, atropurpurea,* and *variegata* are used with many plant names to connote colors. Table 2-1 defines some common specific epithets.

Writing and Pronouncing Botanical Names

The writing of botanical names follows a prescribed pattern. The genus is always capitalized, followed by the specific epithet beginning with a lowercase letter and both are italicized or underlined. Any variety name follows and may be set off from the species by "v.," or "var.,". The abbreviation "cv.," is used to designate a cultivar. Single quotes around the variety or cultivar name can substitute in either case. For example, the variety of ash tree known as Modesto ash may be written as:

> *Fraxinus velutina* 'Modesto'
>
> *F. v.* v. Modesto
>
> *F. v.* var. Modesto

Note that in a listing, the genus or specific epithet is written out once and abbreviated thereafter.

Another botanical abbreviation is the use of the genus name followed by the word "species," "sp.," or "spp." Species or sp. indicates an unknown species in that genus. Spp. is plural and is used to refer collec-

Table 2-1 *English Translations of Common Specific Epithets and Cultivars*

Latin epithet	English translation
alba	white
atropurpurea	dark purple
aureum	golden
carnosa	fleshy
compacta	compact
esculentus	edible
fastigiata	erect
floribunda	free-flowering
glauca	with white or gray coating
grandiflora	with large or showy flowers
horizontalis	horizontal
nanus	dwarf
nidus	nest
occidentalis	from the Western Hemisphere
officinalis	medicinal
pendula	hanging
rotundifolia	round-leaved
rubrum	red
sativus	cultivated
semperflorens	ever (semper)-flowering (florens)
stellata	starlike
sylvestris	of the forest
tuberosum	bearing tubers
variegata	variegated
vulgaris	common

tively to all species of that genus. The latter abbreviation technique is useful when discussing a broad topic such as the diseases of oaks (*Quercus* spp.).

Unlike the standardized method of assigning and writing botanical names, pronouncing the names is somewhat arbitrary. Botanists themselves disagree on certain pronunciations. Common usage has become the basis for pronunciation.

Botanical Classification of Plants

Seed- and Spore-bearing Plants

Ninety percent of plants cultivated are flowering plants that reproduce by seed, but a few common ones are not. These *Pteridophytes* have a completely different reproductive system based on spores (see Chapter 3). Ferns are the most widely known Pteridophytes, and they belong to a branch of the plant kingdom that emerged early in plant evolution (Figure 2-1).

The seed plants are further divided into two groups. Gymnosperms are the smaller of these two groups, consisting primarily of evergreen cone-bearing plants such as pines, spruces, junipers, and yews. Their foliage is generally needlelike, and they do not produce showy flowers. Of the indoor plants, cycad,

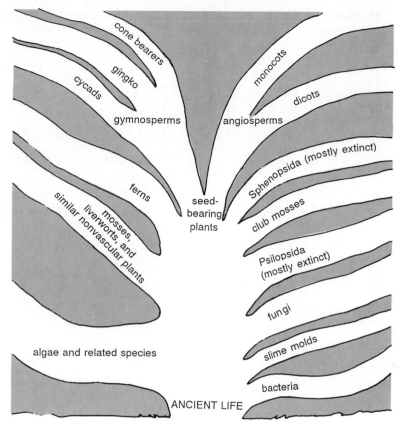

Figure 2-1 Evolutionary history of plants.

podocarpus, and Norfolk island pine are the common gymnosperms.

Angiosperms compose the majority of cultivated plants, including such diverse plants as rose, cabbage, and palm. It is within the angiosperm group that all ornamental flowering plants and nearly all food plants are found. The primary identifying characteristic of angiosperms is the growth of seeds completely within the plant ovary, which swells to become the fruit (see Chapter 3).

Monocots and Dicots

The angiosperms are further separated into the divisions *Monocotyledoneae* and *Dicotyledoneae,* or monocot and dicot for short. The monocots are the smaller of the two groups and are usually nonwoody plants that have a short stem and overlapping leaves arranged in a whorl, a form called a *rosette.* Their leaves are frequently long and narrow, with parallel veins running the length. Other identifying characteristics include flower petals fused at the base to form a bell and often, a bulb or fleshy root system (see "Modified Roots"). Included among the monocots are all grasses, lilies, irises, onions, cattails, and most flowering bulbs.

The dicots, on the other hand, are frequently woody plants which may grow to a large size. Their leaves have a branching vein pattern, and their flowers have a variety of shapes. Most trees and shrubs are dicots, as well as most fruits and garden vegetables.

Being able to recognize the difference between monocots and dicots is not of academic interest only. Whether a plant is a monocot or dicot can also determine its method of propagation and its susceptibility to weed killers.

Plant Families

The classification of plants leads ultimately to the smallest division, variety or cultivar (Figure 2-2). For the most part the intermediate subdivisions are of interest only to botanists, with the exception of plant families, the division prior to genus. Each family groups a number of genera having like characteristics together. These families have both Latin and common names. Some families are listed in Table 2-2 with a sampling of the genera each includes as well as a key identifying characteristic.

Basic knowledge of plant families can help identify an unknown plant. For example, by remembering that all pea family members have pod-type fruits, you could deduce that such diverse plants as mimosa trees and snapbeans are both in that family.

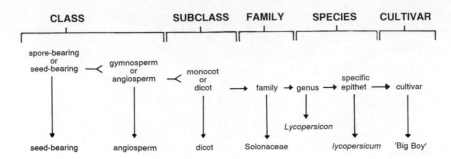

Figure 2-2 A botanical classification of the tomato cultivar 'Big Boy.'

In addition, plants in the same family are often susceptible to the same diseases. If a gardener has had a recurring problem with a disease, he could avoid buying plants with similar susceptibilities through familiarity with the family characteristics.

Horticultural Classification of Plants

Botanists are interested primarily in classifying plants by their evolutionary relationship to each other. Horticulturists frequently classify plants additionally according to use.

Most cultivated plants are valued either for their ornamental or for their edible value, with a few multipurpose plants such as herbs and ornamental vegetables. Within the edible group, plants are commonly classified as the following.

Tree fruits. apple, peach, plum, orange, avocado, and so on

Nuts. pecans, walnuts, filberts, almonds, and so on

Small fruits. blackberries, raspberries, strawberries, figs, grapes, and so on

Table 2-2 *Some Common Plant Families and Their Key Identifying Characteristics*

Family name		Key identifying characteristics	Some included genera	Some included sample plants
Latin	*Common*			
Cactaceae	cactus family	fleshy growth habit and spines	*Mammillaria, Schlumbergera, Opuntia*	pincushion cactus, Christmas cactus, prickly pear
Compositae (Asteraceae)	aster family	daisy or buttonlike flower heads composed of many tiny flowers which resemble petals	*Senecio Chrysanthemum, Rudbeckia*	florist cineraria, chrysanthemum, blackeyed susan
Cruciferae (Brassicaceae)	mustard family	flowers with four petals arranged crosswise	*Brassica Alyssum*	cabbage, alyssum
Gramineae (Poaceae)	grass family	narrow, parallel-veined leaves encircling the stem at the base	*Poa Avena*	bluegrass, oats
Labiatae (Lamiaceae)	mint family	square stems; flowers have two lips	*Coleus, Mentha, Plectranthus*	coleus, mint, creeping charlie
Leguminosae (Fabaceae)	pea family	long, podlike fruits, characteristic flowers	*Lathryus, Phaseolus Gleditsia*	sweet pea, green bean, honey locust
Liliaceae (Amaryllidaceae)	lily family	thin, grasslike leaves; bell-shaped flowers with six petals	*Chlorophytum Lilium, Yucca, Allium*	spider plant, Easter lily, yucca, onion
Rosaceae	rose family	flower petals in multiples of five; flowers have many stamens	*Malus Prunus, Pyracantha*	apple, flowering cherry, firethorn
Umbelliferae (Apiaceae)	carrot family	flowers in an umbrellalike cluster	*Daucus, Anethum, Petroselinum*	carrot, dill, parsley

Vegetables. cauliflower, tomatoes, sweet corn, peppers, squash, and so on

Herbs. rosemary, basil, dill, thyme, and so on

Grains. wheat, rice, corn, oats, and so on

The last category, grains, is generally not covered under horticulture. Grains are classified as field crops and fall under the study of crop science or agronomy.

Within ornamentals, the classification system links the plant to its form and use.

Trees: oak, magnolia, pine, eucalyptus, and so on

Shrubs: juniper, yew, lilac, azalea, and so on

Vines: clematis, ivy, and so on

Groundcovers: bugleweed, periwinkle, and so on

Garden flowers: tulip, marigold, daylily, and so on

Turfgrasses and lawn substitutes: bluegrass, St. Augustine grass, dichondra, and so on

Houseplants: umbrella plant, African violet, spider plant, and so on

Plant Identification

"What plant is that?" is one of the most frequently asked questions among gardeners. Determining the identity of a plant may be relatively easy if the plant is commonly grown, but it becomes difficult and time consuming if the plant is a wildflower or other uncultivated species. As a first step, cut off a stem with foliage and preferably flowers and/or fruits. Without flowers or fruits, identification may be difficult, unless the plant is a foliage houseplant. Enclose the sample in a plastic bag to lessen wilting, having sprinkled it previously with water if possible. Make note of the relative size of the plant, whether it is woody or nonwoody, and where it was growing (sun, shade, marsh, landscape, and any other relevant conditions). All this information can be helpful.

To have the unknown plant identified, begin by showing the sample to a knowledgeable local nurseryman. If he or she is unfamiliar with it, consult mail-order plant catalogs with color pictures or library books with photographs. If the plant was growing wild, botanical references may be necessary. However, these are usually complicated to use (unless you have a background in botany) because they include a considerable amount of technical terminology.

As a last resort, dry the sample by pressing it flat between several layers of newspaper and weighting it with a book. When the sample is completely dry (several weeks), it can be mounted on cardboard with tape and mailed to a university department of botany or horticulture along with the information on size, growing location, and so on, previously noted.

Plant Anatomy

Understanding how the parts of a plant form the whole, as well as the structure of those parts, is necessary if you are to grow plants successfully.

Vegetative Plant Organs and Their Function

Nearly all cultivated plants are composed of a limited number of basic parts, or organs: leaves, stems, buds, and roots (Figure 2-3). Collectively these are termed *vegetative organs* because they are not part of the sexual reproductive system of the plant. In its simplest form, a young plant would have the organs discussed in the next section.

Leaves

Leaves are the most obvious part of most plants. They are usually made up of two main parts: the wide section termed the *blade,* and the *petiole,* or leaf stem (Figure 2-4).

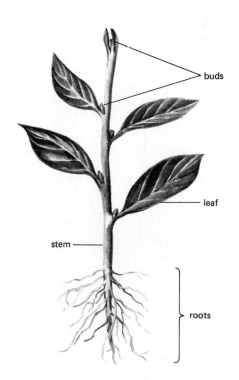

Figure 2-3   Vegetative parts of a typical plant.

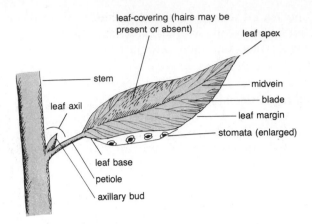

Figure 2-4   Parts of a simple leaf.

Figure 2-6 Leaves of gymnosperm plants. (*Photo by Kirk Zirion*)

The angle formed between the petiole and its supporting stem is the *leaf axil*, and in that axil will be a *bud*.

Figure 2-4 is a diagram of what is called a simple leaf. In contrast to this simple leaf, Figure 2-5 shows several compound leaves. A compound leaf is composed of individual leaflets but will only have a bud at the base of the entire leaf.

Leaves of the types shown in Figures 2-4 and 2-5 are common leaf forms on the angiosperm plants. But the leaves of cone-bearing gymnosperms are often scalelike or needlelike. Figure 2-6 shows sample leaves of gymnosperms.

Parts of the Leaf Blade. Looking closely at a simple leaf, the visible parts are as follows.

Margin. The leaf margin is its outside edge. The margin may be toothed, barbed, lobed, or in any other way different from smooth (Figure 2-7). Leaf margin characteristics are a means of identifying plants. Barbed or spined margins can also protect a plant from being eaten by grazing animals.

Veins. The patterns of veins in leaves can be seen by looking closely or holding the leaf up to the light. Dicot plants have a central vein (midvein) with many branch veins. Monocots have parallel venation running the length of the leaf (Figure 2-8).

Figure 2-5   Types of compound leaves. (*Photo by Rick Smith*)

barbed

toothed

smooth

Figure 2-7 Types of leaf margins. (*Photo by Kirk Zirion*)

(a)

(b)

(c)

Figure 2-8 Venation types: (a) elm leaf with midvein and branch veins, (b) maple leaf with three main veins and numerous branch veins, (c) amaryllis leaf with parallel veins.

Leaf apex. The apex is the tip of the leaf. The apex may be pointed, blunt, notched, or a number of other shapes.

Leaf base. This is the part of the blade that attaches to the petiole or directly to the stem (if the leaf has no petiole).

Leaf covering. Any hair, scales, or film on the leaf blade can be considered a leaf covering. Almost all leaves have an invisible wax layer, called the *cuticle,* which prevents water loss from the leaf surface. A leaf may have hair or scales in addition to this cuticle.

Stomata. Stomata are minute openings found primarily on the undersides of leaves (Figure 2-4). They are flanked on either side by *guard cells* which open and close the opening according to environmental conditions. Stomata regulate the flow of gases and water vapor in and out of the leaves.

Modified Leaves. Modified leaves have evolved to perform functions not customarily associated with leaves. The "petals" of poinsettia are modified leaves called *bracts.* Twining tendrils and spines are also modified leaves.

Stems

Growth of stems increases plant height or width. The stem transports water and other substances. It is also the site of leaf and flower attachment.

Modified Stems. Not all stems grow aboveground. Underground bulbs contain short stems and attached scales (onion layers, for example) which are modified leaves. The potato is an underground stem thickened to serve as a storage site. Frequent use is made of modified stems in plant reproduction, discussed in more detail in Chapter 4.

Shortened stems found in the aboveground portions of plants create the rosette plant form (Figure 2-9). Rosette plants are easily recognized because of the shortened stem; the leaves appear to grow from one central point and radiate outward like the overlapping petals of a rose. Many plants grow in rosette form, including African violets, cabbages, strawberries, daylilies, and all grasses.

Arrangement of Leaves on the Stem. Leaves can be arranged in an opposite or an alternate pattern on a stem. The pattern of leaf arrangement is important and is used in identifying plant species. As the names imply, leaves arranged oppositely grow in pairs, and those arranged alternately are attached stepwise down a stem (Figure 2-10). The site at which a leaf is or was once attached is called a *node,* and the section of stem between nodes is the *internode.* A bud will be present at each node, although it can be undeveloped and difficult to distinguish. Knowing the locations of nodes and internodes on a stem is important in pruning and propagation.

Internal Stem Structure. The life of a plant is dependent on the correct functioning of its vascular system, the network of pathways that moves carbohydrates, minerals, water, and other substances within the plant. These pathways, which can be thought of as a plant's circulatory system, are divided into xylem and phloem.

African violet

strawberry

cabbage

Figure 2-9 Typical rosette-form plants.

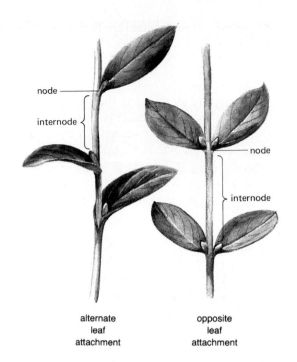

node

internode

node

internode

alternate
leaf
attachment

opposite
leaf
attachment

Figure 2-10　Types of leaf attachment and node locations.

Water and nutrients are moved up in the xylem from the roots, through the stem, and out to the leaves. Carbohydrate (the plant's "food") is manufactured in the leaves and moves down within the phloem to the roots or to other parts of the plant where it is needed.

Dicots have their xylem and phloem arranged in concentric rings inside the stem (Figure 2-11). The phloem is always found outside the xylem. Monocots, however, have xylem and phloem together in *vascular bundles* scattered throughout the stem. Both xylem and phloem extend not only through the stem of a plant but also up the petiole and into the leaf blade.

The vascular cambium is the third component of a dicot stem. The cambium manufactures new xylem and phloem cells to increase the transporting abilities of the vascular system. The cambium is a thin layer of rapidly dividing cells found between the xylem and phloem areas in the stem (Figure 2-11).

A second type of cambium is found in woody stems. This cambium, called the *cork cambium*, produces bark cells and is found directly beneath the bark surface.

Buds

A bud contains immature plant parts and can be of three types: *vegetative, flower,* or *mixed* bud. A vegetative bud contains a leaf or leaves and sometimes an embryonic shoot. A flower bud includes the rudiments of one or more flowers and is frequently larger than a vegetative bud. A mixed bud contains the potential to produce both a shoot and a flower.

The arrangement of buds on a stem places them in either *axillary* or *terminal* positions (Figure 2-12). Axillary buds are found in the leaf axils, usually directly above the point at which the petiole attaches to the stem. Terminal buds are located on the tips of stems.

In some cases buds form at other sites, such as along the veins of a leaf or at the petiole and leaf blade junction. Buds in these unusual places are called *adventitious* buds.

Roots

Roots are the fourth vegetative plant organ, and the point at which they attach to the aboveground portions of the plant is called the *crown*. Roots anchor the plant in its growing position. They also absorb and transport water and nutrients for use in photosynthesis. This absorption takes place mainly through the young tips of roots, rather than through the older larger parts of the root system. In addition, carbohydrates are stored in the older parts of the root system. This storage capacity of roots is very important because carbohydrates are the source of energy for generating new shoots if the top of the plant is damaged.

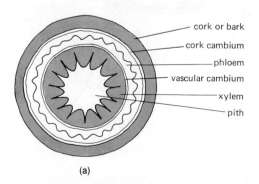

cork or bark
cork cambium
phloem
vascular cambium
xylem
pith

(a)

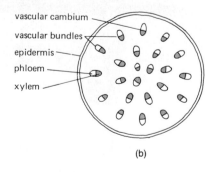

vascular cambium
vascular bundles
epidermis
phloem
xylem

(b)

Figure 2-11　Cross sections showing arrangements of vascular tissue inside dicot (a) and monocot (b) stems.

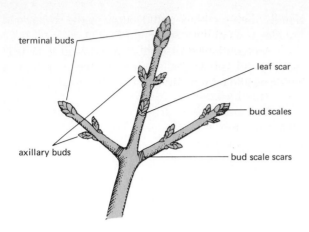

Figure 2-12   Stem with axillary and terminal buds.

Root Types. In the development of a plant from seed, a single primary root is first produced. In many plants this primary root does not branch substantially and remains the primary site of anchorage and absorption. Such roots, called *taproots,* are found typically on citrus, dandelions, and carrot plants (Figure 2-13).

In most plants the primary root branches to form a netlike mass called a *fibrous root system* (Figure 2-13). Grasses, African violets, and squash are examples of plants with fibrous roots.

A third type of root system has *fleshy roots* (Figure 2-13) which are thick like taproots but branch like fibrous roots.

Parts of the Root Tip. Regardless of whether a plant has tap, fibrous, or fleshy roots, the structure of the root tips will be the same. At the tip of every root is the root cap. The cap is made up of a layer of cells that prevents damage to the rest of the root as it pushes through the soil (Figure 2-14). Directly behind is the root *meristem,* an area that produces new cells to replace cells scraped off as the cap pushes against the soil.

Beyond the meristem is the zone of elongation. Cells produced in the meristem are also deposited here, grow, and lengthen the root. Cell lengthening in this zone pushes the root through the soil.

Water and nutrients are absorbed in the next zone, the zone of maturation and absorption. Cells in this area form fragile root hairs (Figure 2-15) about 1/16 inch (1–2 millimeters) long through which water and nutrients enter the plant.

Root hairs often live only a few weeks, and are replaced by hairs developing on younger cells. In this way root hairs constantly contact fresh soil, assuring a steady supply of nutrients.

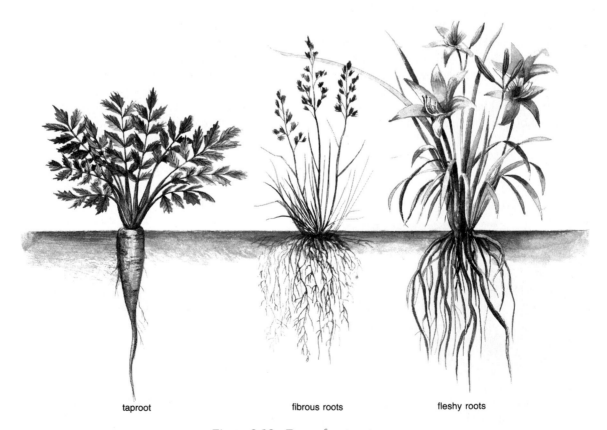

taproot   fibrous roots   fleshy roots

Figure 2-13   Types of root systems.

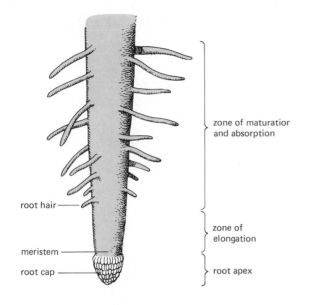

zone of maturation
and absorption

root hair —

meristem —

root cap —

zone of
elongation

root apex

Figure 2-15　Root hairs of a radish seedling. (*From* Botany
*by Carl E. Wilson and Walter E. Loomis. Copyright 1952, 1957,
1962, 1967, 1971 by Holt, Rinehart and Winston. Reprinted by
permission of Holt, Rinehart and Winston*)

Modified Roots. Although most roots resemble those in
Figure 2-13, modified roots exist, much the same as do
modified leaves and stems. Roots fashioned for storing
large quantities of carbohydrates are the most common
modified roots. Sweet potatoes and beets are examples
of this modification.

　Adventitious roots grow from a stem instead of
from other root tissues. The adventitious roots of ivy
and orchids (Figure 2-16) attach them to trees. Adventitious prop roots of corn (Figure 2-16) anchor it
against uprooting by wind.

Sexual Plant Organs

Flowers and fruits are found on most cultivated plants
at some time during the life cycle. They may appear
continuously, seasonally, or only once, depending on
the species and its growing conditions. Although the
biological function of flowers is seed production, flowers are often not essential for the continuation of the
species because man reproduces plants from the vegetative parts. Consequently flowers are frequently of
only ornamental interest to a gardener.

Flower Types

Flowers are classified by their sexual parts. A flower with
both male and female parts is called *perfect;* most species
produce perfect flowers. However, a flower can be exclusively male or female, with blooms of opposite sexes
needed for seed formation. Some species have male and
female blooms on one plant and are called *monoecious.*
Examples of monoecious species are cucumber, sweet
corn, and pecan. Others have male flowers on one plant
and female on another, with plants of opposite sexes
needed to produce seed on the female. These *dioecious*
species include holly, date palm, bittersweet, and ginkgo.

　Occasionally flowers are sterile, that is, they have
neither male nor female parts. Sterile flowered plants
are frequently prized by gardeners because they often
have double sets of petals. The doubleness is caused by
sexual parts mutating into petals.

Flower Structure

A typical flower is made up of a central female structure, the pistil, and numerous surrounding male parts
called stamens (Figure 2-17).

　The pistil contains three main parts: the knoblike
top called the *stigma,* the thin vertical shaft called the
style, and the bulblike base called the *ovary.* It is the ovary
that contains the eggs which will develop into seeds.

　Stamens are composed of the *anther* at the top,
which produces pollen (functionally similar to sperm),
and the supporting stalk, called the *filament.*

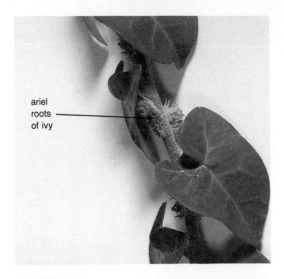

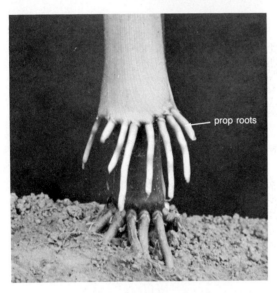

Figure 2-16   Modified roots. (*Top photo by Kirk Zirion, San Luis Obispo, CA. Bottom Photo by Rick Smith, Los Osos, CA. Middle photo from* Botany, 5th ed. *by Carl D. Wilson and Walter E. Loomis, Copyright 1952, 1957, 1962, 1967, and 1971 by Holt, Rinehart and Winston. Reprinted by permission of Holt, Rinehart and Winston.*

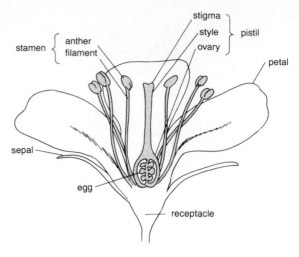

Figure 2-17   Parts of a perfect flower.

Surrounding or beneath the sexual parts are the petals and the *sepals* (the green, leaflike base beneath the petals). Under the sepals is the *receptacle*, the enlarged base on which the flower rests. The colored petals attract insects, which transfer the pollen to the female flower parts. They also shield the pistil and stamens from the weather. The sepals form the covering for the developing flower bud, protecting it from damage.

Figure 2-17 shows a typical form for a flower, but not all flowers have such easily distinguished parts. Instead the parts will be in various forms due to the divergent evolution of plant families and genera. Figure 2-18 illustrates other typical flower forms.

Moreover, not all flowers are produced singly. Often they occur in clusters, and even though each flower may be tiny, when grouped together they form a showy head. Flower clusters are named on the basis of the positioning of the blooms (Figure 2-19).

Fruit Production and Types

After flowering is complete, the flower parts shrivel and drop, with the exception of the ovary. The ovary continues developing, producing a swollen fruit with seeds.

Fruits are either fleshy or dry, depending on whether the ovary is juicy and plump at maturity or dry and shrunken. Fruits of cucumber, orange, and crabapple are fleshy, whereas pea pods, ears of corn, and pine cones are dry.

Physiological Processes

Photosynthesis, respiration, translocation, absorption, and transpiration are complex processes that occur constantly in plants. Understanding these processes

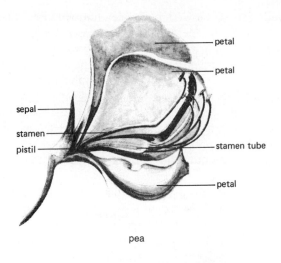

(a)

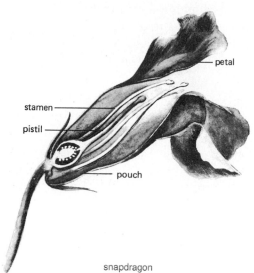

Figure 2-18   Two flower forms.

(b)

will equip the gardener with the information needed to make decisions on proper plant care.

Photosynthesis

Photosynthesis is the process by which plants manufacture the carbohydrates they require to live and grow. The process is made possible by the green chemical chlorophyll. In the presence of water, light and carbon dioxide, chlorophyll produces carbohydrates (starches and sugars) and, as a by-product, oxygen gas:

water + light + carbon dioxide gas

in the presence ↓ of chlorophyll

carbohydrates + oxygen gas

Light, the first essential for the photosynthetic process, generally comes from the sun. With the ex-

(c)

Figure 2-19   Three types of inflorescences. (a) Head. (b) Spike. (c) Umbel.

ception of plants that naturally grow in shady habitats, bright light intensity and long light duration maximize photosynthesis and increase growth.

Water, the second essential, is drawn from the soil by a process called *absorption.* Any lack of water will slow photosynthesis, and a severe deficiency will stop it completely.

Carbon dioxide, the third ingredient, is taken in from the air through leaf stomata. It is through this same pathway that the excess oxygen by-product is released into the air. Lack of carbon dioxide seldom causes insufficient photosynthesis outside. However, photosynthesis can be increased through the addition of supplemental carbon dioxide, which is occasionally used in commercial greenhouses.

Respiration

Respiration can be thought of as an opposite reaction to photosynthesis. Whereas photosynthesis makes carbohydrates out of the sun's energy, respiration breaks carbohydrates down into energy.

Respiration requires two ingredients—carbohydrates and oxygen—and yields three products—energy, water, and carbon dioxide gas:

carbohydrate + oxygen gas →

energy + water + carbon dioxide gas

The carbohydrates are already present in the plant. Oxygen is taken from the air primarily through the stomata on the leaves and also through the roots. Of the three products resulting from respiration, energy is the primary one used by the plant. It is needed for activities such as absorption of water and nutrients, chlorophyll formation, and development of flowers or fruits. However, the carbon dioxide and water released in the reaction can also be reused in photosynthesis.

The photosynthetic process requires light but respiration does not. Respiration continues constantly in every cell of the plant, even at the same time photosynthesis is taking place. It does, however, generally occur at a slower rate than photosynthesis so that more carbohydrates will not be respired than are being photosynthesized, and the excess can be stored in the plant, for example in the roots or developing fruits.

The relative rates of respiration and photosynthesis govern the health and growth rate of plants. To grow a successful crop of vegetables, it is estimated that the photosynthesis rate should be 8 to 10 times higher than the respiration rate. This assures that the majority of the carbohydrates will accumulate in the plant or be used for new growth.

Correct environmental conditions are crucial for obtaining a high photosynthesis/low respiration ratio. Maximum photosynthesis can be encouraged by moderate to bright light intensity, sufficient water in the soil, and warm temperatures. Respiration, on the other hand, can be slowed by cooler temperatures. Since lowering the temperature also decreases photosynthesis, the best growing conditions are warm days with cool nights. Under these conditions the respiration rate is decreased at night, but daytime photosynthesis is not affected.

When photosynthesis and respiration rates are equal, there is no new plant growth. The carbohydrates being photosynthesized are entirely used up for the respiration of existing cells, and the plant merely subsists.

Occasionally a plant will have a higher rate of respiration than of photosynthesis. An example would be a small tree in a forest of larger trees. Shading from the larger trees reduces photosynthesis, while respiration remains moderate and steady. All the carbohydrate being produced, plus part of the stored reserve, is used for respiration. As a result, the tree stops growing and slowly dies because it is not producing enough carbohydrate even to keep its existing cells alive.

Translocation

The movement of carbohydrates, minerals, and water through the plant is called *translocation*. As discussed earlier, the main function of the xylem is water and mineral movement, and that of the phloem, sugar movement throughout the plant.

The forces that cause upward movement of water and minerals through the xylem are complex and beyond the scope of this text. It is sufficient to think of the movement as caused by upward pressure from the roots combined with pressure deficit when water vaporizes from the leaves.

The rate of water movement in the xylem is surprisingly rapid. This can be demonstrated by noting the time required to revive a wilted flower after it is placed in water. Complete revival of a flower on a 6-inch (15-centimeter) stem in 15 to 30 minutes is not unusual.

Movement of sugars in the phloem is called a "source to sink" movement because the sugars are taken from their site of manufacture (source) to a storage or utilization site (sink). One of the primary sinks for carbohydrate is the roots. It is stored there as starch compounds made up of many sugar molecules combined. The starches make compact, energy-concentrated packages which can be efficiently stored.

A second sink is developing flowers, fruits, and seeds. In order to survive, seeds must be well supplied with stored carbohydrate to provide energy for respiration after separation from the parent plant. The constant sugar flow to these sites also accounts for the continual increasing in size of fruits from bloom until maturity and explains why fruits are often sweet.

A third sink for photosynthesized sugars is the growing regions (called *meristems*) at the stem and root tips. They require the constant supply of sugars to grow and to manufacture compounds such as pigments for flowers and fruits, waxes for cuticle formation, and cellulose fibers to strengthen the stems.

Absorption

Water and mineral absorption is the fourth internal plant process. It is primarily a function of the roots; however, it can also occur through the leaves by a process called *foliar absorption.*

Absorption of materials through root cells is a complex topic, which is still being heavily researched today. Research has explained absorption as consisting of two processes: active absorption and passive absorption. Active absorption is an energy- and oxygen-requiring process, the primary process responsible for the absorption of mineral compounds. Passive transport through a process called *osmosis*, on the other hand, is a result of natural physical forces and is the main way in which water is absorbed.

Transpiration

Transpiration, the fifth process, is the opposite of absorption in the sense that it involves the loss of water from the plant. Water in liquid form in the plant is released into the air as vapor from the leaves, stems, and flowers. Of all the water taken from the soil by absorption, a relatively small amount is used in photosynthesis; the rest is lost through transpiration.

As mentioned earlier, the main parts of the leaf through which water is lost are the stomata. These are opened and closed by the bordering cells called *guard cells*. An increase in the amount of water in the leaf and guard cells forces the stomata open and transpiration occurs. As the leaf loses water, guard cells go limp and the stomata close, decreasing transpiration.

Transpiration can be increased or decreased by environmental factors such as temperature, humidity, and wind. High temperature increases transpiration. High humidity decreases transpiration because humid air contains almost its capacity of water vapor, thus suppressing the transpiration of more water from the leaf. Wind usually increases transpiration by drawing newly transpired water vapor from the leaf. The dry air surrounding the leaf then causes rapid transpiration again.

Summary

These five processes—photosynthesis, respiration, translocation, absorption, and transpiration—are basic concepts in plant science. Knowledge of each and how each is related to the others is crucial to a thorough understanding of plant growth.

Selected References for Additional Information

*Coastline Community College. *Photosynthesis and Respiration*. Fountain Valley, CA: Coastline Community College.

*Coastline Community College. *Plant Movement and Transport*. Fountain Valley, CA: Coastline Community College.

Devlin, R. M. *Plant Physiology*. 4th ed. New York: Wadsworth, 1983.

Els, David, Editor. *The National Gardening Association Dictionary of Horticulture*. New York: Viking, 1994.

Esau, K. *Anatomy of Seed Plants*. 2nd ed. New York: Wiley, 1977.

Gledhill, D. *The Names of Plants*. 2nd ed. New York: Cambridge University Press, 1989.

Noggle, G. R., and G. J. Fritz. *Introductory Plant Physiology*. 3rd. ed. Belmont, CA: Wadsworth, 1985.

Radford, Albert E. *Fundamentals of Plant Systematics*. New York: Harper and Row, 1986.

Salisbury, F. B., and C. Ross (eds.). *Plant Physiology*. 4th ed. Belmont, CA: Wadsworth, 1992.

*Indicates a video.

3

Plant Growth and Development

Plant Life Cycles

In a typical life cycle a plant will pass through two states, growth and rest. The growth phase occurs with suitable environmental conditions (temperature, rainfall, and the like) and is characterized by activities such as flowering, shoot lengthening, and leaf production. The rest phase, often called *dormancy*, also reflects the influence of environmental conditions such as cold, drought, or inadequate light. It is characterized by slowed or stopped growth, leaf drop, and/or death of the entire aboveground parts of the plant.

Dormancy should not be thought of solely as an outdoor plant phase. Many indoor plants also have dormant periods, induced by the shorter days of winter or by the plant adhering to its normal outdoor growth cycle although living indoors.

To understand the importance of growth and dormancy in the life of a plant, let us first classify plants according to the length of time they live (Figure 3-1). Plants fall into one of four botanical groups: annual, biennial, perennial, and monocarp.

Annual

An annual plant grows to maturity, flowers, produces seeds, and dies during one growing season. It experiences dormancy only as a seed.

To most gardeners an *annual* is a frost-tender flowering or vegetable plant grown in summer. In mild climates there are also *winter annuals* which are planted in fall to bloom through the winter. These common-usage definitions usually do not clash with the botanical definition of an annual. However, exception must be made for plants that would live more than one year in their warm native climate but are not able to survive winter in cold regions. These plants are not true annuals but function as annuals because of climate. Included in this group are vegetables such as tomatoes and flowers such as lantana and impatiens when grown in cold-winter areas.

Biennial

A biennial plant grows for one season, becomes dormant during winter, resumes growth in spring, flowers, and dies. The life cycle is completed in two years, hence the term "biennial." Biennials are relatively uncommon among cultivated plants. Some examples are hollyhocks, parsley, and sweet William.

Perennial

Perennials, or plants that live for more than two growing seasons, make up the third and largest botanical division. Perennials are further subdivided into *woody* and *herbaceous* types. Woody perennials have sturdy, bark-covered stems which remain alive over the winter. The plants increase in size each year, producing additional new wood, and may grow to a large size. Trees and shrubs are examples of woody perennials.

The herbaceous perennial flower is the type of plant most gardeners refer to as "perennial." Herbaceous perennials only produce tender leaves and stems.

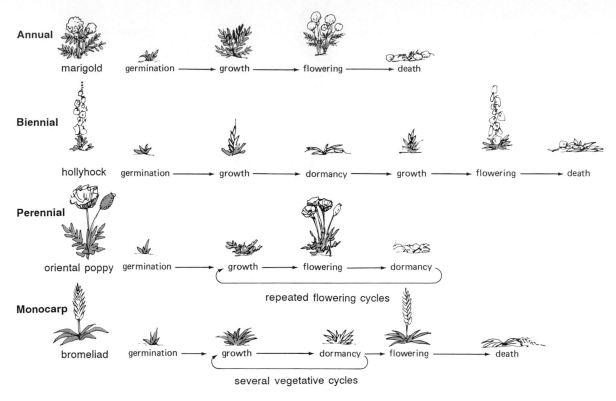

Figure 3-1 The four types of plant life cycles.

In most climates their aboveground portions die in fall, and only the belowground portions survive the winter. But in frost-free areas they may continue active growth all year.

Because they do not produce woody aboveground parts capable of overwintering, herbaceous perennials seldom are taller than about 5 feet (1.5 meters). They produce flowers and seeds yearly after attaining a minimum size or age. Common examples of herbaceous perennial plants are oriental poppies, rhubarb, and strawberries.

Monocarp

A *monocarp* can live for many years but will flower only once in its lifetime and die afterward. Bromeliads and century plants (*Agave* spp.) are two examples of monocarpic plants. These plants bloom after several years of active growth. The tops then die, and new plants are produced by the root system of the old plant.

Stages of Plant Maturation

Regardless of how long a plant lives, it will usually pass through four stages of maturation: germination, juvenility, maturity, and senescence. These stages and the role each has in the life of the plant are discussed in the following sections.

Germination

Many plants begin life as germinating seeds. Germination starts when the seed absorbs water and ends when the primary root emerges (Figure 3-2). After germination, the seedling goes through a period called *establishment*, which lasts until the seedling is independent and photosynthesizing. Both these stages are exceedingly important in the life of a plant, since it is during these phases that plants are most susceptible to adverse environmental conditions and the greatest number of plants die.

Parts of the Seed

A seed must contain an *embryo* which develops into the new plant and a source of carbohydrates to supply energy to the embryo during germination and establishment. In addition, a covering called the *seed coat* helps prevent injury and drying (Figure 3-2).

In many seeds the carbohydrates required for the seed to germinate are stored in two *cotyledons*. These

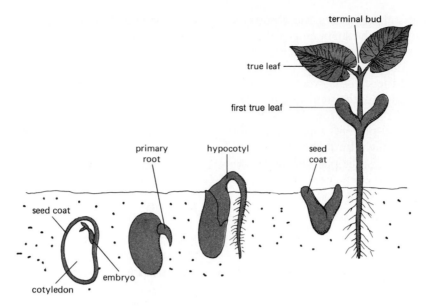

terminal bud

true leaf

first true leaf

primary root

hypocotyl

seed coat

seed coat

embryo

cotyledon

Figure 3-2 Germination of a bean seedling.

cotyledons resemble leaves and are attached to the stem when the seedling emerges from the ground. They are frequently called *seed leaves*, as opposed to the *true leaves* of the plant which are produced later (Figure 3-2).

In other species carbohydrates are stored in an *endosperm*. Whereas the endosperm is a single organ (Figure 3-3) used only for carbohydrate storage, coty-

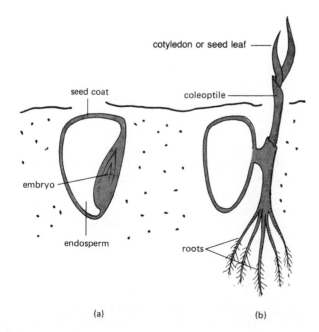

cotyledon or seed leaf

seed coat

coleoptile

embryo

endosperm

roots

(a)   (b)

Figure 3-3 (a) A corn grain showing the enlarged endosperm. (b) The germinated corn grain. Note the absence of the cotyledon leaves found on the bean.

ledons may serve not only as storage sites but also as photosynthesizing organs.

Environmental Requirements for Germination

For a seed to germinate and establish a new plant, correct environmental conditions are necessary. First, water must be present. Second, oxygen must be available. Third, the temperature must be within the range acceptable to that species.

Uptake of water is the first process in germination. The uptake causes the seed to swell and triggers many chemical reactions, including an increase in the respiration rate. As carbohydrates are respired, energy is freed for the growth of the embryo.

Oxygen is necessary for the respiratory process; without it, respiration will not begin and the seed will not germinate. This fact can be demonstrated by the following. Seeds placed in a glass of water will absorb the water and swell, a sign that the germination process is beginning. If taken out of the water a day later and planted, germination will continue. But if the seeds are left in the water without access to the oxygen in the air, the process will stop after the swelling stage.

Suitable temperature is the third essential for germination. Temperatures either too high or too low will kill the embryo or prevent germination from proceeding. For most seeds, 70 to 80°F (21–27°C) temperatures promote vigorous germination, although plants native to cold climates will germinate at considerably cooler temperatures.

After sufficient swelling and respiration have taken place, the *radicle,* or primary root, will emerge. It grows down into the soil, beginning the absorption process, and soon afterward the shoot emerges.

Many times the shoot will arise from the soil as an arch, with the stem of the young plant forming a loop (Figure 3-2). The tip of the shoot is eventually pulled out of the soil by the growth of the stem, and the plant then straightens to an upright position. The pulling of the shoot tip from the soil prevents injury that could occur if it were pushed straight through the soil.

After the shoot emerges, the plant begins photosynthesis. The shoot tip develops leaves, and the cotyledons usually wither and drop from the stem.

Juvenility

After germination, most plants enter a period called juvenility which may last from several weeks to years. The plant will grow rapidly vegetatively but will not begin its reproductive process.

The exact causes of juvenility are not completely known. Pruning, temperature alteration, and applications of growth-regulating chemicals can change a plant from the juvenile to the mature state and back again.

Traits of Juvenility

In most cases the juvenile appearance of a plant resembles the mature form. However, some plants have recognizable traits which signal that the plant is in a juvenile state.

Leaf Form. Leaf form different from that found on the mature plant is a common juvenile trait. The leaves of some acacia species, for example, are frequently compound in the juvenile stage but simple on mature trees. Other plants including sassafras have deeply lobed and divided leaves when they are young, but predominately unlobed leaves when they mature.

Growth Form. Variation in form also distinguishes juvenile from mature plants. The most well-known example is ivy (*Hedera* spp.). Juvenile ivy forms are trailing or climbing, and the leaves are deeply lobed. Mature ivy, however, will produce upright shoots, which grow without support, and have leaves without lobes.

Juvenile branch form is sometimes found on young fruit trees as whiplike vertical shoots. Mature trees occasionally produce such juvenile branches. They usually arise from the base of the trunk and are called *suckers,* or *water sprouts* (see Chapter 11, "Pruning Trees").

Thorns. Thorns are a third common characteristic of juvenility. Citrus and locust trees are frequently thorny when young, but branches produced after the tree reaches maturity are thornless.

Leaf Retention. A tendency for the juvenile parts of a tree to hold leaves throughout the winter is the fourth characteristic of juvenility. Many trees demonstrate this phenomenon (Figure 3-4).

Plants may indicate juvenility other than by visible characteristics. In propagation, juvenile parts almost always root faster than mature portions.

Maturity

Maturity is the third stage in plant growth. It is during this stage that sexual reproduction takes place. The phases of this process are discussed here.

Flower Induction

Flower *induction* is the first indication of maturity and the first of the three phases of flowering. It is an initial chemical reaction that begins the flowering process and occurs before any physical changes take place in the plant. The factors that trigger blooming are well known for some species but unknown for others. Generally as a plant becomes older and larger, its likelihood of flowering increases.

Causes of Flower Induction. Induction of flowering is sometimes caused by environmental factors such as temperature or night length. In nature these factors program flowering into a seasonal pattern so that it occurs at the same time each year. Conditions that cause flowering can also be simulated by man to induce flowering at a preselected time. For example, controlling environmental conditions is a common practice in commercial greenhouses for the timing of flowering of such plants as poinsettias and Easter lilies for holiday sales.

Cool temperature is a crucial environmental factor which causes flowering of many plants. Some plants that grow and flower quite profusely in northern climates either die or grow only vegetatively in warmer locales. They must be *vernalized* (subjected to cold temperatures) before they can be successfully grown. It is because of insufficient vernalization that fruits such as apples, cherries, and pears are not grown in most parts of Florida.

In addition to certain fruit trees, biennial plants may also require vernalization for flowering. The cold received during the winter following the first season of growth induces flowering and completes the life cycle of the plant. Cabbages, carrots, and beets are all biennial plants that require vernalization. However,

Figure 3-4 An oak tree retaining its dead leaves in winter, a characteristic of juvenility. (*Photo courtesy of Dr. Claud L. Brown, University of Georgia*)

since they are raised for their roots and leaves and harvested the first year, the vernalization requirement is not important.

The vernalization needs of bulbs such as tulips, hyacinths, and crocuses are important since they are grown for their flowers. In warm climates the bulbs of these plants are sold prechilled; that is, they have already been vernalized by refrigerated storage. They will flower the following spring regardless of the climate they are grown in. However, they will not bloom more than one year unless they are growing in a climate where they will receive yearly natural vernalization.

Daily night duration is a second environmental factor that can control flowering in plants. It is common knowledge that the nights are longest in winter, shortest in summer, and intermediate in spring and fall. The further north or south of the equator, the more extreme is the difference between the shortest night of summer and the longest of winter. Plants that respond to these changes in the night length are called

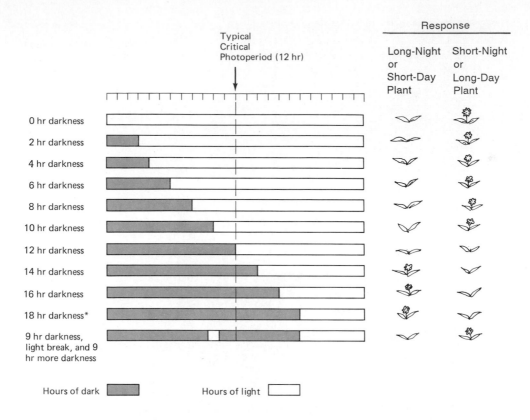

Figure 3-5 The flowering responses of long-day and short-day plants to varying dark/light ratios.

photoperiodic and are common in all but tropical climates near the equator.

Two main types of photoperiodic plants have been studied intensively: *short-day* and *long-day* plants.[1] Short-day plants flower only when the daily light period is *less* than their critical photoperiod of, for example, 12 hours. Long-day plants will flower only if the light period is *more* than their critical photoperiod. They can also flower under continuous light or during a long dark period interrupted at some point with a short light period of an hour or so. In an otherwise long night, a short period of light has the effect of cutting the dark period into two shorter periods. Provided each half is less than the critical period, the plant will respond as if it were exposed to only short nights, and flowering will take place (Figure 3-5).

The way in which night length triggers flowering involves a pigment called *phytochrome,* which alternates between two different forms, each with sensitiv-

ity to differing wavelengths of light in the red to far-red range (see Chapter 16 for a discussion of the wavelengths of light).

Light intensity can also control flowering. Plants that are day neutral and not affected by night length frequently respond to intensity. Generally, greater light intensity is required to induce flowering than is sufficient to keep a plant growing vegetatively. Consequently, many houseplants bought for their blooms never flower again after being moved from a greenhouse to indoors. The light intensity is simply too low to induce flowering.

Stress, such as from drought or the crowding of a plant's roots, can also stimulate flowering. One possible explanation is that the natural production of chemical compounds that trigger flowering of a mature plant is intensified when the plant is dying, and the remainder of the stored carbohydrates is used for flowering and continuation of the species through seed.

Flower Initiation

The second phase of flowering is flower initiation. Like induction, it is a phase of flowering invisible to the naked eye, but changes are taking place in the micro-

---

[1]The confusing emphasis on day length rather than night length stems from the first research on photoperiod. At that time it was mistakenly believed that the length of light rather than darkness caused flowering.

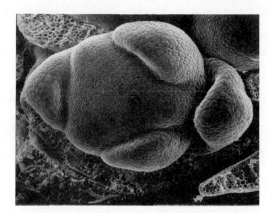

Figure 3-6 Electron microscope photograph of a developing flower meristem of an Easter lily. (*Photo courtesy of Dr. H. Paul Rasmussen, Logan, UT*)

scopic parts of the plant. It is during this phase that the vegetative meristems found at the stem tips or leaf axils change to flower meristems capable of developing into blossoms. This radical changeover from a vegetative to a reproductive structure takes place over several days or weeks. The center of the vegetative meristem, which was formerly mounded, becomes flattened. Small knobs of cells emerge in a spiral around the center (Figure 3-6); these are the beginnings of petals, stamens, and other flower parts. The meristem stops lengthening at this time but development continues.

Flower Development

The length of time from induction to bloom may vary from several weeks to six or eight months. Cold-climate woody plants usually have long developmental periods. The flower buds of these plants are partially formed in late summer, are vernalized over the winter, and finish developing and open in early spring. The opening of the flower is its final stage of development.

Pollination and Fertilization

After a flower opens, it is generally receptive to pollination. Whether it pollinates itself (*self-pollination*) or is *cross-pollinated* by another plant will affect the resulting seeds.

During pollination dust-size pollen grains found on the anthers are deposited by wind or insects on the tip of the stigma. Like seeds, they germinate and grow downward toward the ovary where the eggs are located (Figure 3-7). After reaching the ovary, the uniting of sperm (contained in the pollen) and eggs (in the ovary) occurs in the process of *fertilization*.

In some instances pollen produced by a flower may be unable to fertilize the eggs produced by that

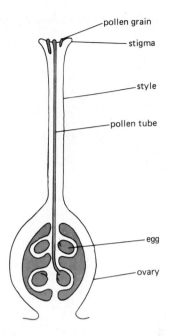

Figure 3-7 Germinating pollen grains growing toward the ovary.

same flower. This *self-incompatibility* is found in about 40 percent of cultivated plants (see Chapter 7). In some cases the pollen grain simply never germinates. In others it germinates but grows poorly and never reaches the egg.

Cross-pollination is thus quite common. Separate fruiting and pollinating tree cultivars are frequently planted in fruit orchards to overcome self-incompatibility problems, since flowers that are not pollinated *and* fertilized usually will not develop fruit.

Cross-pollination can be a hindrance as well as an advantage. Sweet corn and field corn cross-pollinate if planted close together, making the sweet corn taste like corn raised for grain. Sweet and hot peppers will also cross-pollinate. The fruit, which develops from the ovary, will still be edible. But the seeds are the sexual union of two parents, and a mild pepper with hot-tasting seeds could be expected.

Although fertilization of the egg by pollen usually must occur before a fruit will develop, there are cases in which it is not necessary. For example, pollen does not fertilize the egg but the fruit still develops, resulting in a seedless fruit. Seedless fruits formed in this way are called *parthenocarpic*. Parthenocarpy can also be caused by applying growth-regulating chemicals and by specific environmental conditions. For example, excessively warm temperatures during pollination can cause seedless tomatoes to form.

Another way in which seedless fruits form is through the spontaneous death of the seed embryo in its early stages of development. Fertilization is needed

Figure 3-8 A greenhouse-grown cucumber that was not fully pollinated and failed to develop normally. (*Photo courtesy of the Cooperative Extension Service of the University of California*)

to start fruit growth, but after the death of the embryo, the fruit will continue to mature to full size. Seedless fruits formed in this way are also parthenocarpic.

For the majority of cultivated fruits, living seeds are necessary for the fruit to develop normally. For example, if only half the eggs inside an apple are fertilized, only one side will grow; a misshapen fruit is thus produced. Cucumbers sometimes have this problem, with a portion staying small while the remainder develops normally (Figure 3-8). In cases where the fruit contains only one seed, the fruits will drop prematurely if the embryo dies. This happens in the case of apricots, peaches, and cherries.

Fruit Development and Enlargement

During fruit development sugars photosynthesized in the leaves constantly flow into the fruit. They supply energy for the developing ovary which will become the fruit.

Ripening

When a fruit reaches the end of its enlargement period, ripening begins. The fruit may become soft or change color, and its flavor can change from sour to sweet.

Softening of fruits is the result of the breaking down of compounds called *pectic substances* which strengthen the cell walls and cement cells together. In overripening, mushiness occurs because too much of the pectic substance has been lost. Mealiness in apples results from such overripening. The cells are so easily separated from each other that they are chewed without breaking and releasing the flavor of the apple.

Color changes result from the depletion of chlorophyll and the accumulation of other pigments. The chlorophyll continually decreases while the other pigments increase in intensity until coloration is complete. *Carotene* is a pigment which gives orange color to fruits such as oranges and persimmons. *Anthocyanin* is another pigment, responsible for the red color of ripe strawberries and apples.

Senescence

Senescence is the aging past its prime of a plant or any of its parts. It occurs as a part of the natural life cycle of the plant or as a result of environmental factors.

The natural life span of the plant often determines when senescence begins. Annual plants begin senescence at the flowering stage and will die soon after seed formation is complete. Senescence after reproduction is also the rule for monocarpic plants which live several years before flowering.

Senescence of perennial plants is frequently called *decline*. Asparagus is a perennial vegetable that suffers from decline. The productive life of asparagus plants is generally 20 to 25 years. Once the plants reach this age, the yield drops off considerably, and the bed is replanted.

An entire plant need not die for the term senescence to be applied. The top portions of herbaceous perennials and bulbs senesce annually, but the root system remains alive. The lower branches of trees and leaves of houseplants such as red-margin dracaena (*Dracaena marginata*) and bamboo palm (*Chamaedorea erumpens*) frequently die while the top continues growing rapidly.

A colorful show of leaf senescence occurs during the fall when trees turn color. The coloring is due to the same pigment changes that occur with the ripening of fruit. However, environmental conditions trigger this senescence, in this case the shortening of day length in fall and the cooler temperature.

Abscission

The dropping of leaves, flowers, fruits, or other plant parts is called *abscission*. This process involves the manufacture of plant hormones and the formation of a zone of abscission. In simple leaves, this zone is formed at the point where the petiole connects to the

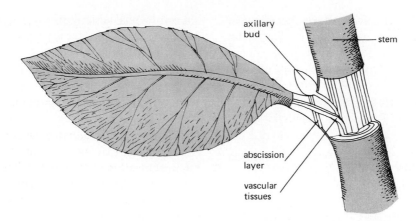

Labels: axillary bud, stem, abscission layer, vascular tissues

Figure 3-9 A leaf just prior to abscission.

stem (Figure 3-9). In compound leaves, the main abscission zone is also at the point where the petiole reaches the stem, although individual leaflets may also form abscission layers and drop.

Flower abscission is often related to pollination, and again a hormone (auxin) is involved. The pollen contains auxin, which it carries to the pistil, triggering abscission of the flower parts. Since the petals, stamens, stigma, and style are no longer needed to attract pollinating insects, their function is complete and their abscission diverts more carbohydrate to the developing fruit.

Fruit abscission can occur at any point in development, but is most common after ripening. Dropping of ripe fruit is a type of seed dispersal. As with leaf drop, an abscission layer generally forms before a fruit drops.

Dormancy

Dormancy is a stage of plant development in which growth slows or stops. It affects all life stages from seed to maturity and plays an important part in adapting plants to their environment and ensuring their survival.

Dormancy occurs during periods not suitable for plant growth and is usually seasonally related. Winter dormancy is found among plants in cold-winter areas. Dry-season dormancy is found in other areas of the country that experience distinct wet and dry seasons, such as California.

Leaf abscission is a common indicator of dormancy, signaling the onset of winter dormancy in many deciduous (leaf-dropping) trees. Although photosynthesis can take place at temperatures below freezing (such as in evergreens like spruce and pine), the leaves of most plants are not adapted to withstand subfreezing temperatures. Instead, they abscise in autumn, and new leaves survive as buds.

The changeover from dormancy to active growth is called *breaking dormancy*. It normally results from changing environmental conditions—from those that induced the dormancy to those that encourage photosynthesis and growth.

Breaking Seed Dormancy

Seeds usually enter dormancy just prior to abscission of the fruit or senescence of the parent plant. The conditions required to break this dormancy and start germination are specifically geared to ensure the survival of the seedling.

For example, many seeds require winter vernalization before breaking dormancy and sprouting. This assures that germination will only take place in the spring, when the plant will have adequate time and suitable temperature to establish itself. If germination occurred as soon as the seed touched the ground in the fall, the young plant would be unlikely to survive the approaching winter. Simulated vernalization is sometimes used on seeds to break their dormancy in commercial nursery crop production. This process, called *stratification,* involves storing moist seeds at temperatures near freezing for one or more months. Peach and apple trees are examples of plants whose seeds are stratified before planting.

Another environmental condition that breaks dormancy in some seeds is heat. Forest fires destroy existing trees but also trigger the germination of dormant pine seeds. Because the seed coats of some pine seeds are woody and thick, only the intense heat from a forest fire will weaken them enough to allow water to be absorbed and the embryos to emerge.

Most seeds with hard seed coats require less drastic measures to break dormancy. Usually the seed coat erodes enough by weather to permit germination. It can, however, be weathered artificially by a process called *scarification*. This involves cutting, scraping, or otherwise injuring the seed coat enough to allow water absorption and subsequent germination.

Cold, heat, and weathering are all ways used to break seed dormancy but are seldom normal practices for the home gardener. The application of water is usually all that is required to start the germination process.

Plant Hormones and Growth-Regulating Chemicals

Plant hormones are chemicals made within the plant body that produce changes in growth. They are effective in extremely small quantities and cause such responses as root formation, seedless fruit formation, leaf drop, and stem lengthening. Plant growth regulators are manufactured chemicals like plant hormones. They have the same effects and may be chemically quite similar to hormones, but they do not occur naturally.

At least nine groups of plant hormones have been identified, and it is likely that others will eventually be discovered. Scientists researching plant hormones encounter a number of difficulties. First, the effects of hormones are not the same from species to species. Second, slight changes in hormone concentrations can completely alter their effects. Third, two or more hormones frequently are found together, and it is difficult to determine which chemical is responsible for a particular effect.

Auxins

Auxins were the first hormones discovered. In varying concentrations auxins have such diverse effects as promotion of rooting, formation of underground tubers and bulbs, prevention of fruit formation, defoliation, and prevention of abscission of leaves and fruits. Auxin manufactured in the terminal bud is also responsible for suppressing the sprouting of axillary buds further down the stem of a plant (see Chapter 14, "Pinching").

Synthetic auxins, which can be purchased in garden centers, have a number of practical home uses. Powders sold to encourage rooting of cuttings are composed of manufactured auxins in a talcum powder base. 2,4-D sold to control lawn weeds is also an auxin. Applied at high concentrations, it kills many plants, but at very low concentrations it is a growth enhancer.

High auxin concentrations can also be used to prevent fruit on ornamental trees when the fruit is an undesirable feature. Ginkgo and horse chestnut trees both have large bothersome fruits that can be prevented from forming by using an auxin spray in spring.

In commercial agriculture, synthetic auxins are used to defoliate plants before harvest, prevent sprouting of potatoes in storage, and prevent premature fruit drop in orchards.

Gibberellins

Gibberellins are a second group of hormones. The first plant activity that was associated with gibberellins was stem elongation. Additional experiments showed other effects including breaking dormancy of seeds, buds, and tubers and inducing flowering in plants that normally require vernalization or a specific photo-period. Increases in flower, leaf, and fruit size were also found to be caused by gibberellins, leading to speculation that they would be wonder chemicals for agriculture, but this did not occur. However, gibberellins are used in greenhouses to form tall tree-form fuchsias and geraniums from cuttings because of their stem-elongation ability. They are also used to increase the size of grapes and to substitute for vernalizing azaleas and fruit trees in the South.

Cytokinins

Cytokinins are the third plant hormones; they have less pronounced effects on growth than auxins or gibberellins. Cytokinins are believed to work in conjunction with light to increase cell division and enlargement. This leads to stimulation of leaf growth, lengthening of stems, and formation of buds. They have also been shown to prevent chlorophyll degeneration and to break axillary bud dormancy.

Growth Retardants

Synthetic and natural growth retardants make up the fourth classification of growth-regulating chemicals. They are sold under the trade names Phosphon, B-Nine, Bonzi, and Cyocel for use on florist crops such as poinsettia and chrysanthemum. They slow elongation of the stems, making the plant sturdier and fuller. On fruit crops they are used to improve color and firmness and prolong storage life. Growth retardants can sometimes be used on hedges and lawns to slow growth and decrease maintenance. They can also be used to maintain bedding plants in a compact size which gives them a neater appearance in formal landscapes.

Abscissic acid is a growth retardant that induces abscission and dormancy and inhibits seed germination. However, its action can be counteracted by growth-inducing chemicals like auxins, gibberellins, and cytokinins.

Ethylene gas is another retardant, whose use dates back to the Chinese practice of ripening fruits in incense-filled rooms. It was later determined that it was the ethylene gas given off in the fumes which caused the accelerated ripening. Ethylene is also produced by the ripening fruits themselves and by cut flowers, or it can be manufactured.

In addition to ripening fruits, ethylene also ages plant parts such as flowers, causing them to abscise and induces flowering in a limited number of species. Pineapples and other members of the bromeliad family (many of which are used as houseplants) will bloom when treated with ethylene. The home practice is to enclose a ripening apple (a source of ethylene) in a plastic bag with the plant for about a week. If the plant is mature enough, flowers will appear in 1 or 2 months.

Vitamins

Vitamins, particularly B vitamins, are occasionally sold as stimulants for plant growth and for use after transplanting. They are the same vitamins sold for human use but act more as hormones in plants.

The effectiveness of vitamins on improving plant growth has not been fully determined. When vitamins are used on bean seeds, limited experiments have shown that they improve germination rates and decrease the time from seed sowing to harvest. Increases in yield have also been reported on some species. When vitamins are used with auxins, improved rooting of cuttings has been found, although it is questionable whether this is due to the vitamin or the auxin. Overall, it is very possible that vitamins do improve plant growth in some cases. However, they should never be used in place of fertilizers or proven-effective chemicals sold for use on plants.

Selected References for Additional Information

Galston, A. W. *Life of the Green Plant*. 3rd ed. Englewood Cliffs, NJ: Prentice-Hall, 1980.

Nickell, L. G. *Plant Growth Regulators: Agricultural Uses*. New York: Springer Verlag, 1982.

Wareing, P. F., and I. D. J. Phillips. *Growth and Differentiation in Plants*. New York: Pergamon Press, 1981.

Wilkins, Malcomb. *Plant Watching: How Plants Remember, Tell Time, Form Partnerships and More*. New York: Facts on File, 1988.

Zimmerman, M. H., and C. L. Brown. *Trees, Structure and Function*. 2nd ed. New York: Springer Verlag, 1985.

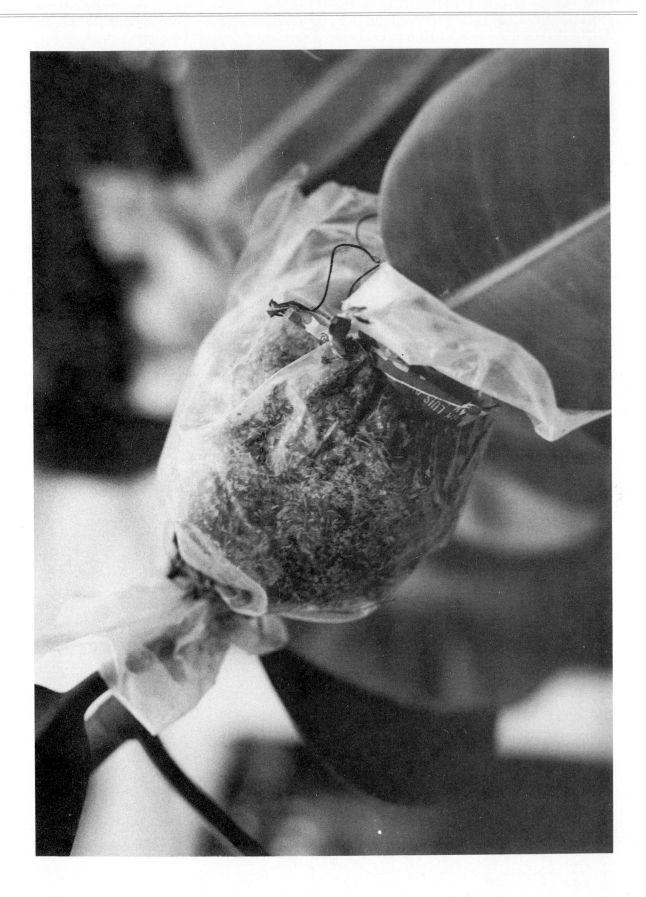

4

Plant Propagation

Sexual and Vegetative Propagation

Plants are reproduced, or *propagated,* either sexually or vegetatively. Sexual propagation involves seeds or spores which result from the union of male and female cells. Vegetative propagation is based on using parts of an existing stock to generate new plants. These *daughter plants* will have traits identical to those of the plants from which they were derived.

Propagation by seed is possible in almost all cultivated plants, and is used extensively. However, vegetative propagation is equally, if not more, important in horticulture for several reasons. First, it is often faster than seed propagation. Most importantly, exact genetic duplicates of the parent plant are obtained. These duplicates are called *clones* and are essential for the preservation of desirable characteristics (such as fast growth or high yield) which could be lost in sexual reproduction. Seedlings may resemble the parent plant on which they were produced, but they will seldom be exactly like it. This is a result of genetic recombination of male- and female-carried genes. Just as brothers and sisters produced by the same parents are not all alike, seeds produced by a plant will not grow into identical plants. The gene mixing will result in offspring with a variety of different traits.

Whether a plant is reproduced sexually or vegetatively depends on many factors: the ease of germinating the seed, the number of plants to be grown, and the importance of preserving a special trait possessed by a parent plant. Table 4-1 groups plants by their uses and lists the ways in which they are normally propagated.

Seed Formation

Seeds can be formed as a result of a plant fertilizing itself (self-pollination) or fertilization by another plant (cross-pollination). Offspring resulting from cross-pollination are called *hybrids* and carry traits of both parents. Although the term "hybrid" can apply to any seed resulting from cross-pollination, a seed package bearing the word "hybrid" indicates that the included seeds are the result of special breeding and are generally superior to other cultivars. Selected hybrid seeds produce healthier, faster-growing plants, a phenomenon called *hybrid vigor.*

Hybrid seed results from the controlled crossing of two groups of plants, one designated to be the *female line* and the other designated the *male line,* of known genetic makeup. The crossing of the two "genetically pure lines" produces seed that will have the best traits of both parents (Table 4-2).

Seed saved from hybrid plants will not grow good plants the following year. Only the original hybrid seed bears the desirable traits. The seeds included in its fruit result from random cross- or self-pollination. They will not produce the same superior plants.

Plant Breeding

Home plant breeding is an interesting activity, but it seldom yields very spectacular plants. Desirable characteristics are the result of luck in most cases. For example, a person who crosses red and white petunias may assume that pink offspring will be produced due to the mixing of red and white. But plant genetics is a complex science, so the result is more likely to be a combination of reds, whites, and shades between.

When breeding plants, a few basic rules of genetics must be followed. First, the two plants to be crossed nearly always must belong to the same genus and often to the same species. Cultivars within a species will cross-pollinate in most cases.

Table 4-1 *Primary Propagation Methods for Cultivated Plants*

Plant category	Primary method of propagation
Annual flowers	Seed
Annual vegetables	Seed
Fruit trees	Seed for rootstock, followed by grafting of vegetatively propagated scion
Groundcovers	Vegetatively by cutting
Houseplants	Vegetatively by cuttings
Ornamental trees	Seed
Perennial flowers and bulbs	Vegetatively by cuttings, underground storage organs, or division
Perennial vegetables	Vegetatively by division
Turfgrass	Seed
Vines	Vegetatively by cuttings

Only in rare instances do plants cross between genera. One example was the cross of English ivy (*Hedera helix*) and Japanese fatsia (*Fatsia japonica*) to produce the popular plant *Fatshedera*, a semivining shrub with ivylike leaves intermediate in size between the parents.

In nature, seeds from cross-pollination make up only 4 percent of all seeds produced. Understandably so, since the pollen-bearing anthers within a flower are close to the female style, making self-pollination most likely.

Among plants that never self-pollinate are species that produce male and female flowers on separate plants (holly and bittersweet, for example). Clearly, self-pollination of these species is impossible.

The mechanics of plant breeding are not difficult. Self-pollination can be achieved by sealing a flower in a paper bag just before it opens (usually in the early-morning). The flower will open within the bag and, having no other source of pollen, will self-pollinate. Tapping the bag will aid in releasing the pollen from the anthers so that it can land on the stigma.

For cross-pollination, flowers that are ready to open should be selected from two parent plants. The petals and anthers of the female parent should then be removed with small scissors and the blossom enclosed in a paper bag. Proceeding to the male parent, the flower should be opened by hand and a small paintbrush inserted to pick up some of the pollen. The pollen should then be taken to the female parent, brushed lightly over the style (Figure 4-1), and the bag replaced.

In several days the pollen will fertilize the eggs, and the pistil will shrivel and fall. The bag can then be removed and the flower tagged with the names of the parents.

Seed Harvesting

When seed is mature and ready to be harvested, usually either the seed pod splits or the fruit drops to the ground. The color of the mature seeds will darken as they ripen.

Cleaning removes the ovary tissues that surround the seed. Dry fruits such as the pods of beans or fruits of many trees usually split open when mature, and the seeds can be shelled out with the fingers. With fleshy fruits such as peach or squash, the fruit should be cut apart and the seeds extracted and dried on a paper towel for several days.

Storing Seed

The best conditions for storing seed depend on the species. However, for general storage of vegetable and flower seeds, a sealed container kept in the refrigerator is sufficient. Glass jars, plastic containers, and freezer bags all are suitable because they maintain constant humidity. The average refrigerator temperature of about 40°F (4°C) slows respiration and keeps the seed viable for the greatest possible time.

The life span of seeds in storage varies from days to years among species. Excess vegetable seeds are

Table 4-2 *Sample Cucumber Hybridization*

Inbred parents	Traits of parents	Traits of resulting hybrid offspring
Cucumber A line (female line)	Long, thin shape (desirable trait)	Long, thin shape
	Pale color (undesirable trait)	Deep green color
	Matures early in season (desirable trait)	Matures early
	Large, tough seeds (undesirable trait)	Small, tender seeds
Cucumber B line (male line)	Short, thick shape (undesirable trait)	
	Deep green color (desirable trait)	
	Matures late in season (undesirable trait)	
	Small, tender seeds (desirable trait)	

Figure 4-1 Pollination of a hibiscus bloom by brushing the pollen on with a paintbrush. (*USDA photo*)

commonly stored for use the following year. Chapter 6 gives the maximum time seeds of common vegetables can be stored under recommended storage conditions and still produce satisfactory plants.

Even under ideal storage conditions, the longer the seeds are stored, the fewer seeds will germinate. The seedlings that do emerge will be weaker. Therefore when planting seed that has been stored, it is advisable to sow the seed thickly to compensate for decreased germination.

Growing Plants from Seed

Seed Growing Outdoors

Sowing seed outdoors is called *direct seeding* and is used for many vegetables and a few flowers and herbs. Since the plants are seeded where they will mature, it is important to choose the site carefully. It should provide the correct amount of light for the species and should also be well drained (water should not stand after a rain). The soil must be loose and crumble easily in the hand. Soils that crust or pack should be conditioned as described in Chapter 5 and spaded and worked to a depth of about 1 foot (30 centimeters) to assure suitable soil conditions for root growth. Clods should be broken and any rocks or debris removed.

The seeds should be sown at a depth and spacing recommended on the package. If sowing directions are not given, the general rule is that seeds should be planted one and one-half times as deep as their diameter. Spacing between seeds can be estimated from the mature size of the plant. Thicker sowing is permissible and even recommended, with the understanding that weaker seedlings will be pulled out to achieve proper spacing for mature plants.

Soil used to cover seeds should be free of even small clods to enable the emerging shoots to reach the soil surface easily. Rubbing the soil between the palms of the hands will pulverize it to the proper degree.

Timing is important when planting seeds outdoors. If planted too early in spring, the seeds will fail to germinate or germinate slowly because of cold weather. Planted too late, they may not mature within the growing season. Package directions are the best source of planting information.

Seeds planted in moist soil do not require immediate watering. They are able to absorb moisture from the soil to start germination. In dry soil, watering will supply moisture and settle the soil around the seed so that the emerging root will have immediate contact with a source of water and nutrients.

Maintaining adequate soil moisture during the germination and establishment parts of a plant's life is essential. A drying of the soil at these times will be fatal to the seedlings.

Soil crusting can occur after rain or watering; it can be recognized by the smooth, glazed appearance of the soil surface. Air and water entry into the soil is inhibited and germination reduced. If soil begins to crust, the surface should be broken carefully with a light rake or hand tool so emerging seedlings are not injured.

A liquid fertilizer "starter solution" (Chapter 5) can be applied to seedlings after germination is complete. Although this is not essential, research has shown that a starter fertilizer will increase the growth rate of seedlings substantially. Seedlings are considered established after true leaves appear (see Chapter 3). True leaves indicate that the root system is absorbing water and nutrients from the soil, and the plant is no longer dependent on carbohydrate stored in the seed.

Natural Reseeding of Outdoor Plants

A number of cultivated plants produce seed that germinates naturally the following season. Among these plants are tomatoes, corn, and petunias. Seedlings that grow untended are called *volunteers* and generally do not produce plants of the same quality as the parents.

Seed Growing Indoors and in the Greenhouse

Seedlings of vegetables, flowers, and herbs can be raised indoors or in a greenhouse. They should be started six to eight weeks ahead of when they will be required for the garden.

A container for raising seedlings need not be elaborate. The bottom of a milk carton or plastic dairy product container is sufficient, provided holes are made at the base for draining of excess water. For large numbers of seedlings, plastic, styrofoam, or wood nursery flats can be purchased (Figure 4-2). Compressed peat moss disks (Figure 4-3) that swell with water and form a plantable container for one seedling are also available in garden centers. Cells of plastic egg cartons can also be used for individual seedlings. Plastic pony or pot packs from plants bought in previous years can also be used.

The material in which seeds are planted is called the *growing medium*, and it must be fast-draining and not prone to packing. Commercial potting soils sold for houseplants fit these requirements and are convenient for raising seedlings. They are not actually soils at all but mixtures of natural treated ingredients. Pure garden soil should not be used, since it packs when restricted in a container and becomes unsuitable for seedling growth.

Sterilization of the growing medium is an important precaution when growing seed indoors because it kills fungi in the medium which cause plant diseases. To sterilize the medium, it should be dampened, placed in a baking dish, and covered with foil or a lid. The oven temperature should be set at 180°F (82°C) and the medium baked for about an hour or until it is heated throughout. If all the growing medium is not used, the excess should be kept in a sealed plastic bag to prevent recontamination.

Houseplant potting soil seldom requires any further preparation than sterilization. After it has cooled, it can be poured into a clean seeding container and the seeds planted immediately. Spacing between seeds can be close provided the young plants will be transplanted soon after they emerge.

Water can be supplied to container-grown seedlings by sprinkling or misting whenever the growing medium begins to dry. An easier way is to water the seeds after planting and then enclose the seed container in a clear plastic bag (Figure 4-2). The plastic prevents moisture from evaporating. No further watering will be required until after the seeds have germinated and the plastic is removed. If the container is made of plastic or other waterproof material, it is not necessary to enclose the entire container. A sheet of plastic food wrap can be laid across the top of the container to hold in moisture.

For most rapid germination, seeded containers should be placed at a temperature of 70 to 80°F (21–27°C). A sunny area is not required for germination and, in fact, should not be used for containers covered or enclosed in plastic. The heat from the sunlight will build up under the plastic and kill the seeds.

Though not essential for germination, bright light or direct sunlight is needed after seedling emergence. The limited light entering through a window, even a bright south window, is often not sufficient to raise healthy transplants. If the seedlings become pale and "stringy" (Figure 4-4) instead of compact, poor garden plants will result. For raising most transplants, artificial light (Chapter 16), a greenhouse, or a greenhouse window is advisable. Exception can be made for seedlings of houseplants (such as palms and Norfolk Island pines) and for bright-light climates listed in Chapter 16.

Figure 4-2   A wooden nursery flat used for growing large numbers of seedlings. The plastic is used to retain moisture around the germinating seeds. (*USDA photo*)

Figure 4-3   A compressed peat moss disk (Jiffy 7 peat pellet) before water is added to it, after water is added to it, and with a seedling. (*Photo courtesy of Jiffy Products of America, Inc., Batavia, IL*)

Damping Off

The period from seedling emergence until several true leaves have formed is the primary time that seedlings are susceptible to damping-off disease. If a fungus is present in unsterilized growing media, the organism rots the seedling stem, which discolors and shrivels at soil level. Within a few days the plant collapses and dies (Figure 4-5).

Once damping-off disease is detected, it can spread rapidly and kill the majority of the seedlings. Fungicidal soil drenches (see Chapter 18) are moderately effective in stopping the spread, but the best prevention is sterilizing the growing medium before

Figure 4-4   Stringy tomato seedlings raised in insufficient light. (*Photo by Kirk Zirion*)

planting. Use of an inorganic material such as vermiculite or perlite to cover the seeds is also helpful.

Transplanting Seedlings

It is advisable to transplant seedlings to larger growing containers as they begin to crowd each other. As leaves of neighboring plants overlap, less light will be available to each seedling, and growth will slow accordingly. The competition of crowded roots for nutrients also causes stunting. At this point, usually corresponding with emergence of true leaves, the seedlings should be transplanted into separate growing containers.

Seedlings are very delicate, and transplanting must be done carefully to avoid injury. A pencil or knife can be used to lift up groups of seedlings (Figure 4-6), after which they can be gently separated with the fingers. Seedlings should always be picked up by a leaf rather than by the stem, since any injury to the stem will kill the seedling.

When transplanting, the growing medium should be firmed lightly around the roots with the fingertips, and each plant should be watered to assure contact between the growing medium and the roots. The seedlings should be kept in a shaded location for two to three days to minimize transpiration water loss while the roots damaged during transplanting resume water absorption.

After this period of reestablishment, the seedlings can be replaced in bright light and fertilized weekly with dilute liquid fertilizer (Chapter 5).

Hardening Off

Seedlings raised on a windowsill, in a greenhouse, or under artificial light grow under less strenuous conditions than they will be exposed to in a garden. To acclimatize them to outdoor conditions before transplanting, they should be subjected to a *hardening-off* period. When the weather is suitable, the transplants should be moved outdoors for several hours each day. Gradually, the time can be increased until they are brought indoors only if night temperatures are predicted to drop below the range acceptable for the

Figure 4-5   A damping-off infection in seedlings. (*Photo courtesy of Dr. R. E. Partyka, Columbus, OH*)

Figure 4-6 Lifting out seedlings with a knife for transplanting. (*USDA photo*)

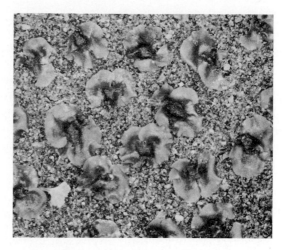

Figure 4-7 Fern prothalli. (*From* Botany, *5th ed. by Carl E. Wilson and Walter E. Loomis. Copyright 1952, 1957, 1962, 1967, 1971 by Holt, Rinehart and Winston. Reprinted by permission of Holt, Rinehart and Winston*)

species. After about a week the transplants will be hardened and less succulent appearing. Hardened plants are less likely to wilt or shock at transplanting.

Transplanting Outdoors

If possible, an overcast day should be chosen for transplanting outdoors to minimize wilting. If this is not possible, transplanting should be done late in the afternoon to allow the roots to recuperate overnight before the plant is exposed to bright sunlight and other drying conditions the following day. The roots should be disturbed as little as possible and the plant watered afterward. Some wilting should be expected, but if it is severe or lasts more than one to two days, the plant will need protection during reestablishment. Shade should be provided by a cardboard tent or newspapers. The soil should be moist at all times so the roots can easily replace water lost through transpiration.

Growing Plants from Spores

Of all the commonly cultivated plants, only one type, ferns, bears spores. In seed plants, the egg (ovule) and sperm (pollen) unite and mature into seeds on the parent plant. In ferns, a spore drops on the ground and grows into a flat plant called a *prothallus* (Figure 4-7). The prothallus then develops the sperm and egg, which unite and form the new plant.

Mature, healthy ferns (with the exception of highly ruffled types which are sterile) develop hundreds of organs called *spore cases* on the undersides of their fronds (Figure 4-8). The spore cases resemble small brown dots or lines, and protect and release the spores.

Several months are required for a fern to develop and ripen spores. When the spores are mature, the spore cases open and release them into the air. The spores themselves look very much like a brown powder.

To capture spores before they are released, the maturity of the spore cases can be tested by lightly tapping the frond over a piece of white paper. When specks appear on the paper, the frond should be cut off and enclosed in an envelope for several days. The spore cases will then open in the envelope, yielding a tiny amount of spores in the bottom.

Spores have the same growing requirements as seeds: warmth, moisture, air, and a growing medium. A small pot or flat should be filled with damp houseplant

Figure 4-8 Spore cases on the underside of a holly fern leaf. (*Photo by Rick Smith*)

soil, then misted, and the spores scattered over the surface. The pot should be enclosed in plastic and set in a shaded location. In several weeks the spores will begin to germinate and grow into flat, green *prothalli* (plural of prothallus). As they grow, the prothalli should be misted daily because the egg and sperm they form can only unite in water. Eventually fronds will grow through the center of the prothalli, which will slowly die. The young ferns can then be transplanted to individual pots as soon as they are large enough to handle.

Vegetative Propagation

Propagating plants vegetatively involves the use of nonsexual plant organs such as leaves, stems, and roots. Whereas a plant propagated from seed may not develop identically to its parent, vegetative propagated plants almost always do. Vegetatively propagation thereby preserves the work of plant breeders and natural mutations which have yielded valuable plants.

A second reason is to reproduce plants that seldom flower or produce only sterile flowers. Foliage plants and navel oranges fall in this group.

A third reason is the relatively short time vegetative propagation requires to produce a mature plant. A cutting can be rooted in as little as two weeks. To raise the same size plant from seed would require a month or more.

Cuttings

Cuttings or "slips" are the most widespread vegetative propagation method. Cuttings are vegetative plant parts such as leaves, stems, and roots that regenerate their missing parts to form new plants. They are cut from parent plants called *stock plants*. The environment required for growing cuttings is the same as for germinating seeds: warmth, moisture, and a growing medium.

Rooting Media

Most parts used for vegetative propagation are taken from aboveground portions of the plant and must regenerate roots. The growing medium that supports and surrounds the rooting area will determine whether roots will form and their quality. The main requirements are that it should drain quickly to admit air to the rooting area yet should retain some moisture.

There is not one superior rooting medium. Many combinations of materials are used, including the following five: two parts (by volume) sand and one part peat moss, one part perlite and one part vermiculite, pure vermiculite, pure perlite, and pure sand. The in-

gredients listed in these media are discussed in more detail in Chapter 15. As with seed media, media used for rooting should always be sterilized.

Cuttings from Outdoor Plants

Shrubs, vines, herbs, groundcovers, and a few perennial flowers are propagated by cuttings taken at different times of the year. Cuttings from woody outdoor plants are classified as hardwood, semihardwood, or softwood, depending on the degree of woodiness of the cutting. A fourth classification, herbaceous, includes cuttings from nonwoody plants such as many groundcovers.

Hardwood Cuttings. Hardwood cuttings (Figure 4-9) should be taken in late fall through early spring from the matured wood of the past season's growth. They are generally about 6 to 10 inches (15–25 centimeters) long and may come from either deciduous or evergreen plants such as juniper, viburnum, and apple.

The specific procedures for rooting cuttings vary among species, but generally the cuttings are wrapped in plastic and stored over winter in the refrigerator. In spring they are unwrapped, and the lower 1 to 2 inches (2.5–5 centimeters) is pushed into damp rooting medium. The cuttings should be enclosed in a plastic bag to prevent moisture loss and placed in a 70 to 80°F (21–27°C) location.

Roots should form in less than 6 weeks, and dormant buds should begin growing shortly thereafter. The cuttings can be hardened off and transplanted several weeks later.

Semihardwood Cuttings. Semihardwood cuttings are taken in summer from the partially matured new

Figure 4-9   Hardwood cuttings of juniper (left), sycamore (middle), and boxwood (right). (*Photo by Kirk Zirion*)

growth of woody plants. The cuttings are made 3 to 6 inches (8–15 centimeters) long and, since they lose water through their leaves very rapidly, should be placed as they are cut in a premoistened plastic bag.

The procedure for rooting is the same as for hardwood cuttings, but rooting is often more rapid.

Softwood Cuttings. Softwood cuttings are taken in late spring from succulent new growth produced that season. They generally root more consistently and quickly than either hardwood or semihardwood cuttings. Picking the cuttings at the right stage of maturity is crucial to success. Shoots that are growing rapidly are too tender and prone to rotting. Ideally, the shoots should be slightly flexible but should snap when bent to a 90° angle. They should be harvested into wet plastic bags and rooted using the same directions as for hard- and softwood cuttings.

Because the younger parts of plants generally root more easily than more mature parts, softwood cuttings are the most reliable way to vegetatively propagate most woody outdoor plants.

Herbaceous Cuttings. Herbaceous cuttings are the equivalent of softwood cuttings but are taken from herbaceous plants such as coleus and impatiens which never become woody. Most houseplants are also herbaceous and are discussed in the next section.

Herbaceous cuttings can be taken and rooted at any time during the growing season. They are generally 2 to 4 inches (5–10 centimeters) long and taken from the tip of a stem. As with all the types of cuttings previously discussed, the bases should be planted in damp rooting medium and the cuttings enclosed in a plastic bag to increase humidity. Keeping the cuttings wet from the time they are cut until they are stuck in the medium will reduce wilting.

Root Cuttings. Root cuttings are sections of thickened roots without leaves or stems attached. Whereas most outdoor plants can be propagated by one or more types of top cuttings, only a few will grow from root cuttings. These species have the ability to generate an adventitious bud on the root section.

Root cuttings of most outdoor plants are best taken in early spring just before new growth begins. At this time carbohydrate reserves are maximum, and the plants are ready to emerge from dormancy. Like other cuttings, root cuttings should be protected from drying by a plastic bag and can be harvested by simply digging out medium-sized roots from around the base of an established plant. The sections can be cut into 3-inch (8-centimeter) lengths and planted horizontally ½ inch (1–2 centimeters) deep in rooting medium. The

medium should be kept damp, and shoots should appear in four to six weeks.

Indoor Plant Cuttings

Cuttings of indoor plants can be of the following four types, depending on the part of the plant from which they are taken. The procedure for rooting the cuttings is the same as for cuttings of outdoor plants.

Stem Tip Cuttings. Stem tip cuttings are the most common type (Figure 4-10) and consist of the top 2 to 4 inches (5–10 centimeters) of a growing stem.

Leaf Bud Cuttings. Leaf bud cuttings (Figures 4-10 and 4-11) are taken from a plant after the stem tip has been used. Each cutting includes a short section of

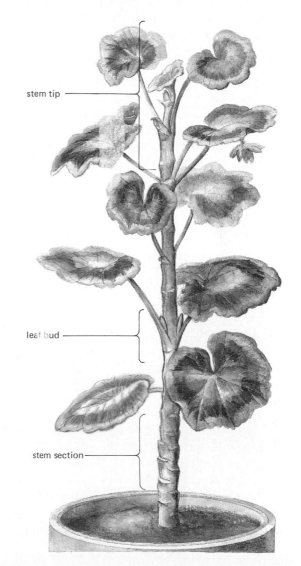

Figure 4-10 A typical dicot plant, showing where stem tip, leaf bud, and stem section cuttings would be found.

Figure 4-11 Rooting stem tip (left), stem section (middle), and leaf bud (right) cuttings. (*Photo by Rick Smith*)

stem, one leaf, and its corresponding axillary bud. When the base of the cutting roots, the axillary bud breaks dormancy and forms the new stem of the plant. Any plant that can be propagated by stem tip cuttings can be propagated by leaf bud cuttings, although the latter method is slower.

Stem Section Cuttings. Stem sections or cane cuttings (Figures 4-10 and 4-11) are short pieces of thickened, leafless stem. They are used primarily for obtaining large numbers of offspring from cane-forming plants such as dumbcanes (*Dieffenbachia* spp.) and Hawaiian ti plants (*Cordyline* spp.). First, the top of the plant is removed for a stem tip cutting. The remaining stem is then cut into pieces containing two to three nodes each. The sections are laid horizontally half-buried in the rooting medium. A plastic covering is not essential provided the medium remains moist, and roots and shoots will be generated in one to three months.

Leaf Cuttings. Leaf cuttings are a single leaf and sometimes its petiole. Most indoor plants cannot be propagated using only a leaf; only certain species generate buds where the leaf blade joins the petiole or along the leaf veins or margins (Table 4-3). The propagation procedures vary among genera. Leaves of African violet and peperomia are picked with the petiole attached. The leaf is buried in the rooting medium up to the blade, and new plants form at the soil line (Figure 4-12).

With *Bryophyllum* and other succulents such as jade plant, burro tail, and echeveria, leaves are picked directly off the plant and laid on the surface of the medium (Figure 4-12). Since they are succulents, a medium largely composed of sand is best; enclosing them in a plastic bag is not advisable. With *Bryophyllum*, up to 20 new plants will form around the leaf margin. With others, single plants will appear where the leaf was broken from the plant.

Fibrous-rooted begonias such as rex begonia produce new plants along the veins of mature leaves and also where the petiole connects to the leaf blade. Whole leaves or pieces of leaves should be laid on a damp rooting medium so that the veins make contact. Cutting the veins with a knife stimulates bud formation, and three to five cuts can be made on each leaf.

Another houseplant grown from a leaf is the snake plant (*Sansevieria* spp.). A mature leaf can be cut crosswise in 1-inch (3-centimeter)-long sections and the base of each stuck in rooting medium (Figure 4-12). Remember that the orientation of the cuttings is critical. The end of the section that was originally on

Table 4-3 *Selected Plants Propagated by Leaf Cuttings*

Common name	Botanical name
Rex begonia	Begonia × *rex-cultorum*
Jade plant	*Crassula argentea*
Echeveria	*Echeveria* spp.
Mother of millions, Bryophyllum	*Kalanchoe beauverdii, laxiflora, and pinnata*
Peperomia	*Peperomia* spp.
African violet	*Saintpaulia ionantha*
Snake plant	*Sansevieria* spp.
Burro tail	*Sedum morganianum*
Christmas cheer	*Sedum × rubrotinctum*

Figure 4-12 Leaf cuttings of a snake plant (*Sansevieria*) (left), begonia (middle), and peperomia (right). Note the locations of the new plants that formed. (*Photo by Rick Smith*)

top must be up, and the part that was lower then planted in the rooting medium. Sections inserted upside down do not root. After the leaf section roots, a new plant will be produced at the base and grow up beside the leaf section (Figure 4-12).

It should be noted that variegated or yellow-banded sansevierias propagated by this method produce green offspring. To maintain variegation, the variety must be propagated by division.

Propagation from leaf cuttings is slow. The generation of an adventitious bud and roots is an energy-consuming process. Several months will be required before the young plants are large enough to be transplanted.

Suggestions for Successful Rooting of Cuttings

Attempts to propagate plants by cuttings meet with varying degrees of success, depending on the care one takes and the readiness with which the species generates new roots. Following the suggestions given here will maximize chances of success with most species. But for plants that are difficult to propagate, consulting one of the references listed at the end of the chapter may be necessary.

First, select cuttings from shoots without flowers or flower buds, or if this is not possible, pick off the flowers. A cutting with blossoms will channel energy into the reproductive parts and will be less likely or slower to root.

Also note the positions of nodes on the stem of the cutting. Roots are often generated first at these sites, so cuttings should be made with at least one node near the base.

Also, leafy cuttings wilt easily and once severely wilted are less likely to root. Therefore cuttings should always be kept moist after cutting and before being stuck in the medium to slow water loss from the leaves.

Any leaves that will be covered after the cutting is stuck into the rooting medium should be removed. If left on, they will rot and provide a breeding ground for disease organisms.

Likewise, during the rooting period leaves that die and drop from the cuttings should be removed along with whole cuttings that appear dead.

Finally, use of a rooting hormone can increase rooting speed and success. The auxins discussed in Chapter 3 are available in powder form in a talc base, as dissolvable powders for mixing with water, or as liquids, and are applied to the stems of cuttings before insertion into the rooting medium.

Gentle tugging of rooting cuttings about once per week is the standard test for determining whether rooting has occurred. If the cutting slips out easily, no anchoring roots have formed. The cutting should be inspected for signs of rotting and, if still healthy, can be reinserted into the medium. If the cutting does not pull out with gentle tugging, it already has roots. The plastic bag can then be opened to accustom the plants to normal humidity and be removed entirely after several days. After one week the cuttings can be transplanted to pots.

Rooting Cuttings in Water

A few outdoor plants, such as willow and pussy willow, and many tropical indoor plants (Table 4-4) can be rooted and grown in water. The main reason more species will not root in water is lack of oxygen in the water which is, in this case, the rooting medium. If air can be bubbled in (for example by an aquarium pump) many more species can be rooted this way.

Stem tip cuttings are usually used, with the lower leaves removed and the cutting absorbing water through the cut end until roots are formed. One of the easiest methods is to fill a glass with water and cover the mouth with foil. Holes can then be punched in the foil and the cuttings inserted and held upright by the foil. To minimize transpiration the cuttings should be kept out of direct sun.

Table 4-4 *Selected Cuttings for Rooting In Water*

Common name	Botanical name
Chinese evergreen	*Aglaonema* spp.
wax begonia	*Begonia semperflorens*
angel wing begonia	*B.* spp.
grape ivy	*Cissus rhombifolia*
coleus	*Coleus × hybridus*
dumbcane	*Dieffenbachia* spp.
dracaena	*Dracaena* spp.
pothos	*Epipremnum aureum*
botanical wonder	*× Fatshedera lizei*
rubber plant	*Ficus elastica*
nerve plant	*Fittonia vershaffeltii*
Tahitian bridal veil	*Gibasis geniculata*
English ivy	*Hedera helix*
polka dot plant	*Hypoestes phyllostachya*
impatiens	*Impatiens wallerana*
bloodleaf	*Iresine herbstii*
prayer plant	*Maranta leuconeura*
philodendron	*Philodendron* spp.
Swedish ivy	*Plectranthus* spp.
African violet	*Saintpaulia ionantha*
German ivy	*Senecio macroglossus*
arrowhead	*Syngonium podophyllum*
wandering Jew	*Zebrina* spp. and *Tradescantia* spp.

Crown Division

Crown division is probably the most common and reliable propagation method. It is used for many herbaceous perennials, a few shrubs, and many houseplants such as ferns, asparagus ferns, African violets, and spider plants.

In division, one plant is separated into two or more pieces, each containing a portion of roots and crown (Figure 4-13).

The division can be done at any time in the growing season, however, plants in active growth must be treated carefully to minimize water loss through the leaves. Therefore division of dormant plants in early spring is common in cold-winter areas of the country.

Shrubs to be divided must be mutli-stemmed, that is, have a number of stems growing out of the ground instead of branches arising from one single short truck. They should be divided when dormant if possible, by cutting through the crown with a spade so that it is broken down into several sections. Each section is then dug, keeping as many roots as possible, transplanted to the new location, and pruned to lessen shock. A part of the parent crown can be left to rejuvenate the shrub. With herbaceous perennials, the entire plant is dug up and cut apart. The sections are then replanted and watered.

Figure 4-13 Division of a plantain lily into several smaller crowns. (*Photo by George Taloumis*)

When dividing houseplants, the parent can be removed from its pot, cut or pulled apart, and the sections repotted. Pruning is normally unnecessary. If wilting occurs, the newly potted sections can be left in plastic bags for several weeks until the roots are reestablished.

Layering

Layering is another method for propagating vines and also shrubs that have a trailing growth habit or flexible branches. A low, flexible shoot is bent to the ground during any time the plant is in active growth and held in place by a bent wire or heavy stone. It is then mounded with soil a short distance back from the tip. Within a few months, roots should have formed on the covered portion of stem, and the layer can be cut from the parent plant and transplanted.

Runners, Rhizomes, and Stolons

These three structures are types of horizontal stems produced as a natural means of vegetative reproduction. Runners (Figure 4-14) grow aboveground and are found on such plants as strawberries, ferns, spider plants, and strawberry begonias. Rhizomes and stolons grow at ground level or below ground and are typically found on bamboo, grasses, and some irises. There are subtle botanical differences among the three, but basically they are handled the same in propagation.

As modified stems, these organs have buds, nodes, and internodes. A runner will have only one bud at the tip, whereas stolons and rhizomes may have several. The nodes are the sites where new plants will form. With a runner on an outdoor plant, the stem can be positioned as it begins to form the new plant. This is a common practice to avoid overcrowding in strawberries (Chapter 8). The runner will root without additional attention and can be severed from the parent and moved after rooting if desired.

Because houseplants are grown in pots, the runner plants may never make contact with a rooting medium without help and can live attached to the parent indefinitely. When a new plant is wanted, a small pot of growing medium can be placed under the runner plant. Roots will form in less than a month, and the runner can then be detached. When this method is impractical (such as with hanging plants), the runner can be detached and treated as an unrooted stem-tip cutting.

Propagating plants from rhizomes and stolons is done much the same way. The stem connecting the young plant to its parent is cut after a moderate amount of roots has formed, and the young plant is transplanted to the new location.

Figure 4-14   Rooting a runner of a spider plant. (*Photo by Kirk Zirion*)

Suckers and Offsets

Suckers and offsets (Figure 4-15 and 4-16) are young shoots that grow from the roots or stems of mature plants. They are functionally similar to rhizomes and stolons and are found in many shrubs and houseplants such as bromeliads, succulents, and cacti. Offsets on cacti (Figure 4-16) are frequently produced on top of the plant and can be broken off and rooted without difficulty. Suckers from the bases of plants may or may not have developed root systems independent from the parent. If so, they can be transplanted directly. If not, they are treated as cuttings.

Storage Organs: Bulbs, Corms, and Tubers

These underground organs are produced by some herbaceous perennials such as lilies, gladiolas, and amaryllis as an underground repository of stored car-

Figure 4-16 A pincushion cactus with offsets. (*Photo by Rick Smith*)

bohydrate. Botanically, these are modified stems with nodes, buds, and modified leaves. Their natural means of vegetative reproduction is the formation of smaller replicas of themselves (called *bulbils, cormels,* or *tubers*) around the base of the parent (Figure 4-17). These can be broken off (preferably while the plant is dormant) and planted in new locations. Blooming of storage organs may take two to three years after the year they are produced, since a minimum size must be reached before flowering will occur.

Air Layering

Air layering is a technique used primarily on large houseplants such as rubber plants. It involves inducing stem rooting of a plant that still has its own root sys-

Figure 4-15 A snake plant with two young offsets. (*Photo by Kirk Zirion*)

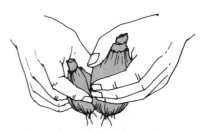

Figure 4-17 Removing a daughter bulb from the mother.

tem. It can be used to obtain new plants from a large branched specimen, to shorten a plant that has grown too tall, or to make a plant that has become tall and leafless short and bushy again. New plants formed by air layering can be very large, up to 3 feet (1 meter) tall if desired.

Many plants having woody stems can be air-layered (Figure 4-18). As the first step, choose the section of the stem where the new root system of the plant is desired. Then a strip of bark about 1 inch (2 centimeters) wide should be cut around the stem and the bark pulled off. This girdling removes the phloem and cambium but not the xylem, which will still translocate water to the top of the plant.

Next, place a handful or two of damp sphagnum moss over the girdled area and wrap with plastic (Figure 4-18). Twist-ties or tape secure the moss at both ends and seal in moisture.

Within two to three months, roots should be growing in the sphagnum. During the waiting period the sphagnum should be checked every one to two weeks for dampness and rewet if necessary since roots will not form if the moss is dry. When several roots 2 to 3 inches (5–8 cm) long have formed, the air layer can be cut from the parent and transplanted to its own pot. If it wilts severely, humid conditions should be provided for about a week.

Budding and Grafting

Budding and grafting are fairly complex methods of propagation used for reproducing valuable fruit and ornamental cultivars in nurseries; they are discussed only briefly here. A gardener who wishes to try either budding or grafting should plan the project ahead of

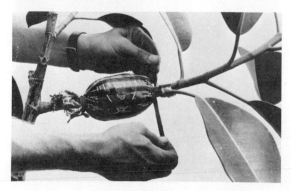

Figure 4-18 Steps in air layering: (1) girdling the branch, (2) applying damp peat moss to the wound, (3) wrapping the peat moss in plastic and sealing the ends, (4) root formation is evident. (*Photo from* Plant Propagation: Principles and Practices *by Hartmann and Kester, 1990. Reprinted by permission of Prentice Hall*)

time and consult several of the reference books listed at the end of the chapter for more in-depth information.

Essentially, budding and grafting unite two genetically different plants so that they heal together and function as a single plant. Budding transfers a bud of one plant to another plant that will function as the root system, whereas grafting attaches a small branch to another plant.

Most frequently, budding or grafting combines two cultivars of a species into one plant that exhibits the best features of each. For example, a plant with a vigorous root system but little ornamental value (called the *rootstock*) could be grafted onto a plant having good ornamental qualities but weak roots (called the *scion*).

Grafting can also serve other purposes: (1) repair of girdled trees that would otherwise die; (2) creation of unusual plant forms such as tree roses or trees with weeping heads atop strong, straight trunks; and (3) changeover of fruit trees in an old orchard to a new cultivar.

Grafting and budding rely on the activity of cambium cells and must be done when these cells are actively dividing and will heal the grafted area quickly. To determine the correct stage for grafting, bark is examined for *slippage*. If it can be separated easily from the underlying wood, the cambium cells are active and the grafting or budding is feasible using the most common techniques. This is usually in early spring before new growth starts, since this greatly reduces water loss through transpiration and improves chances that the cambia of the two plants will grow and unite them. Other grafting techniques can be used even on dormant plants.

Grafting

There are many styles of grafting. The style used will depend on the species and the reason the graft is being made. General rules for grafting are:

1. The rootstock diameter must be equal to or larger than that of the scion, but the scion wood is usually a pencil thickness or slightly larger.
2. The cambium of the stock and scion must be in contact, and preferably over as great an area as possible. If cambium contact is not made, the graft will not heal, and the scion will die.
3. The stock and scion must fit tightly together, and the joint must be protected from drying. A tight graft union is achieved by wrapping the area with special rubber ties or waxed string. Drying is then prevented by a coating of wax.

Figure 4-19 shows the basic steps involved in a style of graft called *whip grafting*.

Budding

Budding (Figure 4-20) involves removing a patch of bark with a bud from a scion and laying it directly against the cambium of the stock. It is a less risky process than grafting because the stock is only slightly damaged if the union does not heal. As with grafting, the area of the union must be protected from drying with rubber ties.

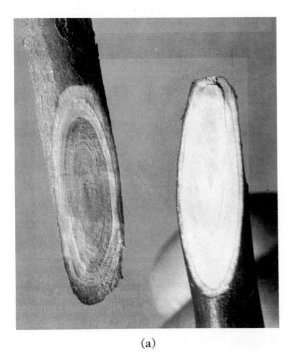

(a)

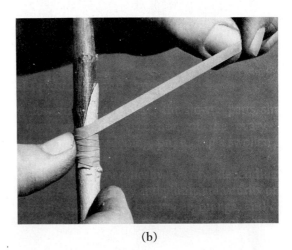

(b)

Figure 4-19 Steps in performing a whip graft: (a) stock and scion cut to equal size and ready for grafting, (b) stock and scion joined and held together with a rubber strip. (*Photo courtesy of Vocational Education Productions, California Polytechnic State University, San Luis Obispo, CA*)

Figure 4-20 Bud patches to be used in budding (above) and a bud patch in place on a branch (bottom). (*Photo courtesy of Dr. Robert F. Carlson*)

Tissue Culture

Tissue culture, also called *micropropagation*, is the propagation of plants from nearly microscopic portions of parent plants (Figure 4-21). The technique has undergone a growth explosion in the past decade and progressed from being primarily a university labo-

Figure 4-21 A black currant plant tissue cultured in a flask. (*Photo from* Small Fruit Crop Management *by G. J. Galletta and D. G. Himelrick (eds.), 1990. Reprinted by permission of Prentice Hall*)

ratory procedure to a commercial propagation method of significant economic consequence. The technique has two distinct advantages over traditional propagation: (1) It enables mass production of a cultivar from an extremely limited amount of *parent stock* and in a relatively small area. (2) It enables the propagator to eliminate disease-causing viruses from the parent material, an achievement unattainable through the use of pesticides, and to propagate numerous virus-free offspring which are healthy and vigorous. This importance of propagation from virus-free parent stock has only recently come to be appreciated as the detrimental effects of unrecognized virus infection have become known.

In its simplest form, tissue culture involves the following steps:

1. Removal of the previously disinfected meristem of a shoot tip using a surgical knife and tweezers under sterile conditions.

2. Placing of this *explant* in a test tube of gel-type growing medium containing plant hormones, again under sterile conditions. The test tube is plugged with sterile cotton to prevent entry of disease-causing organisms.

3. Waiting during a growth and establishment period of several weeks or months during which the explant multiplies many times in size.

4. Division under sterile conditions of the cell mass into numerous near-microscopic pieces which are placed in more gel-filled test tubes containing hormones and plugged with cotton.

5. Waiting for the development of roots and shoots induced by the hormones in the medium.

6. Transferring of the small plantlets from test tubes to pots under greenhouse growing conditions for establishment in normal growing medium.

7. Transferring of plants to outdoor growing conditions.

Tissue culture is not normally considered a home gardener's activity, primarily because it is nearly impossible to achieve the completely sterile conditions necessary. Under nonsterile conditions the gel medium on which the explants and plantlets grow becomes host to microorganisms that rapidly kill the vulnerable young plants. However tissue-cultured plants still in test tubes are sold in nurseries occasionally as novelty items, particularly orchids, which were the main plant's tissue cultured for many years. The test tube is left sealed and treated as a miniature terrarium. When the plant outgrows the tube it can sometimes be transplanted to a pot, although the process is not always successful.

Genetic Engineering[1]

Genetic engineering of plants is a branch of biotechnology, the harnessing of the biological machinery of bacteria to manufacture otherwise hard-to-obtain products, combat genetically caused diseases, improve the tolerance of plants to adverse environmental conditions, and attain other similar laudable goals. Genetic engineering as a means of plant improvement changes the genetic makeup of plants without either breeding or selection, the conventional routes of improvement. Its main advantage is that while plant breeding and selection normally involve only the genes within a species, genetic engineering makes possible the transfer of genes between completely unrelated plants or bacteria, and, in rare cases, even from animals to plants. As an example of the usefulness of this ability, most flowers are not available in blue cultivars, and it is impossible to find or breed blue cultivars because no members of the species carry the genes for this color. However using genetic engineering, it is possible for blueness from, for example, bachelor buttons to be transferred to roses, and in fact blue roses have already been developed by genetic engineering in Australia.

[1]The author gratefully expresses appreciation to Dr. Rick Ward of Michigan State University for his help with this topic. This section could not have been written without his counsel.

The obvious advantage then is that when the possibilities for plant improvement through breeding and selection have been exhausted because of the limited *gene pool* (the genes the plants in the species naturally carry), a species can continue to be improved by bringing in outside genes. The procedure is, of course, complicated, but a simplified version is outlined as follows:

1. The first step is to identify the characteristic that is to be incorporated, for example, disease resistance, frost resistance, or color.

2. The second step is then to either create or isolate from a donor plant the genetic material that will give this characteristic. To isolate the material the donor organism is ground up and then all extraneous cell contents are eliminated chemically until only the genes remain.

3. The third step is to take this extract of genetic material and incorporate it into a *vector*, that will convey the genes to the target plant to be genetically engineered. The first vector is a *plasmid*, a microscopic ring of genetic material found in bacterial cells and available in quantity from scientific supply companies. *Enzymes* (chemicals that cause or accelerate chemical reactions without being affected themselves) are used to open the plasmid rings to admit the new genetic material and then to close them. The treated plasmids, not all of which will have accepted the new genetic material, are then taken up naturally by specialized bacteria in a lab procedure.

4. The next step is to check that *transformation* has taken place, that the plasmids are in the bacteria. This is done by growing the bacteria individually into colonies on a gel medium that has been specifically designed to react through color or another visible sign when in the presence of a transformed bacterium.

5. Those bacteria that are seen to have been transformed are then transferred to new gel medium, grown, and then checked for the presence of the specific genes wanted. This is necessary because although it will have been ascertained that genetic material was transferred, it is uncertain without further checking whether the specifically desired genes are there. One way to accomplish such checking is through chemical analysis. If the trait to be transferred were disease resistance conveyed by a gene that triggers a plant to produce a poison that kills the disease organism, for example, one could check through chemical analysis for the presence of the poison in the bacteria.

6. The next step is to transfer the genetic material in the bacteria into a second plasmid vector within an organism that will put the genes into the target

plant. The organism normally selected for this in genetic engineering of plants is the bacterium that causes the disease crown gall (see Chapter 13). This bacterium has the astounding and rare natural ability to transfer genes to its host plant as a part of its life cycle, and this is the property that makes it invaluable because it can transfer other genes as well.

Crown gall naturally produces tumors in plants by producing cytokinen and indole acetic acid hormone. To eliminate the tumor-causing characteristic and take advantage only of its infective potential, the bacterium is first *disarmed*, that is, rendered incapable of causing galls.

7. The final step is to infect the target plant with the crown gall bacteria which now contain the genetic material to be incorporated. Pieces of root, leaf sections or other plant parts are *inoculated* (touched with the bacteria), and those that become infected (again checked by chemicals in the medium that show a reaction) can then be grown as tissue-cultured plants (see previous section) to regenerate a whole plant. The new plants are now called *transgenic* because they contain genes from outside its own genus. The new genetic material has become an integral part of the plants' genetic makeup and will reproduce itself in every new cell produced.

While the above-described method is the most common, two others are used as well. The first uses a device called a particle gun which actually fires 22 caliber cartridges containing tungsten particles coated with the plasmids carrying the foreign genes. The plant to be genetically altered is literally shot in a vacuum, and the genetic material is propelled through the cell walls and becomes a part of the genetic makeup of the plant.

The second method is *protoplast* transformation. Cells from the target plant have their cell walls removed chemically (thus becoming protoplasts), and are then soaked in plasmids that have had the new genetic material inserted. The new plasmids incorporate into the protoplasts, and the protoplasts are then regenerated by tissue culture into whole plants.

Selected References for Additional Information

*Ball, Jeff. *How to Grow and Nurture Seedlings*. Irwindale, CA: KVC Entertainment.

Clarke, Graham, and Alan Toogood. *The Complete Book of Plant Propagation*. London: Wardlock, 1992.

Coastline Community College. *Plant Reproduction*. Fountain Valley, CA: Coastline Community College.

Garner, R. J. *The Grafter's Handbook*. London: Cassell, 1988.

Hannings, David. *Plant Tissue Culture Series, Parts I and II*. San Luis Obispo, CA: California Polytechnic State University Vocational Education Productions.

*Harrigan, Jim. *Plant Propagation Series, Vols. I and II*. San Luis Obispo, CA: San Luis Video Productions.

Hartmann, H., and D. Kester. *Plant Propagation: Principles and Practices*. 5th ed. Englewood Cliffs, NJ: Prentice-Hall, 1990.

Kyte, Lydiane. *Plants from Test Tubes: An Introduction to Micropropagation*, revised ed. Portland, OR: Timber Press, 1987.

*Labor, Elizabeth. *Budding and Grafting*. San Luis Obispo, CA: California Polytechnic State University, Vocational Education Productions.

*University of Florida. *Handling Tissue-Cultured Plants in the Nursery*. Video VT 233 Extension Video Series. Gainesville, FL: Food and Agriculture Sciences, University of Florida.

*University of Florida. *Lab Procedures for Tissue Culture: A Beginners' Guide*. Video VT-234 Extension Video Series. Gainesville, FL: Food and Agriculture Sciences, University of Florida.

*Indicates a video.

Part II

Growing Plants Outdoors

5

Outdoor Soils and Fertility

The soil is a blanket of loose material that covers the majority of the Earth's land surface. It is a dynamic mixture of minerals, decaying plants and animals (organic matter), microorganisms, water, and air. Relative to plants, the soil serves two primary purposes: first, it provides mechanical support for the plant by permitting roots to grow through it; second, it provides a reservoir for water, air, and nutrients essential for healthy plant growth. How well a particular soil performs each of these functions largely determines the fertility of the soil, that is, how well plants will grow in it.

Classification

Mineral soils are classified primarily by the size and quantity of mineral particles present in the soil. Soil particles range from clay particles, which are so small that they require an electron microscope to be seen, to silt, and then to sand, which can be seen readily by the unaided eye. Each type of soil particle possesses unique properties that influence plant growth. However, the properties of silt are in many respects intermediate between those of clay and sand, and silt is present in relatively small quantities in many soils. Therefore, a basic understanding of soils can be obtained by considering mainly the properties of clay and sand particles. Awareness of these properties will enable the gardener to adapt cultural practices so that the soil environment will be the most conducive to healthy plant growth.

Organic soils (including muck and peat) contain predominately decaying organic matter with smaller amounts of mineral components. They are generally found in small pockets in the north central and northeastern sections of the United States and Canada as well as in the Everglades region of Florida. Organic soils, while very fertile, require careful management;

they do not, in many respects, respond in the same manner as mineral soils. Since organic soils are not important as home garden soils, they are not discussed in this text. Persons gardening in organic soils should consult a soils text or their county extension agent for further information.

Soils seldom contain only one type of particle. Instead they contain several types in varying percentages, and the soil is classified according to the type of particle that predominates (Table 5-1). For example, if a soil contains mostly clay particles but also a small amount of sand, it will be classified as a clay soil. When the percentage of sand increases to a point where it nearly equals the percentage of clay, the soil is classified as a loam. When the percentage of sand further increases to the point where it greatly surpasses the percentage of clay, the soil is designated as a sandy soil. Thus the term "loam," which is often used on seed packages to describe the ideal soil in which seeds should be planted, refers to a natural soil having nearly equal percentages of sand and clay particles.

As mentioned previously, clay and sand particles each possess unique characteristics. If a soil is classified as a loam, its properties will be intermediate between those of clay and sand particles. The higher the percentage of one type of particle, the more closely the soil properties will resemble those of that particle group. A loamy clay soil, for example, mostly possesses the characteristics of clay particles but is modified somewhat by the sand it contains. A clay loam, on the other hand, is basically a loam with a slight extra amount of clay.

Characteristics of Clay Soils

Clay soils (Table 5-2) are often referred to as heavy soils, composed primarily of platelike clay particles. These

Table 5-1 *Sample Classifications of Soil by Particle Types*

Soil	Average particle size	Suitability as garden soil
Clay	Smallest	Fair
Loamy clay	Small	Fair
Clay loam	Medium	Good
Loam	Medium	Good
Sandy loam	Medium	Good
Loamy sand	Large	Good
Sand	Largest	Fair

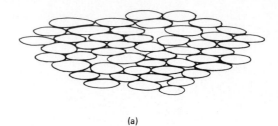

(a)

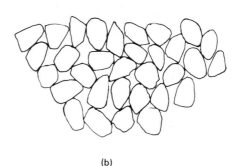

(b)

Figure 5-1 Clay particles (a) lie flat and have smaller pore spaces between them than sand particles (b).

particles are flat and disk-shaped, and their small size and flattened shape are responsible to a large extent for the properties of clay soils. Because the particles are flat, they are able to lie in close contact with one another. This closeness makes the spaces between individual clay particles relatively small compared to the spaces between more rounded particles such as sand. This might be likened to pennies (clay particles) being stacked in a jar as compared to marbles (sand particles). The spaces between the pennies are much smaller than the spaces between the marbles (Figure 5-1). Conversely, the jar filled with marbles, which is likened to a sandy soil, contains fewer particles but larger spaces. These small spaces between clay particles (called *pore spaces*) account for one of the main properties of clay soils: slow water and air movement into and out of the soil.

Water movement through pore spaces causes a soil to be either fast- or slow-draining and determines whether its drainage is good (fast) or poor (slow). Thus soils composed mostly of clay are considered poor-draining. They hold water for long periods of time, and this can be detrimental to plant growth.

Another property of clay soils attributable to the flattened shape of the particles is *plasticity*. This is the property of clay responsible for its being sticky and moldable when wet (Figure 5-2) and hard and cloddy when dry. Since clays are highly plastic when wet, they are also subject to compaction, a packing down of the

soil particles which practically closes up all the pore spaces (Figure 5-3). Compaction is the result of pressure such as walking or driving on the soil when it is wet. It makes water penetration and drainage poor, resulting in poor plant growth due to a lack of air in the soil. Another result of the plastic properties of clay is that it expands when water is added and contracts when it dries out. This tendency to expand and contract results in the surface cracking of the soil that is often evident when clay soils become very dry.

Clay soils also have a great capacity to attract and hold nutrients. This property (called *cation-exchange capacity*) will be discussed in more detail later. Basically, it means that nutrients tend to be stored in clay soils rather than washed away by rainfall or irrigation.

Table 5-2 *Properties of Sand and Clay Soils*

Clay	Sand
Ability to hold a large amount of water	Inability to hold water
Slow movement of air and water in the soil (poor drainage)	Rapid movement of air and water in the soil (good drainage)
Small, numerous pore spaces	Large and fewer pore spaces
Tendency to expand when wet and contract and crack when dry	Does not expand or contract as a result of water
Good ability to attract and hold nutrients for plant growth (high cation-exchange capacity)	Poor ability to attract and hold nutrients for plant growth (low cation-exchange capacity)
Tendency to compact under pressure when wet	No tendency toward compaction
Plastic when wet and highly moldable	Not plastic when wet

Figure 5-2 The plasticity of clay allows it to be formed into ribbons when wet. (*Photo courtesy of Vocational Education Productions, California Polytechnic State University, San Luis Obispo, CA*)

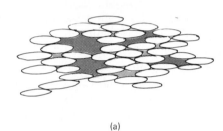

(a)

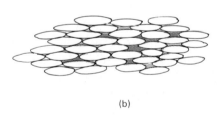

(b)

Figure 5-3 (a) Aggregated clay soil, (b) compacted clay soil. Note the decreased pore space in the compacted soil.

Characteristics of Sandy Soils

Sandy soils (Table 5-2), or *light soils*, can be thought of as possessing the opposite properties of clay soils. Whereas water moves through clay soils slowly via the small pore spaces, in sandy soils it moves freely and rapidly through large pore spaces. Sandy soils do not retain moisture well; nor do they expand and contract, become sticky or cloddy, or compact readily. The cation-exchange capacity of sandy soils is quite low, so nutrients wash away readily when water drains through.

The difference between the properties of clay and sand particles is due mainly to the larger size of sand particles and their rounded or irregular shape. Being irregularly shaped, they do not pack tightly (compact). Nor do they slide easily past one another when wet, as clay particles do. Consequently they lack the plastic properties inherent in clay soils.

Characteristics of Loam Soils

Loam soils possess qualities of both sand and clay proportional to the relative amount of clay or sand particles in the loam soil. They are considered to be ideal for most plants because they do not have the extreme properties of either sands or clays. For example, clay soils may become waterlogged and unworkable for many days after rain, and sandy soils hold so little water that irrigation must be applied frequently. However, a loam soil will absorb a large quantity of water and not require frequent irrigation, yet it will dry enough to be workable in a much shorter time than will a clay soil. Loams also hold more nutrients than do sandy soils; however, they are not as prone to becoming cloddy or compacted as clay soils.

Organic Matter in the Soil

In addition to the mineral components of soil such as sand and clay, organic matter is also a valuable component. In fact plants can be grown in pure organic matter quite satisfactorily without any inclusion of sand or clay. Organic matter in the soil is the result of vegetative matter such as leaves and roots and animal products such as manure returning to the soil and decomposing. Though organic matter is present in most soils in relatively small quantities (a few percent), it greatly influences the properties of the soil. It commonly, for example, is responsible for half of the cation-exchange capacity of a soil. In sandy soils organic matter increases water- and nutrient-holding capacity; in clay soils it improves drainage and air movement and decreases compaction and plasticity. Its net effect, then, is improvement of any soil to which it is added.

One of the most important effects of organic matter on clay soils is to cause *aggregation*. Clay particles have a tendency to lay very close together and exclude air. When organic matter decomposes, the resulting *humic acid* acts as a glue and causes the clay particles to clump together loosely (Figure 5-3). Since the small clay particles are then clumped together, they form much larger particles which function more like sand. The effect of this aggregation is that since the soil has larger pore spaces, water and air movement is improved, resulting in better plant growth. When clay soils are worked while wet, these aggregates are broken apart, resulting in a poorer soil. For this reason, clay soils should never be disturbed until they are dry enough to let the soil crumble easily in the hand.

Another beneficial effect of soil organic matter is that it decreases the tendency of soils to crust. *Crusting* refers to the formation of a thin, hard soil layer at the soil surface which impedes the movement of water and air into and out of the soil and may physically impede the germination of seed. Crusting is most common in soils that are deficient in organic matter and is seldom a problem where careful soil management has regularly returned organic matter to the soil.

Cation-Exchange Capacity

The cation-exchange capacity (CEC) of a soil is the relative capacity of the soil to attract and hold nutrients (cations) on the surface of the soil particles. Soil and organic matter particles have negative charges; since particles with opposite charges attract each other, they are able to attract positively charged ions from the soil water and hold them against the surface of the soil particle (Figure 5-4). This process is called *adsorption*. In soils with a poor CEC such as sand, the number of cations each sand particle can attract is quite low, due largely to the small surface area of each particle. Thus, when nutrients are added to the soil, only a small amount can be held by the soil particles. The rest remain dissolved in the soil water and are washed away readily with rainfall or irrigation.

In soils with a high cation-exchange capacity such as clays or those high in organic matter, a very large quantity of nutrients can be held by each soil particle. Thus, when a fertilizer such as potassium is added, a large amount of the nutrient will be held in the soil. These adsorbed nutrients cannot be leached away by watering; instead, they are stored in the soil for future use by plants. This *buffering capacity* makes the soil easier to manage and means that chemicals added to the soil will affect the soil less readily than in soils with a poor CEC.

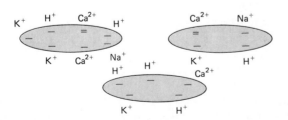

Figure 5-4 Negatively charged clay particles attract and hold cations, providing a reservoir of nutrients.

Understanding cation-exchange capacity is important in fertilizer application. Fertilizers need to be applied less frequently to soils high in cation-exchange capacity because of their nutrient storage capabilities. A high cation-exchange capacity is therefore a desirable soil quality.

Of the three commonly applied fertilizer elements, nitrogen, phosphorus, and potassium, only nitrogen and potassium leach readily in soils that are low in CEC and thus must be applied frequently. Fertilization can be reduced in high CEC soils if fertilizer containing nitrogen in an ammonia form, such as urea or ammonium nitrate, are used.

Air and Water in the Soil

Varying amounts of air and water are contained in any soil. Most higher plants require that plant roots be in contact with adequate amounts of both water and air. Thus a good soil will provide a balance between water and air. Whenever either water or air is lacking in the soil, the plant will die. In a sandy soil, which tends to provide abundant quantities of air but does not retain water well, lack of water commonly restricts plant growth. On the other hand, in a heavy clay soil where water drainage is poor, plant growth may be restricted by a lack of adequate air for root respiration.

In any soil the pore spaces are filled with either air or water. At any given time the percentage of pores filled by water or air is determined by the size of these pores and the relative abundance of either air or water. Therefore the balance between air and water is partially determined by the soil itself (pore size and number) and partially determined by the quantity of water supplied.

Pore size affects the air/water ratio in the soil because small pores hold water better than do large pores. When water is added to the soil, it has a tendency to move downward in response to gravity. It is a force called *adhesion*, the attraction of solid materials (soil particles) to liquids, which holds water in the soil rather than allowing it to flow through. Adhesion can be demonstrated by dipping a marble in water and observing the thin film of water that adheres to it. At the same time water is adhering to soil particles, it is also *cohering* to other water molecules. *Cohesion* is illustrated by the droplet of water that forms at the bottom of the marble after it is removed from the water (Figure 5-5). This droplet is composed of a layer of water molecules adhering directly to the marble and another layer cohering to the first layer of molecules. When too many layers of water molecules cohere to the first layer, the force holding them together is not strong enough to support the weight of the

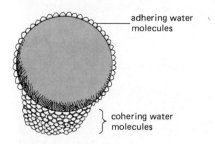

Figure 5-5 Adhesion and cohesion of water molecules around a marble show how water is held in the soil.

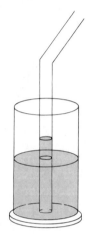

Figure 5-6 Capillary action moves water upward in this straw in the same way that water moves upward in the soil.

droplet and the droplet will fall from the marble. The same phenomenon occurs in soil. If a pore space is small, a greater quantity of water will remain, due to the combined forces of adhesion and cohesion.

Using this information and knowledge of the size of pore spaces found in clay and sand, it is easily seen why clay has a higher water-holding capacity than sand. The presence of a large number of small pores in clay, as compared to the fewer but larger pores in sandy soil, explains the difference.

Water, in addition to moving downward in the soil, may also move sideways or upward. Water moving in these directions is due to the combined forces of adhesion and cohesion and is called *capillary water.* Capillary water is analogous to water that moves upward in a straw placed in a glass of water (Figure 5-6). In this case the water adheres to the sides of the straw, and then the water molecules cohere to each other, thus creating a column of water that extends above the level of water in the glass. In the soil, water moves upward in a column through the pores to a root through which the plant absorbs it or to the surface, where it evaporates. Through the process of capillary water movement, water that has not been utilized by the plant or

has evaporated is constantly replaced by water from below. Capillary action illustrates the importance of watering deeply rather than applying only a small quantity of water to the soil surface.

If water is added to the soil through rainfall or irrigation the soil will eventually become completely saturated to the point that it can hold no more water. When this point is reached, and after any excess has drained away, the soil is said to be at *field capacity.* As time passes and water evaporates or plants absorb the water in the soil, the water content drops until there is no longer any water left that plants can use. At this point, called the *permanent wilting point,* the only water left in the soil is *hygroscopic water,* which is too tightly held to the soil particles for plants to remove. Thus, *available water* for plant growth is that water between field capacity and the permanent wilting point.

Soil Layers

Most soils possess distinct layers that affect plant growth. Soil scientists divide the soil into layers called *horizons.* The topmost horizon is called the A horizon which also corresponds with the *topsoil.* In undisturbed soil the topsoil may extend to a depth of several feet or may be as shallow as a couple of inches. The A horizon is characterized by relatively high levels of organic matter, a darker color than the deeper horizons, and generally higher fertility. The topsoil layer is the one that is usually involved in soil preparation activities and that provides the best substrate for plant growth.

Below the A horizon lie the B and then the C horizons which are collectively referred to as the *subsoil.* These horizons are low in organic matter and relatively infertile. They are progressively lighter in color than the topsoil. Roots may extend into the B horizon but, except for large woody plants, rarely extend into the C horizon.

Unfortunately, home gardeners are often in the position of trying to garden in areas where developers have removed the topsoil during construction leaving only subsoil. Because of the inherently infertile nature of subsoil, the gardener will have to made extensive use of soil amendments and careful soil management practices in order to garden successfully. With time and careful management even the most infertile subsoil can produce a beautiful garden.

For the gardener, the most important layers are the topsoil and the subsoil (Figure 5-7). In addition to the natural topsoil and subsoil layers, another layer called a *hardpan* (Figure 5-8) may occasionally be present. A hardpan is a layer of compacted soil that slows

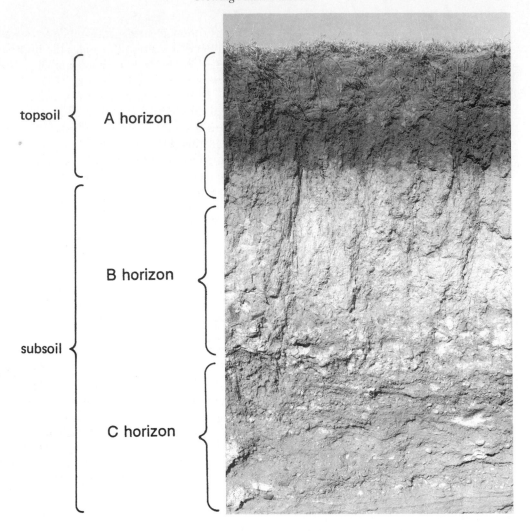

topsoil { A horizon

subsoil { B horizon

C horizon

Figure 5-7   Topsoil and subsoil layers. (*USDA photo*)

or stops the movement of water and air and the growth of roots into lower layers. This results in the total displacement of air in both large and small pores, with consequent damage to plant roots due to the absence of air. When air is lacking, roots become susceptible to

several disease organisms present in the soil and will be weakened or killed.

Hardpans may occur naturally or as a result of heavy equipment moving over the soil. If a hardpan is present, the best solution is to dig through it with a

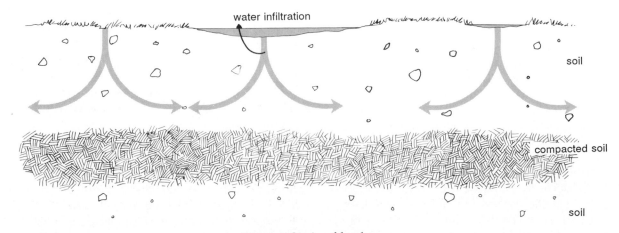

water infiltration

soil

compacted soil

soil

Figure 5-8   A soil hardpan.

posthole digger or, if this fails, to construct raised planters. The presence of a hardpan and accompanying poor drainage can be identified by filling a planting hole with water and waiting for it to drain away. If the water remains after a half-hour, the hardpan is still present.

Erosion

Erosion is the movement of soil particles from one place to another and is a process that is detrimental to the fertility of the soil. Water or wind are usually the agents of the movement with the soil either being washed away by heavy rain or improper irrigation or blown away by wind.

Though erosion is a more serious problem for commercial agriculture, where it results from improper soil management practices, the home gardener should also be aware of possible erosion problems. Erosion is most likely to be a problem when gardening on steep banks where water erosion will result in movement of soil from the top of the bank to the bottom. In order to minimize erosion, the gardener should be sure that appropriate plant materials are used to cover the bank. Plants with extensive, fibrous root systems and spreading growth habits are useful for controlling erosion. When planting a bank, the gardener should disturb the soil as little as possible. Preparing individual planting holes for shrubs or groundcovers is, for example, preferable to rototilling the entire bank. Where preparation of the entire bank is necessary as in the case of seeding a grass, planting should proceed immediately after soil preparation, and the soil surface should be mulched to stabilize the soil until the grass is sufficiently established to hold it.

Even where banks are minimally disturbed during planting, mulches or commercially available materials such as jute netting should be used to minimize erosion. Irrigation systems used on banks should be carefully matched to the water infiltration rate of the soil so that run-off does not occur. This means that systems will have to deliver water slowly so that the water will soak into the soil rather than run down the slope carrying soil with it.

Where extensive bank areas are to be planted, commercial hydroseeding is often employed. *Hydroseeding* is a process in which the seed is mixed with water, fertilizer, and a soil stabilizer and then sprayed on the bank. Hydroseeding is an effective method of establishing erosion-preventing plants on steep banks.

Soil Organisms

The soil is inhabited by a vast and ever-changing population of small, often microscopic, plants and animals. These organisms affect the soil and the higher plants growing in it in a number of ways. They improve drainage and aeration, assist in the decomposition of organic matter, aid in plant nutrient and water uptake, cause diseases, or feed directly upon plants, insects, or pathogens. Although some are harmful, causing plant injuries or diseases, most are beneficial and form an essential part of the soil complex. Included among soil-inhabiting organisms are such animals as earthworms, insects, and nematodes and such microscopic plants as mycorrhizae, algae, fungi, and bacteria.

Perhaps to the home gardener the most interesting, beneficial animal in the garden soil is the common earthworm. These animals feed by passing soil containing organic matter through their bodies, where it is subjected to digestive enzymes and a grinding action. The material that has passed through the earthworm's body is called a *casting*. Castings are much higher in organic matter and nutrients than the surrounding soil. In addition, the earthworm burrows increase soil aeration and drainage.

Earthworms are most abundant in moist clay or loam soils rich in organic matter. In cold climates they are least abundant in the spring, with populations increasing all summer. Since earthworms are killed by early fall freezes in bare soil, they should be protected by surface mulches to maintain high populations in the home garden. After the first cold snap of the year, earthworms begin to acquire cold tolerance and migrate deeper into the soil where the temperature is more favorable.

Soil pH

The pH of a soil is a measure of its acidity or alkalinity. It is measured on a logarithmic scale that runs from 0 to 14, although a soil pH below 4 or above 9 is uncommon (Figure 5-9). A pH of 7 is neutral, whereas a pH above 7 is alkaline and below 7 is acid. The optimum range for the majority of plants is 6.5 to 7.0. Many plants adapted to a lower pH are called "acid-loving" while other plants are tolerant of alkaline soils.

A soil pH that is not optimum may be corrected by adding materials to the soil to change the pH. Since the pH scale is a logarithmic one, a pH of 6.0 is 10 times more acidic than a pH of 7.0; a pH of 5.0 is therefore 100 times more acidic than a soil with a pH of 7.0. Thus relatively large quantities of acid- or alkaline-forming materials must be added to cause even a small change in pH, especially in clay soils.

Lime is one material utilized to raise soil pH. It is available in several different grades and types, all of which will make the soil more alkaline. Ground limestone or calcitic limestone ($CaCO_3$)is the most common and the least expensive type available. Dolomitic limestone ($Ca \cdot MgCO_3$), although slightly more expensive than

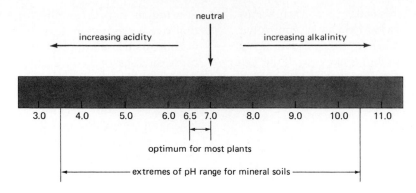

Figure 5-9   The pH scale of soil.

calcitic limestone, performs the added function of providing magnesium, a necessary nutrient lacking in some soils. Whichever type of lime is used, it will take between 3 and 6 months to accomplish the desired pH change. This is due to the slow speed with which lime dissolves. To speed up the pH change as much as possible, buy finely ground lime, and mix it thoroughly into the soil. Surface applications of lime are only minimally effective in raising the pH below the soil surface.

Acidification of the soil is generally accomplished through the use of the chemical sulfur or aluminum sulfate or by using organic amendments such as oak leaves, pine needles, or sphagnum moss which are acid-forming. When using organic amendments, large quantities are needed, and the change is very slow. Many fertilizers such as ammonium sulfate and potassium sulfate are also acidic and can be used to acidify soils, though large amounts cannot be used due to the danger of damage to plants from excess fertilization. Sulfur or aluminum sulfate also changes soil pH but only after the use of relatively large quantities. Due to the high cost of acidifying materials, large-scale soil acidification is generally not done. Instead the soil immediately surrounding an acid-loving plant is adjusted. Agricultural sulfur is the most commonly used acidifying agent and is less costly than aluminum sulfate; however, a greater quantity is required than of the latter, and the effect is slower. Aluminum sulfate is more acid-forming than sulfur but must be used carefully since, if overused, the aluminum can become toxic to plants. Soils with a high buffering capacity will require larger quantities of pH-adjusting materials than those with a low buffering capacity.

Soil Improvement

Soil Testing

The best way to determine the pH and nutrient content of a soil is to have it professionally tested. The results can then be used in the soil improvement program. In most of the United States a soil test will be performed for a small charge by the Cooperative Extension Service. The usual procedure is to submit a soil sample to the county agent, who will forward it to the soil testing laboratory at the university. In other areas commercial soil testing laboratories must be used.

The most important step in soil testing is taking the sample. The sample must be representative of the area being tested or it will be meaningless. To obtain a representative sample, a combined sample is taken from the area. The best technique is to establish a grid pattern over the area being sampled (Figure 5-10) and then take a small sample from each grid intersection. Each individual sample can best be taken by digging a hole 6 to 8 inches (15–20 cm) deep and then skimming off a small slice of soil (Figure 5-11) from the top to the bottom of the hole. At least a dozen samples should be combined in a plastic bucket and mixed thoroughly, and then a 1-pint (0.5 liter) sample should be taken from this.

Soil Improvement through Addition of Amendments

A soil amendment is a material added to soil to improve physical properties such as drainage, CEC, aeration, and water retention. It may or may not supply nutrients, depending on the amendment. The addition of amendments for soil improvement is a fundamental gardening practice and will greatly influence gardening success.

The best amendment is chosen by considering several factors including the problem to be corrected and its seriousness, properties of the amendments available, and cost. Often the best amendment will not be practical because of cost, and a less expensive amendment will have to be used.

The most important rule to remember about the use of any soil amendment is that amendments must be used in large quantities to achieve the desired benefits.

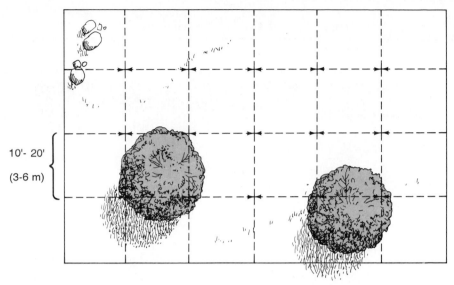

10'- 20'

(3-6 m)

Figure 5-10 Following a grid pattern for soil testing will assure a representative soil sample.

Figure 5-11 Slicing off an undisturbed section of soil for testing.

The addition of a small amount will not make any appreciable difference in soil properties. The amount of amendment required will depend on the amendment chosen and the severity of the soil problem. As a general rule, an amount of amendment equal to 25 percent of the volume of the soil should be applied.

The easiest way to apply an amendment is to spread it over the surface of the soil to a depth of several inches (8–10 centimeters) and then spade or rototill it into the soil. The thickness of the amendment layer needed can be determined by comparing the soil depth to be spaded or rototilled to the amendment depth. For example, if a 2-inch (5-centimeter) amendment layer is used and the soil is spaded 8 inches (20 centimeters)

deep, the amendment added equals approximately 25 percent of the soil volume. If large quantities of amendments are added, the process is best repeated several times, with about a 2-inch (5-centimeter) layer of amendment incorporated each time.

Each type of soil amendment has properties that must be evaluated when determining its suitability. An important consideration is the quantity of nitrogen it contains and how rapidly it decomposes. If the amendment decomposes rapidly and contains little nitrogen, nitrogen may actually be taken from the soil to supply the nitrogen needed for the process of decomposing the amendment. This nitrogen will be released when decomposition is completed, but in the meantime nitrogen deficiency symptoms will often occur in plants growing in the area. This effect will be intensified if the amendment is finely ground and the soil is warm and moist, since these conditions are conductive to the most rapid decomposition of the amendment and thus the greatest nitrogen deficit. To counteract this, a fertilizer containing nitrogen should be added with the amendment. Amendments that take up nitrogen excessively include sawdust and wheat straw.

Other considerations are salinity and ash content. Certain amendments are high in chemicals such as sodium and ammonium salts. This may not be a serious problem in the eastern United States and Canada where abundant rainfall occurs; however, in the drier western parts of the hemisphere, rainfall is often insufficient to wash away the salts, resulting in salt toxicity to plants. Manure often contains high levels of ammonia salts and therefore will "burn" plants if used in large quantities.

Ash content refers to the amounts of mineral matter present in the amendment. In organic amendments

Table 5-3 Common Soil Amendments

Amendment	Uses	Nitrogen requirement	Comments
Inorganic			
Perlite	Work into heavy soils to improve drainage and aeration and reduce compaction	None	Expensive; lasts a long time; best for planters; lightweight floats
Sand (coarse)	Same as perlite	None	Less expensive than perlite but required in large quantities for substantial improvement of clay soils; heavy; do not use beach sand because of salts
Topsoil (loam)	Work into very sandy soils to improve water and nutrient holding	None	Inexpensive but heavy; may be contaminated with weed seed, nematodes, fungi, and other undesirable organisms
Vermiculite	Improves water and nutrient holding in sandy soils	None	Expensive; lightweight; if soil is worked frequently, the structure of the vermiculite will be damaged, reducing its effectiveness
Scoria (lava rock, cinder rock)	Same as perlite	None	Less expensive than perlite; medium weight
Organic			
Bark	Mix with heavy soils to improve drainage and aeration and decrease compaction; some improvement of water- and nutrient-holding capacity in sandy soils	Add about 1 lb (0.5 kg) of actual nitrogen per cubic yard (cu. meter)	Usually inexpensive; breaks down fairly slowly; finely ground bark is best
Compost (also called humus)	Improves both clay and sandy soils; contains some nutrients	None	Variable; may have a high ash content
Leaf mold	Similar to compost but different in that it is derived primarily from leaves	None	May acidify soil slightly
Manure (composted)	Improves both clay and sandy soils; contains some nutrients	None	May be high in both salts and ash; may have a strong odor; inexpensive
Michigan peat (Hypnum peat, muck)	Improves both clay and sandy soils if used in large quantities	None	Inexpensive, but large quantities are needed; high ash content; not readily available west of the Rockies
Redwood compost	Best for clay soils, with some improvement of sandy soils	Add about ¼ lb (0.1 kg) of actual nitrogen per cubic yard	Inexpensive, however, does not hold water or nutrients well
Sawdust	Improves clay soils	Add about 1 lb (0.5 kg) of actual nitrogen per cubic yard (cu. meter)	Very inexpensive. Breaks down so rapidly that nitrogen deficiencies may develop despite the addition of nitrogen; improves soil for a relatively short time. If used, mix with a better amendment. Some toxicity may occur if used fresh.
Sphagnum peat moss	Excellent for clay or sandy soils; high water and nutrient capacity	None	Expensive, but less is required; hard to wet when fully dry; acidifies soil
Sewage sludge (Milorganite)	Improves both sandy and clay soils; adds some nutrients	None	Inexpensive; may have a high ash content and a strong odor; do not use in the vegetable garden if root vegetables are grown
Rice hulls	Improves both clay and sandy soils	Add about ¼ lb (0.1 kg) of actual nitrogen per cubic yard (cu. meter)	Inexpensive, holds water well

the presence of large quantities of ash is usually undesirable, as ash does not improve the soil. Leaf molds, manures, and compost often contain large amounts of ash.

Inorganic amendments such as vermiculite and perlite are usually too expensive to be used as amendments in garden soils, although they are commonly used in potting soils. Sand is sometimes used in garden soils but must be used in large quantities to be effective. When they are used, the effects are often quite permanent, whereas organic amendments must be added periodically. Table 5-3 contains a list of soil amendments, their characteristics, and uses.

Composting

Compost (Figure 5-12) is one soil amendment that can be made rather than bought and which is excellent for soil improvement. It is composed of partially decomposed plants and is made by piling garden refuse such as weeds, grass clippings, fallen leaves, and plant clippings in a pile to rot. The piling helps keep in moisture and heat generated by the decomposition process and speeds the normal rotting process.

A compost pile can be relatively simple or quite complicated. In its most basic form, the plant refuse can be simply piled in an out-of-the-way spot and allowed to decompose slowly without any further attention. With this method, compost will be made very slowly, and three to four months may pass before appreciable compost forms.

To form compost at a faster rate, a compost heap must be built. A wire or wooden bin should be constructed as shown in Figure 5-13. A layer of refuse should then be spread in the bottom 6 to 9 inches (15–23 centimeters) deep, sprinkled with a handful of high-nitrogen fertilizer, covered with a thin layer of soil to supply microorganisms, and wet slightly. The layering and wetting process should be repeated until the bin is full. The composting material should be kept damp by sprinkling with water as necessary and turned and mixed with a pitchfork ever four weeks to ensure even decomposition of the entire pile. If not turned, the pile will generate heat and decompose mainly in the center, with the outer layers rotting only very slowly. Compost can be considered ready for use when the material is evenly dark-colored and decomposed to the point where the materials are not recognizable. It is then used for soil amending or mulching.

Figure 5-12 Compost. (*USDA photo*)

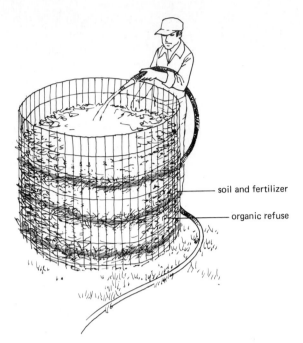

soil and fertilizer

organic refuse

Figure 5-13 Cross section of a compost pile built in layers for rapid decomposition. Keeping the pile moist speeds the process.

Among the other materials that can be composted are vegetable peelings and rotten vegetables, coffee grounds, and other kitchen refuse other than meat scraps. Diseased plants should not be composted since there is a chance the disease-causing organisms will survive despite the heat of decomposition (often up to 150°F) (66°C). Woody materials such as twigs and small branches can be used but will break down very slowly. Their decomposition can be hastened by chopping them into small pieces before adding them to the pile.

Plant Nutrition

All plants obtain a large number of mineral nutrients from the soil. Since many nutrients are known to be utilized by plants, they are grouped for ease of discussion into two groups based on the relative quantities utilized by the plant. The first group, called *macronutrients*, includes those nutrients used in fairly large quantities by plants. The macronutrients include nitrogen (N), phosphorus (P), potassium (K), calcium (Ca), magnesium (Mg), and sulfur (S). The second group, called *micronutrients*, or trace elements, is equally important but is used by plants in small quantities. The most important micronutrients are iron (Fe), copper (Cu), zinc (Z), boron (B), molybdenum (Mb), chlorine (Cl), and cobalt (Co). An example of the difference in

quantity of nutrients used by a plant is that an acre (0.4 hectare) of corn might easily remove 150 pounds (60 kilograms) of nitrogen from the soil while removing less than an ounce (28 grams) of cobalt.

For optimum plant growth to occur, all of the nutrients needed by that plant not only must be present in the soil but must be present in a form a plant can utilize.

Macronutrients

Because macronutrients are those nutrients used by plants in the largest quantities, they are the primary ingredients in most garden fertilizers. Of the six macronutrients plants obtain from the soil, calcium, magnesium, and sulfur are generally present in large enough quantities to be adequate for many years and seldom need to be added when fertilizing. For cases where these nutrients are lacking, a list of fertilizer sources is given in Table 5-4. Phosphorus, potassium, and particularly nitrogen are frequently not present in sufficient quantities for optimum growth; they are therefore the main macronutrients supplied in fertilizers.

Nitrogen

When nitrogen is lacking, plants grow slowly and lose their deep green color. The application of nitrogen will often result in a flush of leafy growth and the return of a deep green color within a week or two. Anyone who has applied nitrogen fertilizers to a lawn can vouch for the resulting fast growth and increased necessity of frequent mowing.

Since nitrogen stimulates rapid vegetative growth, it should be applied frequently to plants grown for their foliage such as turfgrass, leafy vegetables, and indoor foliage plants and to all young plants to encourage the rapid development of leaves which photosynthesize sugars needed to fuel subsequent flowering and fruiting. High levels of nitrogen fertilization do, however, have the disadvantage of stimulating leafy growth often instead of flowers and fruit. A common example is the vegetable gardener who uses too much high-nitrogen lawn fertilizer on tomatoes. The result is a tomato plant of astounding size with healthy, dark green leaves that never produces many flowers or fruit. In this case a lower N fertilizer or smaller amounts of lawn fertilizer should have been used once the plants had produced sufficient vegetative growth.

Another disadvantage of high nitrogen fertility levels is that succulent new growth of shrubs and trees may continue too late in the summer, resulting in cold injury when winter arrives. Thus, when nitrogen fer-

Table 5-4 *Chemical Soil Amendments for pH Adjustment and Micronutrient Addition*

Name	Chemical formula	Speed of effectiveness	Use
Limestone	$CaCO_3$	Slow	Raise soil pH
Hydrated lime	$Ca(OH)_2$	Rapid	Raise soil pH
Gypsum or calcium sulfate	$CaSO_4$	Medium	Add calcium, which displaces sodium in soils and improves drainage
Sulfur	S	Slow	Add sulfur and lower soil pH
Epsom salts	$MgSO_4 \cdot 7H_2O$	Rapid	Add magnesium
Aluminum sulfate	$Al_2(SO_4)_3$	Rapid	Add sulfur and lower soil pH
Calcium nitrate	$Ca(NO_3)_2 \cdot 2H_2O$ (analysis 15-0-0)	Rapid	Add calcium and nitrogen and raise soil pH
Ammonium sulfate	$(NH_4)_2SO_4$ (analysis 20-0-0)	Rapid	Lower pH sharply, add sulfur and nitrogen
Magnesium sulfate	$MnSO_4$		Lower pH and add magnesium
Iron sulfate	$FeSO_4$		Lower pH and add iron
Chelated iron	9–12% Iron		Add iron in absorbable form
Borax	$Na_2B_4O_7 \cdot 10H_2O$		Add boron
Copper sulfate	$CuSO_4$		Add copper and lower pH
General micronutrient fertilizers	Contain a combination of most essential micronutrients		Add deficient micronutrients

tilizers are used, the consequences must be assessed and the quantity and timing of each application regulated to maintain adequate, but not excessive, nitrogen in the soil.

A characteristic of nitrogen that greatly affects fertilizing rate and frequency is that nitrate nitrogen is not bound to soil particles as are many other nutrients. Virtually all nitrate nitrogen added to soil will remain dissolved in the soil water. As a consequence, it readily leaches from the soil. This means that whenever heavy rainfall follows nitrogen applications, much of the nitrogen will be leached away and lost. The drainage characteristics of the soil should thus be assessed when deciding how often to fertilize with nitrogen.

In a sandy soil, nitrogen will be leached when rainfall occurs, so slow-release fertilizers, or frequent small applications of quick-release (nitrate) fertilizers rather than a single large application are best. Slow- and fast-release fertilizers and the forms of nitrogen contained in fertilizers and used by plants are discussed under "Fertilizers." In slower-draining clay soil loss of nitrogen is slower and the benefits from slow-release fertilizers are less valuable.

Another consequence of the lack of nitrate adsorption is that virtually all nitrates will remain in the soil solution. If too much fertilizer is applied, severe root damage may result. With nutrients that are adsorbed, burning is less likely since some nutrients will be removed by adsorption to soil particles.

Nitrogen can be added to the soil in many different forms, each having characteristics that determine its suitability for a particular usage. Consult Table 5-5 for characteristics of a number of common nitrogen sources.

Phosphorus

Phosphorus, the second macronutrient, influences many plant functions including flowering and fruiting, root development, disease resistance, and maturation.

Phosphorus behaves very differently from nitrogen in the soil. It is insoluble in the soil water and, when added to the soil, reacts readily with aluminum, iron, calcium, and other elements, forming various compounds unusable by plants. Consequently, relatively little phosphorus is ever in the soil water at any one time, and there is negligible leaching and little danger of plant damage from overapplication. The problem, then, with phosphorus is to maintain enough of it in the soil water for plant use. This can best be done by maintaining the pH between 6.0 and 7.0, maintaining high amounts of organic matter, and wherever possible, placing the fertilizer in bands next to plants to minimize soil contact.

Since phosphorus is insoluble, it will remain where it is placed and will not leach through the soil. Therefore, it must be supplied in the root zone of the plant. Surface applications of phosphorus will have little effect on deeply rooted plants.

An important, rather specialized use of phosphorus is its use as a transplant or starter fertilizer. (Starter fertilizers are characteristically low in nitrogen and potassium but high in phosphorus.) Starter fertilizers

Table 5-5 *Garden Fertilizers and Their Characteristics*

Name	Analysis	Speed of nutrient release	Effect on pH
Inorganic and chemically manufactured			
Ammonium sulfate	20-0-0	Fast	Very acid
Sodium nitrate	15-0-0	Fast	Alkaline
Potassium nitrate	13-0-44	Fast	Neutral
Ammonium nitrate	33-0-0	Fast	Acid
Urea	46-0-0	Fast	Slightly acid
Monoammonium phosphate	11-48-0	Fast	Acid
Diammonium phosphate	18-46-0	Fast	Acid
Treble superphosphate	0-40-0	Medium	Neutral
Superphosphate	0-20-0	Medium	Neutral
Potassium chloride	0-0-60	Fast	Neutral
General balanced granular	10-10-10, 5-10-10, and so on	Various	Various
Organic			
Bone meal, steamed	2-27-0	Slow	Alkaline
Bone meal, raw	4-22-0	Slow	Alkaline
Hoof and horn meal	13-0-0	Slow	
Dried blood	12-0-0	Medium	Acid
Sewage sludge	2-1-1	Medium	Acid
Activated sewage sludge (microorganisms added)	6-5-0	Medium	Slightly acid
Cattle manure (fresh)	0.5-0.3-0.5	Slow	Slightly acid
Chicken manure (fresh)	0.9-0.5-0.8	Slow	Slightly acid
Horse manure (fresh)	0.6-0.3-0.6	Slow	Slightly acid
Sheep manure (fresh)	0.9-0.5-0.8	Slow	Slightly acid
Swine manure (fresh)	0.6-0.5-0.4	Slow	Slightly acid
Wood ashes	0-2-6	Medium	Alkaline
Rabbit manure (dried)	2.25-1-1	Slow (unless steamed)	Slightly acid
Cattle manure (dried)	2-3-3	Slow (unless steamed)	Slightly acid
Swine manure (dried)[a]	2.25-2-1	Slow (unless steamed)	Slightly acid
Seaweed	1.7-0.75-5	Slow	
Cottonseed meal	6-2-1	Medium	Acid

[a]The nutrient content of the dried, processed manures not listed can be approximated as four times that of the fresh manure.

are applied as a liquid at the time of transplanting and can bring about reduced transplanting shock and a resulting quick resumption of growth. It is not uncommon for tomatoes that received a starter fertilizer to produce fruit several weeks sooner than a similar plant that did not receive starter fertilizer.

Potassium

The third main fertilizer element is potassium, essential for starch formation, movement of sugars in the plant, formation of chlorophyll, and flower and fruit coloring.

Potassium (or potash) is abundant in many soils. However, most of the potassium is not available for plant use, necessitating the addition of potassium in usable forms. Potassium, unlike phosphorus, leaches fairly readily and may be wasted if heavy rainfall occurs after application. The problem of leaching is less severe in clay soils since potassium will be absorbed or held by soil particles and thus prevented from leaching. This potassium can then be drawn upon and used by plants at a later time.

Heavy applications of this nutrient are undesirable because the plants will absorb more potassium than they need if it is available. This results in fertilizer waste with no benefits in plant growth or appearance, or what is called *luxury consumption*. Moreover, overdoses of potassium will injure or "burn" plants (overdoses of nitrogen are even more likely to do so).

Potassium fertilizers can be used most efficiently by being added frequently in small amounts rather than in a single, larger application.

Micronutrients

Although micronutrients are just as important to plant growth as macronutrients, they are needed in such small quantities that soils generally contain enough to supply most plants. Occasionally micronutrients are lacking. Frequently, however, a micronutrient deficiency occurs even though large quantities of the nutrient are in the soil. At certain pH levels, nutrients react chemically with other elements in the soil, forming compounds that cannot be utilized by the plant. Each nutrient has a definite pH range within which it is most available to plants. In most soils, as can be seen in Figure 5-14, the majority of nutrients is readily available between pH 6.5 and 7.0. If pH shifts out of this range, certain elements are likely to become unavailable. If micronutrient deficiencies occur consistently, check soil pH.

Correcting micronutrient deficiencies involves correcting soil pH, if practical, and applying the nutrient to foliage and roots. When applying micronutrients to the soil, it is best to purchase the nutrient in *chelated* form. Although more expensive, it is formulated to be unreactive with other elements in the soil and will remain available to the plant. Foliage applications of micronutrients (as discussed under "Fertilizer Application") will speed the elimination of deficiency symptoms. Optimum soil moisture and warm temperature also decrease chances of microelement deficiencies.

Fertilizers

Fertilizers are compounds which, when properly used, provide the necessary nutrients for plant growth. In common usage, they are divided into organic and inorganic types.

Organic fertilizers are derived from the decomposition of plant and animal products and include blood meal, bone meal, manure, and sewage sludge. Organic fertilizers have the advantage of acting as soil amendments

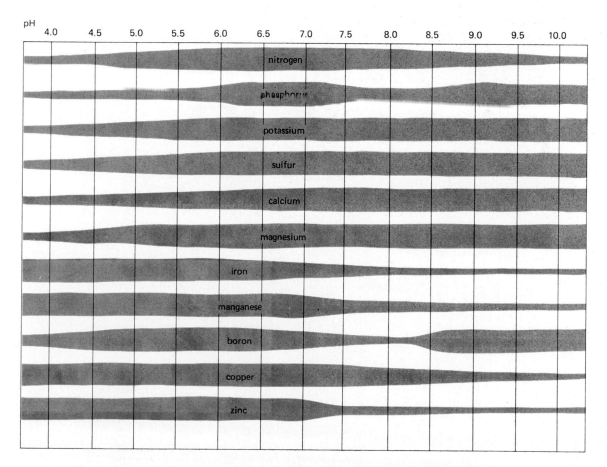

Figure 5-14 The general relation of pH of soil to the availability of nutrients to plants.

in addition to supplying nutrients. Since nutrients from organic fertilizers are released slowly during decomposition, burning of plants caused by too much fertilizer dissolved in the soil solution at one time is less likely. In general, organic fertilizers are much more expensive than chemical fertilizers on the basis of the amount of nutrients supplied. Additionally, since decomposition is slow in cold soils, organic fertilizers are not reliable fertilizers in most areas during winter and early spring.

Inorganic fertilizers are manufactured using raw materials such as natural gas and phosphate rock. They are much more concentrated than organic fertilizers. Some inorganic fertilizers are formulated to release nutrients rapidly so rapid plant response to the fertilizer occurs. Others release nutrients slowly over a period of time. These two types of fertilizers are described as *fast release* and *slow release,* respectively. Synthetic organic fertilizers such as urea generally react the same as inorganic fertilizers though they are more concentrated.

Relatively recently, new products that seek to combine the soil-building attributes of organic fertilizers with the relatively concentrated fertilizer analysis of chemical fertilizers have become available. These products combine a concentrated organic base, rich in hu-mic acid with chemical fertilizers, often in slow-release form. Because they are rich in humic acid, they promote aggregation of clay soils without having to add the quantity of organic matter necessary with other organic materials. Their concentrated fertilizer analysis means that sufficient fertilizer is available to the plants. Thus these products overcome the disadvantages of chemical fertilizers by building the soil and of organic fertilizers by containing sufficient fertilizer.

To understand the difference between organic and chemical fertilizers, it is helpful to understand the nitrogen cycle, defined as the series of transformations that nitrogen undergoes in the environment (Figure 5-15).

Nitrogen comprises about 70 percent of the air. However, plants, with the exception of those in the legume family and a few grasses, cannot use this atmospheric nitrogen. Legumes (beans, peas, peanuts, and others), with the aid of bacteria which live on their roots, are able to enrich the soil by "fixing" atmospheric nitrogen found in the air spaces in the soil.

Most plants can use nitrogen only when it is present in the soil in either the nitrate (NO_3^-) or the ammonium (NH_4^+) chemical form. Though some plants use nitrogen in the ammonium form effectively, most

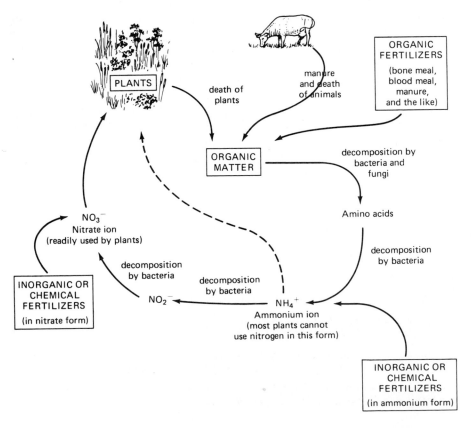

Figure 5-15 The nitrogen cycle, showing the breakdown of nitrogen-containing compounds from organic matter so that nitrogen can be used by plants.

garden plants use nitrogen primarily in the nitrate form. This means that a considerable amount of decomposition must occur before nitrogen (which is added in the form of leaves or blood meal, for example) moves through the decomposition cycle to the point that nitrates are available. This accounts for the slow-release characteristic of organic fertilizers.

If a chemical fertilizer such as ammonium nitrate is added to the soil, part of the nitrogen is already in the nitrate form and part is in the ammonium form. The portion in the nitrate form can be used immediately by the plant. Depending on the plant, the ammonium portion can be either used immediately or converted into nitrates through the decomposition process. This accounts for the rapid plant response to ammonium nitrate and the possibility of burning plants due to high concentrations of usable nitrogen.

It is important to realize that whether the fertilizer is added in an organic form, such as animal manure, or in the form of a chemical fertilizer, such as ammonium nitrate, the plant absorbs the nitrogen primarily as nitrates. The only difference is that the nitrogen is added at a different point in the nitrogen cycle. The main disadvantage of chemical fertilizers is that if organic matter is not added to the soil (by compost, natural plant residues, or other means), the amount of organic matter in the soil will decline, causing a worsening of physical characteristics of the soil.

Fertilizer Forms

Fertilizers are usually available in several different forms—liquids, soluble powders, granules, and tablets.

Liquids are popular primarily because of their ease of application (usually with a hose-end sprayer) and the quick response obtained. These fertilizers often contain micronutrients so that through foliar absorption, micronutrient deficiencies can be alleviated. Since all the nutrients are in a water-soluble form, it is easy to burn plants if too much fertilizer is used. In addition, much of the nitrogen will leach away if heavy rainfall or irrigation occurs after fertilization, and the benefits of the fertilizer will be short-lived.

Soluble powder fertilizers are popular since they dissolve readily in water and can be applied in the same manner as a liquid fertilizer with the same advantages and disadvantages. They are usually a better buy than liquid fertilizers. They often contain a blue dye which will stain hands and clothes but serves to alert the applier who uses a hose-end sprayer that the sprayer is working properly.

The most common and widely used form of fertilizer is the granular form. During the manufacture of a granular fertilizer, the nutrients are compressed into bead-sized particles which are heavy enough to eliminate problems of blowing. Moreover, since the nutrients need not be immediately soluble in water (as in liquid fertilizers), slow-release formulations are available to prolong the release of nutrients and decrease the risk of burning.

Tablet fertilizers, which have long been available for house plants, are becoming increasingly popular for outdoor use. Though more expensive than granular forms, tablets do not have to be measured; one or two tablets are simply dropped in the planting hole. Additionally, tablets release their nutrients slowly over a period of several months up to a year, thus minimizing waste and burning of the root system. Tablets are designed for use in planting holes or pots and do not work well when applied to the soil surface.

Purchasing Fertilizers

Several factors are involved in determining which fertilizer is the best buy: analysis, form, organic or inorganic, slow or fast release, and price.

The analysis of a fertilizer is a statement of the quantity and type of nutrients contained in the fertilizer and is required by law to be printed on every bag of fertilizer. The standard form for the statement of fertilizer analysis is a set of three numbers separated by hyphens, for example, 5-10-20. Each number refers to the percentage of a particular nutrient contained in the fertilizer (Figure 5-16). The first number always refers to the percentage of nitrogen, the second number to the percentage of phosphorus in a chemical form called *phosphorus pentoxide* (P_2O_5), and the third number to the percentage of potassium in a chemical form called *potash* (K_2O). Thus, for the sample analysis above (5-10-20), the fertilizer contains 5 percent nitrogen, 10 percent phosphorus compound, and 20 percent potassium compound. Another way of stating this is that the fertilizer contains 5 pounds of nitrogen, 10 pounds of phosphorus compound, and 20 pounds of potassium compound in every 100 pounds of fertilizer.

Fertilizers are often grouped into three categories based on the relative amounts of nitrogen, phosphorus, and potassium they contain. The first group is called *balanced fertilizers* since all the numbers in the analysis are identical (8-8-8). The second group, called *complete fertilizers,* contains nitrogen, phosphorus, and potassium but in varying amounts (5-10-15). The last category, called *single element fertilizers,* contains only one of the three major nutrients such as urea (46-0-0).

The analysis is the first item to look at in buying any fertilizer since nutrients are the important part of

Figure 5-16 Three bags of fertilizer showing the analysis in large numbers. (*Photo courtesy of Vocational Education Productions, California Polytechnic State University, San Luis Obispo, CA*)

the fertilizer. Other ingredients may be present which make the volume of fertilizer larger and make it appear to be a better bargain. These filler materials may be useful as soil amendments but cannot be considered a good buy when purchased at the price of fertilizer.

Consider both the total quantity of nutrients (by adding the numbers) and the relative amounts of nitrogen, phosphorus, and potassium. If you are buying a lawn fertilizer where the primary nutrient needed is nitrogen, the nitrogen analysis is a prime consideration. For most, a balanced fertilizer that contains about equal proportions of nitrogen, phosphorus, and potassium should be chosen.

Next look at the form of the fertilizer. It is difficult to compare liquid pricewise with either granular or insoluble powder forms due to the weight of the water; however, the latter two can be compared. Powdered fertilizers should be distinctly less expensive than granular forms. If a substantial difference in price does not exist, there is little justification for purchasing a powdered fertilizer.

Whether a fertilizer is organically derived may be an important consideration if you are committed to the exclusive use of organic materials. If an organic fertilizer is selected, you should be prepared to pay a much higher price for the same quantity of nutrients and should not expect a high fertilizer analysis. This is because, by definition, an organic fertilizer will contain bulky organic material in which the nutrients are contained, and shipping costs will often be a substantial component of the price. Either fast- or slow-release fertilizers can be purchased in inorganic form but the slow-release types will be more expensive. Whether or not they are worth the added expense depends on how they are to be used.

Price often becomes the primary consideration in the purchase of fertilizers for home use. In order to determine the best buy, all of the factors discussed above should be weighed relative to the price.

Fertilizer Application

Topdressing

The method of fertilizer application will vary according to the type of plant being fertilized, but the basic rule is that the fertilizer must reach the root zone. For shallow-rooted plants such as turfgrasses, groundcovers, annual flowers, and some shrubs, surface applications, called *topdressings*, followed by rainfall or irrigation will be sufficient to supply nutrients to the root zone. When surface applications are used, the fertilizer should be broadcast around the plant, being careful to remove fertilizer that collects around the stem or upon the leaves.

Preplanting Incorporation

Fertilizers can also be incorporated into the soil along with soil amendments. This *preplant incorporation* method of application has the advantage of getting phosphorus into the root zone of the plants.

Sidedressing

Sidedressing is a form of surface fertilizer application useful on rows of plants such as those found in vegetable gardens and some landscape plantings. Using this method, a narrow band of fertilizer is applied along one or both sides of the row (Figure 5-17) over the root area of the plants but away from the crown, where it could cause burning injury by contact. This banding makes fertilizing long rows of plants relatively easy and avoids the sloppiness and fertilizer waste that can occur with topdressing. It also minimizes contact of the fertilizer granules with the soil and thus helps prevent susceptible nutrients such as phosphorus from reacting with soil components.

For more deeply rooted plants such as large shrubs and trees, surface applications are not effective since the roots are located 18 inches (45 centimeters) or more below the soil surface; most of the phosphorus and a large amount of the potassium will never leach down to the roots. In addition, if surface fertilizer applications are used routinely, surface roots will be encouraged and these will reduce the plant's ability to survive drought.

Needle Feeding

Two methods are commonly employed to place fertilizer in the root zone of trees. The first is the use of a feeding needle, often called a *root feeder*. This apparatus (Figure 5-18) attaches to a garden hose and contains a fertilizer chamber and a pipe with a pointed tip. When the water is turned on, it passes through the fertilizer chamber, dissolving the fertilizer, through the pipe, and out the tip. The water passing through the tip aids the operator in pushing the pipe to the desired depth in the soil. This method is easy and provides both water and available nutrients to the root zone when injected at several points over the root zone of the tree.

Drill Hole

Another effective technique that does not involve the purchase of any special equipment is the drill hole out the root zone of the tree (Figure 5-19). These holes are then filled with a granular fertilizer according to the

Figure 5-17 Sidedressing a row of peas. (*Photo by Rick Smith*)

Figure 5-18 A liquid feed needle. (*Photo by George Taloumis*)

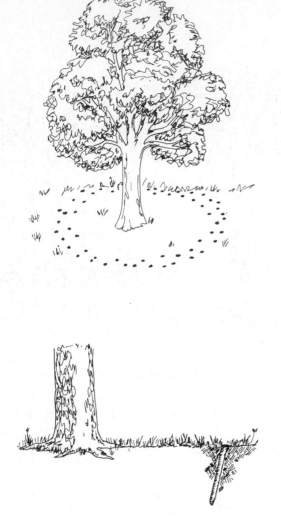

Fertilizer spikes are a modification of the drill hole method. The same results can be obtained using the drill hole method and at less expense.

Foliar Feeding

Another method of supplying small quantities of nutrients to plants is by means of *foliar feeding*. Diluted method, whereby holes are pounded 12 to 18 inches (30–45 centimeters) deep into the ground through-water-soluble fertilizers are sprayed on plant foliage, and the plant then absorbs small quantities through the leaves. Since leaves absorb only small quantities of nutrients, this method is best reserved for micronutrients. The advantage of foliar fertilization is that if deficiency symptoms are obvious, improvement will be evident in a matter of days. In addition, where the micronutrient is likely to become unavailable in the soil, soil application can be supplemented. As a general rule, foliar fertilization does not supply enough nutrients to completely satisfy requirements and should be used in addition to regular fertilizer that is added to the soil, or sufficient liquid fertilizer should be used to wet the soil.

Figure 5-19 The drill hole method for fertilizing trees.

recommended rate and covered with peat moss or topsoil. To be effective, 10 to 20 holes for each inch (2.5 centimeters) of trunk diameter must be made throughout the root zone of the tree. Holes may be made with a crow bar, reinforcing bar, pipe, or auger.

Selected References for Additional Information

Bush-Brown, J., and L. Bush-Brown. *America's Garden Book.* Revised ed. New York: Charles Scribner's Sons, 1980.

*Coastline Community College. *Plant Nutrition.* Fountain Valley, CA: Coastline Community College.

Hausenbuiller, R. L. *Soil Sciences Principles and Practices,* 3rd ed. Dubuque, IA: W. C. Brown, 1985.

Janick, Jules, et al. *Plant Science: An Intro to World Plants,* San Francisco: W. H. Freeman, 1981.

Singer, Michael J., and Donald N. Munns, *Soils: An Introduction.* New York: MacMillan, 1987.

*Indicates a video.

6

Vegetable Gardening

Vegetable gardening is one of the most rewarding activities in home horticulture. It combines the enjoyment of cultivating plants and watching them grow with the pleasure of harvesting fresh produce at the peak of flavor and maturity. Most people will argue that "home grown" tastes much better than "store bought," and there are several reasons why. Vegetable cultivars grown on a commercial scale must have certain characteristics including the property of shipping well without bruising or becoming overripe. Therefore many commercial vegetable cultivars are bred to be less juicy. They must be picked and shipped while not completely ripe in order to reach the consumer undamaged, and this frequently diminishes the ultimate flavor quality. Secondly, from the moment a ripe vegetable is picked, its quality will normally begin to slowly deteriorate. In some cultivars of sweet corn, for example, sugar begins to turn to starch almost immediately and the flavor of the vegetable can be completely altered in less than two hours. Although not always the case, the general rule of "the fresher, the better" is fairly reliable.

The economic benefits of vegetable gardening for saving on food bills have been widely acclaimed. Studies quote that a vegetable plot of a given size will supply all the vegetable needs of a family for a year. Although this is true with intensive, systematic cultivation and conscientious use of all vegetables as they become available, for most families it will not hold true. It is more reasonable to regard a vegetable garden as a money saver with recreational value that provides superior quality vegetables harvested and eaten at their prime.

Types of Vegetables

Just like all other plants, vegetables are either annual, biennial, or perennial. Most fall in the annual group in-

cluding peppers, squash, and beans. Others are technically biennials but are treated as annuals because only the vegetative parts are eaten. These include vegetables such as beets. Still fewer are perennials, including asparagus and rhubarb. Surprising to many people is the fact that the tomato is a perennial, but it is not frost-tolerant and is therefore treated as an annual. Table 6-1 lists the common garden vegetables and their classifications as annuals or perennials.

More important than whether a vegetable is annual or perennial is its classification as either *warm season* or *cool season* (Table 6-1). Warm-season vegetables thrive at daytime temperatures ranging from 65 to 90°F (18–32°C) and at nighttime lows not less than about 55°F (13°C). At lower temperatures they grow slowly. Some, like tomatoes, fail to develop fruit, whereas others, like peppers, develop fruit but produce only small fruits which are not fleshy or full. These vegetables will not live through a frost.

The opposite is true with cool-season vegetables. They will tolerate light frost and grow best at daytime temperatures from 50 to 65°F (10–18°C). In warmer weather, quality is often poor, for example, lettuce may become bitter. Some cool-season vegetables will flower when leaves alone are wanted, and this too is undesirable. The flowering, called *bolting*, is a response to the shortening of the nights and the warmer weather during the summer.

Another factor affecting the culture of vegetables is whether they are grown for roots, leaves, or fruit (Table 6-1). First, fertilizer selection must be considered. Leaf vegetables produce best with nitrogen-rich fertilizer which encourages vegetative growth. Root and fruit vegetables, on the other hand, are best with a fertilizer formula containing less nitrogen. In fact an excess of nitrogen will prevent some vegetables from bearing at all. Second, leaf and many root vegetables can be grown successfully in semishady areas. Fruit-producing

Table 6-1 *Garden Vegetable Cultivation*

Vegetable	Cool/ warm season	Grown as annual, perennial	Edible portion	Moderate planting for family of four	Distance between plants	Distance between rows
Artichoke, globe	Cool	Perennial in zones 9-10, annual in others	Flower buds	3 or 4 plants	4' (1.2 m)	5' (1.5 m)
Artichoke, Jerusalem	N/A	Perennial	Tubers	10-15 plants	1' (0.3 m)	5' (1.5 m)
Asparagus	N/A	Perennial	New shoots	30-40 plants	1' (0.3 m)	5' (1.5 m)
Bean, lima	Warm	Annual	Fruit	15-25' (5-8 m) row	6" (15 cm) (bush) 2' (0.6 m) (pole)	30" (0.5 m)
Bean, snap or green	Warm	Annual	Fruit	15-25' (5-8 m) row	3" (8 cm) (bush) 2' (0.6 m) (pole)	2' (0.6 m) 2' (0.6 m)
Beet	Cool	Annual	Leaves, root	10-15' (3-4.5 m) row	2" (5 cm)	18" (0.5 m)
Broccoli	Cool	Annual	Flower buds	7-10 plants	2' (0.6 m)	3' (1 m)
Brussels sprout	Cool	Annual	Axillary buds	7-10 plants	2' (0.6 m)	3' (1 m)
Cabbage	Cool	Annual	Leaves	10-15 plants	2' (0.6 m)	3' (1 m)
Carrot	Cool	Annual	Root	20-30' (6-9 m) row	2" (5 cm)	18" (0.5 m)
Cauliflower	Cool	Annual	Flower head	10-15 plants	2' (0.6 m)	3' (1 m)
Celery	Cool	Annual	Petioles	20-30' (6-9 m) row	5" (13 cm)	18" (0.5 m)
Chinese cabbage	Cool	Annual	Leaves	10-15' (3-5 m) row	1' (0.3 m)	30" (0.75 m)
Collard	Cool	Annual/ perennial	Leaves	10-15' (3-5 m) row	18" (0.5 m)	3' (1 m)
Corn	Warm	Annual	Seeds	4 rows of 20-30' (6-9 m)	6" (15 cm)	3' (1 m)
Cucumber	Warm	Annual	Fruit	6 plants	2' (0.6 m)	4' (1.2 m)
Eggplant	Warm	Annual	Fruit	4-6 plants	2' (0.6 m)	3' (1 m)
Endive	Cool	Annual	Leaves	10-15' (3-5 m) row	10" (25 cm)	18" (0.5 m)
Kale	Cool	Annual	Leaves	10-15' (3-5 m) row	10" (25 cm)	2' (0.6 m)
Kohlrabi	Cool	Annual	Enlarged stem base	10-15' (3-5 m) row	3" (8 cm)	2' (0.6 m)
Leek	Cool	Annual/ perennial	Enlarged stem base	10' (3 m) row	2" (5 cm)	2' (0.6 m)
Lettuce	Cool	Annual	Leaves	10-15' (3-5 m) row	12" (0.3 m) (head) 6" (15 cm) (leaf)	2' (0.6 m) 2' (0.6 m)
Melon	Warm	Annual	Fruit	5-6 hills	6" (15 cm)	6' (2 m)
Mustard green	Cool	Annual	Leaves	10' (3 m) row	8" (20 cm)	18" (0.45 m)
Okra	Warm	Annual	Fruit	10-20' (3-6 m) row	18" (0.5 m)	3' (1 m)
Onion, green	Cool	Annual	Leaves	10-15' (3-5 m) row	1" (2-3 cm)	18" (0.5 m)
Onion, bulb	Cool	Annual	Bulb	30-40' (9-12 m) row	3" (8 cm)	18" (0.5 m)
Parsnip	Cool	Annual	Root	10-15' (3-5 m) four	3" (8 cm)	18" (0.5 m)
Pea, English	Cool	Annual	Seeds	30-40' (9-12 m) row	2" (5 cm)	3' (1 m)
Pea, edible pod or Chinese	Cool	Annual	Fruit	30-40' (9-12 m) row	2" (5 cm)	4' (1.2 m)
Pepper	Warm	Annual	Fruit	5-10 plants	2' (0.6 m)	3' (1 m)
Potato	Warm	Annual	Tubers	100' (30 m) row	1' (0.3 m)	30" (0.75 m)
Pumpkin	Warm	Annual	Fruit	1-3 plants	4' (1.2 m)	6' (2 m)
Radish	Cool	Annual	Root	4' (1.2 m) row	1" (2-3 cm)	18" (0.5 m)
Rhubarb	N/A	Perennial	Petioles	2-3 plants	3' (1 m)	4' (1.2 m)
Rutabaga	Cool	Annual	Root	10-15' (3-5 m) row	3" (8 cm)	18" (0.5 m)
Spinach	Cool	Annual	Leaves	10-20' (3-6 m) row	3" (8 cm)	18" (0.5 m)
Squash, summer and zucchini	Warm	Annual	Immature fruit	2-4 plants	2' (0.6 m)	4' (1.2 m)
Squash, winter	Warm	Annual	Mature fruit	2-4 plants	4' (1.2 m)	6' (2 m)
Sweet potato	Warm	Annual	Root	50' (15 m) row	1' (0.3 m)	3' (1 m)
Swiss chard	Cool	Annual	Leaves	3-4 plants	1' (0.3 m)	30" (0.75 m)
Tomato	Warm	Annual	Fruit	10-20 plants	2' (0.6 m) (bush) 1' (0.3 m) (staked)	3' (1 m)
Turnip	Cool	Annual	Root	10-15' (3-5 m) row	2" (5 cm)	18" (0.5 m)

vegetables generally require full sun since they are unable to photosynthesize the necessary carbohydrate for flowering and fruiting without bright sunlight and may produce only foliage in shady situations. This fact is important for city dwellers whose gardens are shaded by tall buildings or for other gardeners who find themselves restricted to a shady area. Undue time and effort need not be wasted on fruit crops that will produce no reward. The space can be reallocated to vegetables which will grow more successfully.

Climatic Factors

Vegetable gardening is considered a spring and summer activity by most people, with seed sowing in spring and harvest through summer ending at the first killing frost. But in mild-winter areas of the country, vegetable gardening continues all year. In most other areas it can be extended far beyond the traditional season by choosing the right crops and cultural practices.

Vegetable Gardening in Mild-Winter Areas

The mild-winter areas of North America (see Figure 6-1) are its vegetable bowl. Here most of the fresh vege-

tables are grown for winter consumption. Gardeners located in one of these areas are fortunate, for they can grow fresh vegetables the entire year and can easily supply all their vegetables from the garden.

The climate characteristics of the locale will determine which vegetables can be raised in each season. In the deep South, for example, winter will be cool and nearly frostless and summer will be very hot. The cool-season vegetables can be raised in winter and the warm-season vegetables in summer when the temperatures exceed the acceptable range for the former. In the Florida Keys, the only tropical area of the North America north of Mexico, winter temperatures are moderate (50–75°F, 10–24°C), high enough for warm-season vegetables but not too high for cool-season vegetables. In the summer only warm-season crops could be grown. Along much of the coast of California temperatures are moderated by the Pacific Ocean; there is a minimal difference between winter and summer temperatures. They often fall just short of the acceptable range for warm-season vegetables, and gardeners in this area grow primarily cool-season crops all year.

Vegetable Gardening in Temperate Climates

Most of the United States and Canada have widely varying winter and summer temperatures. Temperatures

Figure 6-1 Areas of the continental United States where vegetables can be grown throughout the winter.

rise slowly in spring, reach a maximum in summer, and cool slowly as winter approaches. In these climates spring and fall are the best seasons for cultivating cool-season crops, whereas summer is most conducive for raising warm-season crops.

The number of days from the last spring frost until the earliest fall frost defines the period of growing warm-season crops. This *frost-free period*, an important concept in vegetable gardening, varies widely with latitude, ranging from 250 days in the deep South to 60 days in parts of North Dakota.

The importance of knowing the number of frost-free days lies in its relevance to cultivar selection of warm-season crops. Modern plant breeding offers the home gardener a wide selection of vegetable cultivars bred for taste, canning suitability, and rate of maturity. The rate of maturity is the number of days from either seed or transplant (depending on the vegetable) to the time when the crop is ready for harvest. For example, there are tomato cultivars that mature in as little as 54 days and others that require 80 or more frost-free days.

The benefits of fast-maturing vegetables to gardeners in the far North are obvious. Even gardeners in more moderate climates value fast-maturing vegetables because they are ready for harvest earlier in the season.

Cool-season vegetables are not restricted to the frost-free period. Not only are they tolerant of light frost, but their flavor may actually be improved by it. Extending the vegetable gardening season in temperate climates is largely a matter of taking advantage of the spring and fall growing periods, which have occasional frosts but are generally favorable for the growth of cool-season vegetables.

Early spring, for example, is fine for sowing peas, lettuce, carrots, and other cool-season vegetables grown from seed. It is also ideal for setting out transplants of broccoli, cabbage, or cauliflower. Although they may appear to grow slowly at first, they will establish their roots and grow faster later in the season. Planting time for cool-season crops starts as soon as the ground is thawed and dry enough to be workable.

Fall Vegetable Gardening

Fall vegetable gardening is a second way to extend the growing season. It involves delaying the planting of cool-season vegetables so that they mature after the rest of the crops have been harvested. Although many of the cool-season vegetables planted in spring are started as transplants, they are usually seeded when grown for a fall crop. This adds six to eight weeks to the "days to maturity" designation of the cultivar and must be factored in when deciding the date to sow seed. Crops such as leaf lettuce and radishes may be started in late

summer or even fall, since they mature quickly, but slower crops should be seeded in early summer. The following vegetables are suitable for summer sowing and fall harvest:

Beets	Kohlrabi
Broccoli	Leeks (sow in spring or over winter seedlings)
Cabbage	Lettuce (leaf and semihead)
Carrots	Mustard greens
Cauliflower	Parsnips (sow in spring)
Chard	Peas, English and edible pod
Chinese cabbage	Radishes
Collards	Spinach
Endive	Turnips
Kale	

Overwintering Vegetables in the Ground for Continuous Harvest

The cold tolerance of some root vegetables comes as a surprise to many gardeners. It is not unusual to discover these vegetables growing through the snow in spring and still good to eat. Occasionally even leaf vegetables such as lettuce and parsley will remain green under the snow cover and renew growth in the spring.

This overwintering phenomenon is due not only to the cold hardiness of the vegetable itself but also to the insulating effects of snow cover and the latent heat of earth (Chapter 20). When these two conditions combine, the soil may remain unfrozen or just at freezing all winter, despite air temperatures reaching far below 32°F (0°C).

It is both possible and practical to make use of these conditions to "store" root vegetables in the ground and extend their harvest through winter. The technique is fairly simple. A section of garden should be set aside for the winter harvest area and planted with root vegetables timed to mature in late summer or fall. These include beets, carrots, Jerusalem artichokes, leeks, parsnips, large radishes (Chinese or icicle type cultivars), rutabagas, and turnips. In fall the area should be mulched with compost, newspapers, leaves, or any other suitable material. The depth of the mulch needed to prevent freezing will depend on the expected winter temperatures. If winter temperatures only reach 20 to 25°F (−7 to −4°C), a layer of mulch 6 to 8 inches (15–20 centimeters) thick would be sufficient. The depth should be increased, ranging up to 2 feet (60 centimeters), if the temperature will drop below 0°F (−18°C). Continual snow cover on the mulch will provide additional insulation. As a precaution, the winter garden can be located near the house in a protected area to preserve the ground heat.

Harvest of the vegetables can begin any time. The mulch should be removed, the roots harvested, and the mulch replaced each time.

Planning a Vegetable Garden

Choosing the Area

A potential vegetable area should have both fast-draining soil for vigorous root growth and full sun-light to encourage maximum growth rate and flowering. Accordingly, the garden should be situated away from trees and buildings. A previously gardened area will probably possess better soil than one that has been planted in grass, but any soil could be improved to a suitable quality by adding amendments, as discussed in Chapter 5, or by green manuring, as described later in this chapter.

Determining the Size of the Garden

For the inexperienced gardener, a small plot is preferable to a large one. After the first gardening season, the amount of work involved and the number of vegetables produced will be known. The gardener can expand or decrease the size accordingly. A 25 × 25-foot (8 × 8-meter) plot will enable a first-time gardener to grow most popular vegetables except for large space consumers like melons and potatoes.

Garden Layout

The first tasks are to decide which vegetables will be grown and to lay out on paper where they will be located in the garden. Rows should be planted running east and west, if possible, to prevent shading. Taller or trellised crops should be placed north with shorter crops to the south. Figure 6-2 shows a typical layout for a plot producing many different kinds of vegetables.

It is often difficult to estimate how much of each vegetable to plant; Table 6-1 gives an average planting for a family of four for each vegetable. When using these estimates, it is best to divide the amount to be planted into two or three parts, and stagger the sowing of the seed so that the harvest is continuous throughout the season. This practice, called *succession cropping,* will assure a steady supply of vegetables instead of an overabundance at one time.

For example, two leaf lettuce sowings made two weeks apart in spring will provide lettuce into summer. A third sowing in late summer will yield a fall crop, giving a total of three successive crops of lettuce. Succession cropping requires careful management of the garden space. It entails more work than the straight "plant and harvest" method, but the crop rewards are greater throughout the season. Almost all vegetables with the exception of the slow-maturing warm-season crops can be succession cropped.

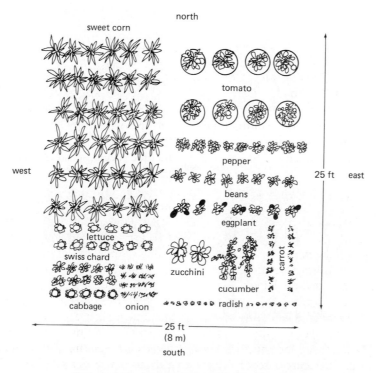

Figure 6-2 A typical vegetable garden plan.

Intensive or Block Gardening

In intensive or block gardening (Figure 6-3), vegetables are grown in blocks rather than rows. It was the original form of gardening practiced before the plow was invented and is still popular in countries where land is scarce. In the United States and Canada block gardening is useful in cities because it gives maximum yield from minimum space.

For block planting, a garden should consist of squares or rectangles no more than 4 feet (1.2 meters) across to keep the plants within reach of the path. Seed is sown by scattering it evenly over the area or in row fashion.

A variation of the original block garden involves growing vegetables in raised beds. The soil in these beds is thoroughly conditioned and fertilized to grow superior-quality crops.

Vegetable Cultivar Selection and Seed Purchasing

The Cooperative Extension Service of most states yearly evaluates new and existing vegetable cultivars to determine the best ones for the climate of that state. A list of the recommended cultivars is then compiled, published, and made available to the public. This list is probably the single most important information source on vegetable cultivars a gardener can have.

Seed catalogs list numerous cultivars and their merits but say little about their drawbacks. The gardener must evaluate the suitability of each cultivar to his needs and climate, noting days to maturity, resistance to disease, and other characteristics. It is wise to use catalogs from companies located in a climate similar to one's own so that the cultivars that are offered are geared to similar climates. For example, southern gardeners should avoid most catalogs from northern companies. It would be unfortunate to pick a cultivar by the picture, only to find it a poor performer under local growing conditions.

Many people buy their seeds from racks in groceries or nurseries. If so, the packages should be examined for the year in which they were packaged for use. If the stamped date is not the current year, the seed is old and may germinate poorly. Moreover, a seed rack may have a limited selection of cultivars, all or none of which may be recommended for the area.

"Seed tape" (Figure 6-4) is an expensive way to purchase seeds but will save labor. The seeds are sandwiched between layers of plasticlike material that dissolves when the row is irrigated.

Overbuying vegetable seed is a common mistake and results in leftover seed at the end of the growing season. Most vegetable seeds can be kept for the fol-

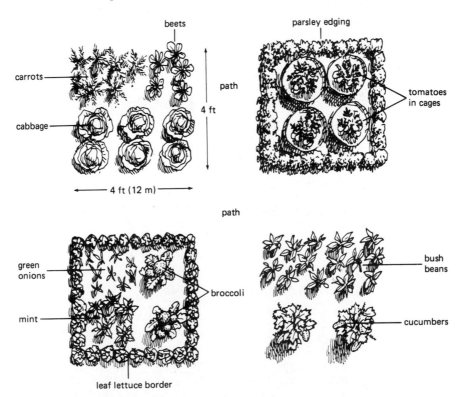

Figure 6-3 A block garden made up of four blocks.

Figure 6-4 Using seed tape to plant a row of vegetables. (*Photo courtesy of W. Atlee Burpee and Co., Warminster, PA*)

Table 6-2 *Approximate Life of Vegetable Seeds Stored Under Cool Conditions*

Vegetable	Years
Asparagus	3
Bean	3
Beet	4
Broccoli	5
Brussels sprout	5
Cabbage	5
Carrot	3
Cauliflower	5
Celery	5
Chard, Swiss	4
Chinese cabbage	5
Collard	5
Corn	1-2
Cucumber	5
Eggplant	5
Endive	5
Kale	5
Kohlrabi	5
Leek	3
Lettuce	5
Muskmelon	5
Mustard	4
Okra	2
Onion	1-2
Parsley	2
Parsnip	1-2
Pea	3
Pepper	4
Pumpkin	4
Radish	5
Rutabaga	5
Spinach	5
Squash	5
Tomato	4
Turnip	5
Watermelon	5

lowing season by storing them in the refrigerator (see "Seed Storage," Chapter 4). Table 6-2 lists the relative keeping quality of various vegetable seeds.

Ornamental Vegetable Cultivars

Vegetables can be used not only for eating but also as ornamentals (Figures 6-5 and 6-6). Leaf lettuce, for example, makes an attractive flower border. Cultivars of vegetables selected for their colorful flowers, foliage, or fruits are available through seed catalogs. A few are listed in Table 6-3.

Patio pot gardens are another ornamental use of vegetables. Ornamental peppers and chard grow well in containers. Cherry tomatoes can be planted in hanging baskets, and runner beans trained on strings to screen a porch or balcony. By using vegetables as ornamental plants, the gardener with space restrictions can also make better use of a limited area of land.

Soil Preparation

The time spent in soil preparation will depend on whether the area was previously gardened or not. If it was and the soil condition is good (not packed down, for example), little preparation will be needed except for pulling or killing weeds and mixing in fertilizer be-

fore planting, called *preplanting incorporation.* Preincorporation replaces nutrients used by the previous season's crop and positions a fresh supply in the root zone. Chapter 5 gives directions on preincorporating fertilizers.

Annual weed seedlings can be controlled by simply *scraping* the soil surface with a hoe, provided the seedlings are still small enough to be cut off at soil level. The area should not be dug to eliminate annual weeds, as this perpetuates the problem by exposing new seeds and encouraging them to germinate. Perennial weeds that have come back from the roots can be dug out individually, if they are few, or treated with a weed killer if they are numerous. A local nursery can recommend

Figure 6-5   Dark green oakleaf lettuce used as a border in a perennial flower bed. (*Photo by George Taloumis*)

Figure 6-6   The cucumber cultivar Salad Bush grown as an ornamental pot plant. (*Photo courtesy of All-America Selections, Downer's Grove, Il*)

a chemical to control existing perennial weeds, but weed samples should be brought to the nursery to be sure of selecting an effective chemical. The nursery also needs to know that the area will be planted in vegetable garden so that a chemical with a short effective period (called its *residual life*) can be selected and not affect the crops. If the area was previously a lawn, a typical situation, the sod must be removed, although as garden preparation is begun in fall it can be simply turned under and allowed to rot in place. In either case eliminating the grass with a weed killer will be necessary if the lawn contains vigorous grass species that spread by rhizomes, since it is nearly impossible to remove all rhizomes by hand and those left in the soil will resprout continually and create a weed problem.

After sod removal it may be necessary to mix in organic matter to improve the structure of the soil, to "loosen it up." Nurseries sell rotted manure, bark products, and various other soild conditioners for this purpose.

Planting the Garden

Sowing Seed

With the exception of block gardening, vegetables grown from seed are planted in rows or hills. When sowing in rows, each row should be marked with a labeled stake and the seed sown thickly and in accordance with package directions. Heavy sowing assures a sufficient stand of plants in case of poor germination.

With slow-germinating seeds such as carrots, a fast-germinating seed is sometimes planted in the same row to mark its location until the other seedlings emerge. In soils that tend to crust, the rapidly germinating seeds also break the surface to ease the emergence of less vigorous crops. Radishes germinate in 3 days and are often used for row making. They can be pulled as soon as the main crop emerges but are usually left in place. Since radishes mature in as little as 21 days, they can be harvested in time to prevent crowding of the main crop. This technique called *interplanting*, saves space and is frequently done with small, fast-maturing crops such as radishes and leaf lettuce. A full crop of leaf lettuce, for example, can be raised by seeding between transplants of broccoli or cauliflower (Figure 6-7).

Directions on packages of squash and vining vegetables like cucumbers normally specify planting in *hills*. Hill planting refers to grouping several plants together, not planting seeds in a mound of soil.

Table 6-3 *Ornamental Vegetable Varieties*

Vegetable	Ornamental species or cultivars	Remarks
Artichoke	*Cardunculus* species (common name, cardoon)	Large gray leaves and thistlelike purple flowers; petioles and roots are edible
	Scolymus species (common artichoke)	Purple flower heads, gray foliage; smaller plant than cardoon
Artichoke, Jerusalem	Any	Mature plants are 6 ft (2 m) or more tall with yellow sunflowers
Asparagus	Any	Feathery foliage and red berries in fall
Bean, bush	'Royalty'	Purple pods
Bean, pole	'Scarlet Runner'	Red flowers; shade- and cool-temperature-tolerant
Cauliflower	'Purple Head' and 'Royal Purple'	Heads tinged lavender at maturity
	'Greenball'	Head green at maturity
Chard	'Rhubarb'	Red petioles
Corn	Indian corn, 'Rainbow'	Red, yellow, and white kernels
	Strawberry corn	Small, strawberry-shaped ear with red kernels
	'Gracillima'	Dwarf size
	'Japonica'	Leaves striped white and pink
Cabbage, ornamental	Park's Super Cole Hybrids	Ruffled leaves are lavender, pink, and green; does not head
Kale, flowering	Park's Super Cole Hybrids	Ruffled leaves, red- or green-tinged
Lettuce, leaf	'Oakleaf'	Dark green lobed leaves
	'Red Salad Bowl'	Red-tinged leaves
	'Ruby'	Red leaves
Pepper, bell	'Golden Calwonder'	Yellow fruits
	'Bell Boy Hybrid'	Red fruits
	'Midway'	Red fruits
Pepper, hot	'Fiesta'	Red, yellow, and green fruits
Eggplant	'Morden Midget'	Deep purple, miniature fruit
Sunflower	No cultivars suggested	Enormous yellow flowers on 6- to 12-ft (1.8- to 3.6-m) stems; edible seeds excellent when toasted
Tomato	'Patio'	Dwarf plant to 1 ft (0.3 m) tall with golf-ball-sized tomatoes
	'Small Fry'	Vining plant producing large quantities of cherry tomatoes; good in a hanging basket
	'Tiny Tim'	6-in. (15-cm)-high plant with tiny tomatoes
	'Goldie'	14-in. (35-cm) tall plant produces 1 in. (3 cm) plum-shaped yellow fruit
	'Minibel'	A cascading growth habit for hanging basket culture and 2 in. (5 cm) diameter red fruit

Figure 6-7 A crop of leaf lettuce planted between broccoli plants.

After seeding it is important to keep the surface of the soil moist. Drying of the soil surrounding the germinating seeds will kill the emerging seedlings and necessitate resowing.

Buying Transplants

Selecting well-grown transplants contributes to the success of the garden, and quality should not be sacrificed in favor of price. The plants should be short, sturdy, and with foliage to the base. Those with yellowed foliage or bare stems should be avoided because they have been growing too long in the transplant container and are likely to be root-bound and slow to begin growth. In some cases they will flower prematurely and there will be no crop because the plants are not large enough to bear fruit.

Setting Out Transplants

If possible, an overcast day or early evening should be selected for transplanting to lessen shock. Watering the plants before removing them from the pack is advisable.

If the plants were grown in individual cells of a plastic pack, there will be very little root shock during transplanting. However, if after removing a transplant, it is found to be root-bound, the root ball should be cut shallowly on each side (Figure 6-8). The cutting will force branching of the roots into the soil. If uncut, the roots may continue growing within the original soil ball, and establishment will be slow.

Transplants grown in pots of compressed peat moss can be transplanted with the pot intact, but the top of the pot should be broken off down to the soil. If the rim is not broken and is exposed to the air, it will wick water from the peat moss pot, restricting the penetration by the roots and slowing growth.

Figure 6-8 Cutting through the wrapped roots of a pot-bound transplant. (*Photo by Rick Smith*)

Figure 6-9 Deep planting of a tomato encourages stem rooting.

Vegetable transplants should be planted slightly deeper than they were originally grown. Tomatoes should be set with much of the stem below the soil surface (Figure 6-9) since they will form adventitious roots along the submerged stem. These stem roots create a larger root system and a more vigorous plant.

After planting, the transplants should be watered with starter fertilizer as discussed in Chapter 5. The watering assures full soil contact with the roots and provides maximum soil moisture to prevent shock.

Vegetable Garden Maintenance

Thinning

Thinning is the removal of excess seedlings which are spaced too closely for best growth (Figure 6-10). Thinning can be done either once or twice for each crop.

One-time thinning should be done as soon as the leaves of neighboring plants touch and should remove as many seedlings as necessary to achieve the recommended final spacing for the vegetable.

Twice-over thinning is practical for vegetables grown for their leaves, such as chard, lettuce, and Chi-

Figure 6-10 A row of beets before thinning (on the left) and after thinning (on the right). The thinnings are in the foreground. (*Photo courtesy of National Garden Bureau*)

nese cabbage. The first thinning is done while the plants are seedlings, but it removes only enough plants to prevent severe overcrowding. The second thinning takes place 2 to 4 weeks later and thins the remaining plants to the final spacing. By the second thinning, plants are large enough to be harvested and eaten.

Weeding

Weed control is very important to a successful vegetable garden. The detrimental effect of weeds on plant growth is masked, showing up as a slowed plant growth rate due to the competition of the crop with the weeds for water, nutrients, and light. Many gardeners incorrectly attribute this sluggish growth to poor soil or adverse weather, not knowing the real cause.

Because seedlings have limited root systems, they are especially vulnerable to competition from vigorous weeds within the row. If unchecked, the weeds will eventually crowd out the desirable plants altogether.

If seedling plants are invaded by weeds, pulling is probably the only control method. Later, hoeing or mulching is effective. Because the roots of vegetable plants are shallow, hoeing for weeds should be a scraping rather than a digging operation. The scraping will cut weed seedlings at the soil line, and if they are annuals, they will not resprout.

Mulching

Mulch is a layer of plant-derived or synthetic material laid on the soil surface over the roots of plants. Mulching can considerably reduce the time spent weeding and will conserve soil moisture. In hot areas of the South and Southwest it helps keep soil from overheating and consequently is beneficial to root growth. Since

vegetable gardens are tilled every year, many people find it easier to use plant-derived mulches which will decay in the soil. Compost, leaves, and grass clippings from lawns that have not been recently treated with weed killer are satisfactory.

Mulch should be applied 2 to 3 inches (5–8 centimeters) thick as a general rule. Transplants should be mulched sparingly about 1 inch (3 centimeters) deep when young, gradually increasing the depth to 3 inches (8 centimeters) as the plants grow taller. For low-growing plants such as lettuce and radishes, mulch should be applied on either side of the row at a depth of 1.5 to 2 inches (4–5 centimeters). Vining crops such as melons should be mulched over the entire area where the vine will grow. Covering the area will help keep the fruits clean and reduce losses from rot diseases which occur when the vegetables lie on the soil.

Newspapers held down with soil are another suitable vegetable garden mulch. Although not attractive, they are readily available and inexpensive.

Plastic mulches (Figure 6-11) are very useful in northern vegetable gardens because they absorb heat and radiate it into the soil underneath. They will speed up the growth of warm-season crops by heating the soil and can mature the crop a week or more ahead of normal. Either clear or black plastic mulch may be used. The former will raise the temperature more than the latter; however, weeds will grow beneath it.

For especially early harvests, a plastic mulch can be used with transplants of cucumbers and melons. The time gained by using transplants combined with the warmth provided by the plastic will encourage rapid growth and fruit set.

Plastic mulches can be applied in two ways. For row crops, strips of plastic are laid on either side of the

Figure 6-11 A black plastic mulch used in a row of tomatoes. (*USDA photo*)

Figure 6-12 Hot caps being used to cover transplants in the garden. (*USDA photo*)

row, with the plants between. For vegetables in hills, the area can be covered and a flap can be cut in the plastic so the shoots can emerge.

Hot Caps and Frost Protection Devices

The use of *hot caps* (Figure 6-12), paper or plastic domes set over plants in early spring, is another method of hastening growth. The caps admit light but insulate against heat loss, keeping the air surrounding the transplants several degrees warmer than the outside air. This slight rise in temperature is often enough to speed growth considerably during early spring. As an added feature, hot caps offer frost protection. They enable gardeners to set out transplants of warm-season vegetables and sow seeds several weeks ahead of the last predicted frost date.

Frost protection of warm-season crops in fall is more difficult because of the size of the plants. If only light frost is expected, a layer of newspapers over the plants will often give adequate protection.

Watering

Watering the vegetable garden should begin as soon as seeds are planted and be done as often as necessary to prevent wilting. Do not assume, however, that every time a plant wilts, the garden needs watering. Many times large-leaved squash or cucumber plants wilt repeatedly during the hottest part of the day due to an inability to absorb enough moisture to compensate for the enormous water loss through the foliage, a phe-

nomenon called *diurnal wilting*. However, if they do not recover at night, water is definitely needed. If, after watering, they still do not recover, the plant may be infected by a wilt disease or nematodes.

Watering the vegetable garden should soak the soil to a depth of about 18 inches (0.5 meter). Perennial vegetables such as asparagus and artichokes will need to be irrigated more deeply.

Fertilizing

If fertilizer is worked into the soil of the vegetable garden during preparation, it will often be sufficient to supply the needs of the crops to maturity. On sandy soils, on soils of low fertility, or if deficiency symptoms are noted, a midsummer sidedressing of balanced granular fertilizer may spur crop growth. Nitrogen fertilizers can stimulate leaf production at the expense of the fruit and should be used sparingly after plants have reached mature size.

Training

Peas, runner beans, tomatoes, and cucumbers are vining plants which can be trained on stakes, strings, or wire. Training keeps the fruits from becoming dirty and can lessen the chances of fruit rot by preventing contact of the fruits with the soil. It also optimizes space usage in the garden and makes harvesting easier.

Teepee Training

Teepee training (Figure 6-13) can be used for a group of several cucumber or bean plants. Ideally the teepee should be built first and then seeds planted at the base of each stake. The vines are tied to the stakes at 1-foot (0.3-meter) intervals.

Staking

A single sturdy stake (Figure 6-14) will support one tomato plant or two to three runner beans. Beans will climb without help, but tomatoes should be tied loosely with pieces of cloth at intervals along the stem.

String Trellises

Runner beans and peas are trained easily on a trellis made of twine strung between poles (Figure 6-15). The poles should be pounded in deeply not more than 10 feet (3 meters) apart to provide a sturdy support for the trellis and vines.

Figure 6-13   Teepee training of cucumbers.

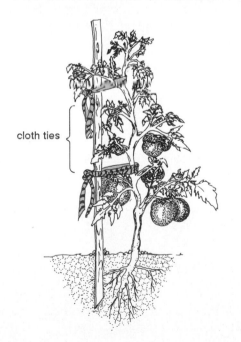

cloth ties

Figure 6-14   Staking is one of the most common ways to train tomatoes.

Tomato Cages

Tomato cages using a wire-mesh cylinder either purchased or made at home (Figure 6-16) are an easy way to support tomatoes. Concrete reinforcing wire has an ideal strength and a good mesh size for tomato cages;

many types of galvanized fencing (not chain link) are also suitable. It is only necessary that the mesh be sufficiently large to reach through to pick ripe fruit. Tying the vines is not required.

Crop Rotation

Crop rotation is the planting of crops in different areas of the garden every year. It involves, for example, switching the locations of the beans and cabbage each year. Crop rotation deters the buildup of soil-carried disease organisms associated with particular crops. Although crop rotation is not essential, it is a sound practice that does not require much effort.

The easiest form of crop rotation is to simply reverse the vegetable garden plan each year, being careful to avoid shading problems due to placement of tall plants. The progression shown in Figure 6-17 is another rotation sequence.

Cover and Green Manure Crops

Cover cropping (Figure 6-18) is used to maintain fertility in established vegetable gardens, but green manure crops are used to improve poor soil before a garden is planted. In most areas of the country, cover crops are planted in fall over the vegetable garden. They grow during fall and winter and are turned under

Figure 6-15　A string trellis used to support runner beans. (*USDA photo*)

Figure 6-16　Tomato cages. (*USDA photo*)

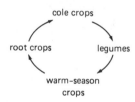

Figure 6-17　Suggested crop rotation for a home vegetable garden.

Figure 6-18　A cover crop in spring before being plowed under. (*Photo courtesy of the Cooperative Extension Service, Michigan State University*)

in spring, supplying organic matter and increasing the fertility of the soil as they decay. Green manure crops perform similarly but are planted at any time of year, grown halfway to maturity, and then turned under to decay.

　　　Green manuring can be used in place of adding organic soil amendments and is much less expensive. The general practice is to sow the green manure crop, wait until it is 8 to 10 inches (20–25 centimeters) high (or until flowering if it is of the pea family), turn it under, wait 10 to 14 days, and then resow another green manure crop within three weeks. The plowing and resowing are repeated as many times as necessary until the soil is of acceptable quality. Members of the pea family such as alfalfa, clover, cowpeas, soybeans, and some cultivars of vetch are among the best green manure and cover crops. They absorb nitrogen from the air and, when turned under to decay, will release that nitrogen into the soil. Winter-growing cover crops include wheat, rye, ryegrass, and buckwheat. Buckwheat is very tolerant of adverse soil conditions and is especially useful in very poor soils.

Culture of Common Garden Vegetables

Asparagus

Asparagus (*Asparagus officinalis*) is a perennial vegetable grown for its tender sprouts. Some gardeners grow asparagus for its tall, fernlike foliage and ornamental red berries. Sprouts appear early in the spring in temperate climates and throughout the year in tropical areas.

Sandy soils high in organic matter are best, but this vegetable will succeed in any soil. If grown in loam or clay soils, crowns should be planted 5 to 6 inches (13–15 centimeters) deep; in lighter soils, crowns should be planted 6 to 8 inches (15–20 centimeters) deep.

Asparagus can be grown from either seed or 1- to 2-year-old crowns. Seed will not produce an appreciable crop until the third year except in zones 10 and 11 where harvesting may begin the second year. Consequently, growing from crowns is the preferred method.

Selection of an asparagus cultivar should, like other vegetables, be based on the adaptability of the cultivar to local growing conditions. In addition to this basic essential, it is useful to know that recently "all-male" genetic lines of asparagus have been developed and can be grown from seed. Normally asparagus is dioecious, and a cultivar is a mixture of male and female plants. However female plants normally produce less yield of asparagus, diverting a part of their energy instead to fruit and seed production. Male plants flower, but don't fruit, so their surplus carbohydrate can be stored instead in the root system to foster additional vegetative yield in the form of spears.

Examples of male hybrid cultivars currently available are Jersey Gem, Jersey General, Jersey King, Jersey Knight, and Greenwich.

Asparagus crowns should be planted 8 to 12 inches (20–30 centimeters) apart in rows in the spring or fall. Harvesting should be delayed until the second year to allow the plants to become established. During this time plants should be watered regularly and fertilized with a balanced fertilizer. During the second year, spears should be harvested for a two-week period and then allowed to leaf out. In the third year spears may be harvested until the plants show weakening by producing small-diameter spears or ones that branch close to ground level. Care should be taken not to harvest for too long a period since this will weaken the crowns and cause yield less the following year.

Asparagus spears should be harvested when they are 6 to 8 inches tall (15–20 centimeters)—before they begin to branch. The best way to harvest is to snap the tips by hand (Figure 6-19). The shoots will break at the point where they become fibrous and inedible. During warm weather, beds can often be harvested daily or even twice daily. If a frost threatens, all emerged sprouts should be harvested regardless of length since they are not frost-tolerant.

Asparagus should be refrigerated immediately after harvest because its quality diminishes rapidly in warm temperatures. Flavor will be retained best under high humidity and at a temperature just above freezing.

Beans

The common bean (*Phaseolus vulgaris*) includes many different garden beans known as snapbeans, stringless or string beans, wax beans, and green beans. Beans may be either climbing vines called *pole types* or *runner beans,* or low-growing, bushlike plants called *bush* or *dwarf types.* The edible pod is eaten while immature and should be picked before the seeds inside enlarge. Since quality declines rapidly after harvesting, beans should be refrigerated and kept in high humidity. For best quality they should be eaten within 24 hours.

Beans are warm-season annuals and will not tolerate cool temperatures or frost. Therefore beans should be planted only after the weather has warmed in the spring. Any well-drained soil is acceptable for growing beans. A pH of between 6.5 and 7.0 is best.

Beans should be planted 1 inch (2–3 centimeters) deep and 2 to 3 inches (5–8 centimeters) apart and do not require thinning. If bush beans are grown, staking is not necessary; however, if pole types are grown, stakes should be provided, preferably at planting time. Beans should be sidedressed with a balanced fertilizer prior to flowering if leaves are not a dark green color.

Summer irrigation is required in much of the South and Southwest to prevent blossom drop. This malady is aggravated by hot, dry winds.

Bush beans normally yield sooner than pole beans, but the harvest period will be shorter. Succession plantings should be made two to three weeks apart to prolong the harvest. Although pole beans take longer to begin producing, they will often bear until killed by frost if all pods are kept picked.

Beets

Beets (*Beta vulgaris* var. *crassa*) are one of the easiest vegetable crops to grow. They can be grown in all climate zones in the summer and in warmer zones as a

Figure 6-19   Picking asparagus by the snap method. (*USDA photo*)

winter crop. The plants are only about 1 foot (30 centimeters) tall, with red-tinged leaves originating from a large red root. The root is the primary edible portion of the plant. The leaves are also edible; in fact they are more nutritious than the roots.

Beet roots can be harvested and eaten at any time. However, the quality is best before they reach 3 inches (8 centimeters) in diameter. Leaves can be harvested either when the plants are thinned or by picking individual leaves from the best plants. No more than two leaves should be removed from a single plant, or root formation will be retarded. Roots can be stored for three to five months in the coldest part of the refrigerator if tops have been removed.

A light, well-drained soil with a pH above 6.5 is preferred; however, beets will grow in most garden soils. Since beets tolerate light freezes, they should be planted as early as possible in the spring or year-round in mild-winter areas. For most cultivars a "seed" will contain several embryos, resulting in several plants at each location. Thinning is thus necessary to space the plants 2 inches (5 centimeters) apart. If greens are de-

sirable, thinning can be delayed until several 3- to 4-inch (8- to 10-centimeter) leaves have formed. The tenderest beets are produced when plants are growing quickly, so a sidedressing of a balanced fertilizer is desirable four to six weeks after planting. Irrigation should also be provided if rainfall is inadequate. When harvesting roots, pull beets starting when they reach one inch (3 centimeters) so that more room is provided for those remaining.

Broccoli

See "Cole Crops."

Brussels Sprouts

See "Cole Crops."

Cabbage

See "Cole Crops."

Cantaloupe

See "Cucurbits."

Carrots

Carrots (*Daucus carota* var. *sativus*) are an easy-to-grow root crop for all climate zones. They are a cool-season crop that can be grown as a summer crop in all zones and as a winter crop in warm-winter regions.

The best quality carrots are produced in deep loose soils with good drainage. In heavy soils roots will often be crooked or forked (Figure 6–20); short-rooted cultivars are likely to produce the best results in such soils.

Seeds should be planted shallowly in soil that has been well prepared to eliminate clods. Where crusting of the soil surface is a problem, organic matter should be mixed with the soil to cover the seeds.

Carrot seeds are small and slow to germinate. To aid emergence and to mark where the carrots are planted, radish seed can be mixed with the carrot seed. The radish seed will germinate rapidly and break the surface crust as well as mark the location of the carrot row. The radishes can then be removed after the carrots germinate.

It is important to thin carrots to provide adequate room for the roots to develop. Thinning can be done several times to ultimately leave 1½ to 2 inches (3–5 centimeters) between carrots. The last thinning can usually be delayed until the thinnings are pencil-size and can be eaten. Harvesting can then continue until they are between 1 and 2 inches (3–5 centimeters) in diameter. After harvest carrots should be refrigerated in high humidity. If fully mature, they will keep for four to five months under these conditions. Younger carrots can generally be stored only four to six weeks.

If carrot leaves are a light green, a sidedressing with a balanced fertilizer will improve yield. As with most root crops, fast growth will produce a more tender and tasty root.

Cauliflower

See "Cole Crops."

Chard

Chard (*Beta vulgaris* var. *cicla*) (Figure 6-21) is probably the easiest to grow and the most useful leaf vegetable for the home garden. As the botanical name indicates, chard is actually a beet but is grown for its spinachlike leaves rather than for its roots. The culture is the same as for beets. Leaves are harvested individually by cutting the outer ones off at the base as they reach edible size. Chard will produce leaves year-round in mild areas and from early spring until late fall in colder areas. It is a vegetable that is practically pest-free and is highly recommended for the home garden.

Cole Crops

The term *cole crop* refers to several cool-season crops related to cabbage and having similar cultural requirements. Included are broccoli, Brussels sprouts, cabbage, cauliflower, collards, and kale as well as several less common vegetables.

All cole crops thrive under cool temperatures and will survive light frosts. In hot-summer areas they should be grown as a spring or fall crop; where winters are warm, they should be grown as a winter crop. Collards alone are tolerant of very hot summer weather.

Any well-drained soil is acceptable. A pH between 5.5 and 6.5 is ideal, but a wide range is satisfactory in practice. All cole crops germinate readily from

Figure 6-20 Carrots deformed like these are caused by rocky, compacted soil and too much moisture. (*Photo courtesy of Stoke's Seeds, Inc., Buffalo, NY*)

Figure 6-21   Swiss chard. (*USDA photo*)

seed planted ¼ inch (6 millimeters) deep or may be started from transplants. Abundant moisture and a sidedressing with a high-nitrogen fertilizer are beneficial. The most serious problem with growing cole crops is the control of caterpillars and aphids.

Individual Requirements

Broccoli. (*Brassica oleracea*, Italica Group). "Sprouting" cultivars which form a large number of small heads rather than one large head should be selected for planting. The edible part is the flower head, which should be harvested before the buds begin to open. If all heads are kept picked, broccoli will continue to produce for many months (Figure 6-22).

Brussels Sprouts. (*Brassica oleracea*, Gemmifera Group). The edible parts are the small axillary buds which form along the main stem. As the plant grows taller, the lower leaves can be removed until leaves remain only at the top of the plant. At this point the plant will somewhat resemble a palm tree with side growths looking like small cabbages.

Sprouts should be harvested when about 1 inch (2–3 centimeters) across but before they begin to open.

Figure 6-22 The main head of this broccoli has been harvested, but it is continuing to produce edible side shoots. (*Photo courtesy of National Garden Bureau*)

Figure 6-23 Blanching cauliflower. (*USDA photo*)

If the weather is warm, they will often be soft and leafy rather than hard. Flavor and quality will improve as cool weather arrives in fall. If long storage of sprouts is desired, the entire plant should be pulled up in fall and replanted in damp sand in a cool cellar. Sprouts will store for several months in this manner.

Cabbage. (*Brassica oleracea,* Capitata Group). Cabbage should be harvested when the heads are hard. If allowed to remain in the garden for too long, the heads will split, especially after a heavy rain. Since cabbage heads tend to mature at the same time, succession planting or the use of several different cultivars with varying harvest dates is desirable. Cabbage will store for a month or more under cold temperatures and high humidity.

Cauliflower. (*Brassica oleracea,* Botrytis Group). This is one of the more difficult cole crops to grow. For best results, the plants should be kept growing vigorously by watering and fertilizing; any slowdown in growth will cause the production of a small head of poor quality. Transplants which are not vigorous should be avoided because these will often produce a head only 1 inch (2.5 centimeters) or so in diameter immediately after transplanting, a phenomenon called *bolting.*

Like broccoli, the edible parts of cauliflower are the immature flower buds. The head is harvested while the buds are still pressed closely together. Quality declines rapidly once the buds begin to separate. To produce white heads similar to those available in markets, the heads must be *blanched.* This is accomplished by tying several of the upper leaves over the head as soon as it begins to form in order to shade it (Figure 6-23). Several "self-blanching" cultivars are also available. These will naturally cover the developing flower buds with leaves, but only if the weather is cool. Failure to blanch cauliflower will result in heads light green in color

which, although more nutritious, have a stronger flavor. Cauliflower plants produce only one head and should be pulled out after the head is harvested.

Collards. (*Brassica oleracea,* Acephala Group). This nonheading cabbage is tolerant of both very cold weather and the heat of summer; as a result it is one of the most reliable plants for the production of greens. Individual leaves can be harvested at any time, although during hot weather the flavor is likely to be strong.

Kale. (*Brassica oleracea,* Acephala Group). Kale is another easy-to-grow plant for greens, but it will not survive hot summers. It will, however, often survive the coldest of winters, producing tender greens as soon as the snow melts. To harvest, remove the outer leaves individually. Kale is less prone to insect problems than other cole crops.

Collards

See "Cole Crops."

Corn

Sweet corn (*Zea mays* var. *saccharata*) is one of the most popular home-garden vegetables. It is a warm-season crop and should be planted only after danger of frost is past.

It will thrive in a wide variety of well-drained soils. A pH between 6.0 and 6.5 is best. It should be planted 1 to 2 inches (2.5–5 centimeters) deep and 8 to 10 inches (20–25 centimeters) apart. Several short rows rather than one long row should be planned to ensure pollination. After corn has germinated, it is necessary to control weeds, provide water, and give one or two sidedressings of high-nitrogen fertilizer. Since corn tends to reach harvesting stage all at one time, several succession plantings should be made.

Since hundreds of cultivars of sweet corn are available, the gardener can pick from midgets less than 3 feet (1 meter) tall to giants 6 to 7 feet (meters) tall. However, regardless which cultivar is selected, popcorn or field corn should not be grown within 100 feet (30 meters) of sweet corn to prevent cross-pollination and poor kernel quality.

Corn is harvested when the silk (female flowers) on the ears turns dark brown and the kernels are plump and filled with milky sap. Since with the exception of "super sweet" cultivars the sugar turns to starch quickly after picking, sweet corn should be cooked immediately after harvest.

Cucumbers

See "Cucurbits."

Cucurbits

The group of vegetables called *cucurbits* includes many warm-season vine crops with similar growth requirements. Members of the group include cucumbers, squash, melons, and pumpkins. In all cases the edible portion of the plant is the fruit.

Cucurbits require a warm climate for optimum growth and fruit development. The best quality of fruits is obtained when summers are long and hot. This does not preclude the growing of these plants in northern areas; however, plantings should be made only after all danger of frost is past and night temperatures are consistently above 45°F (7°C). Since melons require the highest temperatures and the longest growing season, melon cultivars should be selected carefully for adaptability to the local climate. Winter growing of cucurbits is restricted to the warmest areas of California, Florida, and Texas.

Cucurbits can be started either from seed sown ½ inch (1–2 centimeters) deep in late spring or from transplants. Since they do not transplant easily, seedlings should be purchased only if they are grown in peat pots or other plantable containers.

Cucurbits can be planted in rows or hills. If only a few plants are to be grown, the hill method is best because of better space utilization. Vines can be either allowed to run along the ground or trellised. Trellising works very well for cucumbers but is generally not used for melons and squash because of the heavy fruits. Gardeners are often disappointed by the lack of fruit set early in the year on cucurbit plantings. However, with most cultivars this is natural because cucurbits produce

separate male and female flowers, and the normal pattern is for male flowers to be produced first, followed later by female flowers and corresponding fruit set.

Individual Requirements

Cucumbers. Compared with melons, cucumbers (*Cucumis sativus*) require a shorter growing season and less heat and therefore are adapted to a wider range of climates. Two types of cucumbers are commonly grown in the home garden, depending on their proposed use. These are pickling cucumbers and slicers. Pickling cucumbers are bred for their qualities as pickles and are harvested while immature. Slicers are bred for eating fresh, are larger, and have thicker skins and a more attractive, less warty appearance. For the home gardener who desires cucumbers for both pickling and fresh uses, pickling cultivars are best because they are of a good quality for fresh use as well as for pickles.

Bitterness in cucumbers is due to a number of factors including genetic composition, excessively warm temperatures, and lack of water. Though difficult to prevent, adequate water will decrease the chance of bitter cucumbers.

Melons. The most commonly grown melons are cantaloupes (more correctly, muskmelons, *Cucumis melo*, Reticulatus Group), watermelons (*Citrullus lanatus*), and honeydew melons (*Cucumis melo*, Inodorus Group). All require a long growing season, hot days, and plenty of moisture. The best areas for melon growing are warm regions of the country, although cultivars are available that can be grown in most of the United States and Canada. Honeydews require the most heat, followed by watermelons and muskmelons. Since the climate requirements are rather stringent, selection of cultivars adapted to local conditions is imperative. As a general rule, cultivars that produce small or midget fruit perform best where summers are cool.

Often one of the most difficult tasks the home melon grower faces is determining when the fruit is ripe. Watermelons should be harvested when the stem of the melon has shriveled, whereas muskmelons should be picked when the stem slips easily from the fruit with slight pressure from the thumb. Honeydew melons are at their best when they develop a yellowish skin color and sweet smell.

Squash and Pumpkins. Squash and pumpkins (*Cucurbita maxima, C. pepo, C. mixta, or C. moschata*) are the easiest of the cucurbits to grow, and they produce an abundance of fruit. They are divided into two types based on whether the immature or the mature fruit is eaten. Those squash eaten when the fruit is mature and

Figure 6-24 A selection of winter squashes in storage. (*Photo courtesy of National Garden Bureau*)

possessing a hard skin are *winter squash* (Figure 6-24). The name is derived from the fact that these types ripen in late fall and can be stored in a cool place through much of the winter. Pumpkins are essentially a type of winter squash. *Summer squash*, eaten while immature, include zucchini, yellow crookneck, and yellow straightneck squash (Figure 6-25). They are perishable and must be eaten soon after harvest.

Winter squash should be harvested when the rind becomes hard, evident by an inability to puncture it with a thumbnail. Usually a change in color will accompany the ripening process. Summer squash are picked before the skin begins to harden but not necessarily when they are very small. If summer squash be-

Figure 6-25 A selection of summer squash: patty pan on the left, yellow crookneck at bottom right, zucchini in the basket on the left and yellow straightneck in the basket on the right. (*Photo courtesy of the National Garden Bureau*)

come large, they are still edible by scooping the seeds from the center prior to cooking.

When selecting cultivars of squash and pumpkins, consider selecting bush types instead of vine types since they require much less room and produce the same quality of fruit.

Eggplant

Eggplant (*Solanum melongena* var. *esculentum*) is a warm-season vegetable related to tomatoes and peppers. The plant is bushy with heart-shaped leaves and purple star-shaped flowers. The edible portion is the egg-shaped fruit, which is generally large and purple in color and is harvested when it has reached the size appropriate for the cultivar but before the skin has lost its natural sheen. Storage of eggplant should ideally be at 50 to 55°F (10–13°C). They can be damaged by storage in the refrigerator.

Eggplants require a long period of warm weather to set and mature fruit properly. When growing eggplant in northern climates, early maturing cultivars should be planted. They are grown from transplants since the plants are very slow to develop from seed. A starter fertilizer should be used at transplanting, followed by a sidedressing of a balanced fertilizer after the first fruit has set. Since eggplants are subject to several soilborne diseases, they should be rotated in the garden to avoid being grown in the same area where related plants (tomatoes, peppers, potatoes) were planted.

Honeydew Melons

See "Cucurbits."

Kale

See "Cole Crops."

Lettuce

Lettuce (*Lactuca sativa*) is a cool-season vegetable. It is available in a number of types (Figure 6-26), those which form a firm head (called *head lettuce* or *iceberg*), semiheading types such as bibb and Cos (romaine), and a type that produces no head at all, called *leaf lettuce*. Heading and semiheading lettuce is harvested by cutting the head from the roots when it is full grown and, in the case of head lettuce, firm. Leaf lettuce can be harvested by removing leaves individually or by cutting the plant at ground level. Head lettuce and romaine will keep one to two weeks in the coldest part of

Figure 6-26 Four types of lettuce, clockwise from upper left: head or iceberg type, a curly leaf lettuce, romaine or cos type, and butterhead type. (*Photo courtesy of W. Atlee Burpee and Co., Warminster, PA*)

the refrigerator, but leaf and bibb lettuce should not be stored more than one or two days.

Lettuce will grow in most soils, however, emergence of the seedlings in heavy soils may be hindered by soil crusting. To remedy this, mix sieved organic matter into the soil used for covering the seeds. The crop requires cool temperatures and is therefore best grown as either a spring or a fall crop. It can also be used as a winter crop in mild-winter areas, provided no hard freezes occur. During hot weather lettuce will become bitter and tend to bolt. As a preventive measure, plant only during cool weather and choose heat-resistant cultivars.

Leaf and semiheading cultivars are generally seeded shallowly and later thinned to a 10-inch (25-centimeter) spacing. Head and romaine lettuce may be either seeded or transplanted. Transplants should be spaced at least 10 inches (25 centimeters) apart; seedlings should be thinned to the same spacing as soon as they begin to crowd. If not thinned, the lettuce will be prone to rot and may fail to head.

Difficulty in germinating may be due to overly warm soil. If this is suspected, place the seed in water in the refrigerator overnight prior to planting. Lettuce should be kept well watered and should be side-

dressed with nitrogen-rich fertilizer once during its growing period.

Melons

See "Cucurbits."

Muskmelons

See "Cucurbits."

Onions

Onions (*Allium cepa*) are a popular, easy-to-grow vegetable crop. They grow best in a sandy soil rich in humus. When onions are growing in heavy soils, large amounts of organic matter should be added before planting.

Onions are classified as a cool-season crop, but are tolerant of a wide range of temperatures including frost. The best-quality bulbs will be produced when temperatures are cool during the early part of the growing season and moisture is abundant.

Onions can be grown from seed, transplants, or small bulbs called *sets*. Green onions (sometimes mistakenly called shallots) can be easily grown from seed, but bulb types will be ready for harvest sooner if sets or transplants are used. They may be planted 3 to 4 inches (8–10 centimeters) apart or closer together with every other plant pulled for use as a green onion. Regular watering and one or two sidedressings of balanced fertilizer during the growing season are recommended.

The most important factor affecting onion growing is photoperiod. Onions form bulbs only when the night is shorter than the critical photoperiod for that cultivar. If the critical photoperiod is reached too soon after seeding, the onions will bulb before the plants are large enough, and the resulting bulbs will be very small. If the critical photoperiod is never reached, the onions will never form bulbs.

To avoid bulbing problems, select only cultivars recommended for the area. In areas where both winter and summer crops can be raised, a different cultivar must be selected for each, due to the difference in night length between the seasons.

Onions can be harvested at almost any stage of maturity, depending on the proposed use. Young onions, regardless of cultivar, can be harvested before bulbing for use as green or "bunching" onions. In this stage they are perishable and should be refrigerated. If mature onions with bulbs are desired, a cultivar bred for its storage characteristics should be selected, and the onions should be allowed to remain in the garden until the tops fall over. This signals that the bulb is mature. After digging out the bulb, cut the leaves off about 1 inch (2–3 centimeters) above the bulbs. The bulbs should be dried outside if the weather is warm or in slatted crates or mesh bags in a warm indoor location. After several weeks of drying, the onions can be stored in a cool, dark location.

Peas

Peas (*Pisum sativum*) are a cool-season garden crop that shows a spectacular improvement in quality over purchased produce when home grown and picked immediately prior to eating. Peas require cool temperatures and abundant moisture to grow well. They will tolerate moderate freezes but not heat. They are best grown as an early spring or late fall crop or, in mild-winter areas, as a winter crop. Where warm weather makes growing peas difficult, a heat-resistant cultivar such as 'Wando' should be chosen.

Pea seeds are sown about 1 inch (2–3 centimeters) apart and are not thinned. A common practice is to sow peas in double rows about 8 inches (20 centimeters) apart and train two rows on a single string trellis.

It is important to supply water regularly to peas and to pick the pods as soon as they are mature to lengthen the harvest. A side-dressing with balanced fertilizer is recommended when the plants are 4 to 6 inches (10–15 centimeters) tall.

The peas should be harvested when they begin to fill out the pod but before becoming tough. Storage of peas is not recommended, but if unavoidable, they should be kept in the coolest part of the refrigerator.

Peppers

Peppers (*Capsicum annuum*) are known as either "sweet" or "hot." The sweet peppers include bell, pimento, Mexican stuffing ('Anaheim') and sweet banana types. The hot peppers usually have fruits that taper to a point and can be green, red, or yellow. Sweet peppers may be picked any time after the fruit has attained full size and the color characteristic of the particular cultivar. Bell peppers that are not picked soon after they reach full size will eventually turn red; they are still edible and not hot, and some people prefer them at this stage because they are less bitter. Hot peppers can be picked when the full size and proper color have been reached or can be allowed to dry on the plant for winter use. However, picking the fruits as they ripen will increase the total fruit production.

A common problem encountered by home gardeners raising both hot and sweet peppers is cross-pollination. Cultivars of peppers cross freely, and although the fruits will each retain their proper taste, the seeds will bear the characteristics of both parents. The seeds of sweet peppers grown near hot peppers should therefore be removed before eating.

Another problem is poor fruit set which is usually weather-related. Both excessively warm temperatures and cloudy, wet weather will prevent fruit set.

Fresh peppers should be stored in the refrigerator and will retain good quality there for a week or more.

The soil, climate requirements, and culture are the same as for eggplant.

Pumpkins

See "Cucurbits."

Radishes

Radishes (*Raphanus sativus*) are the fastest-growing garden vegetable and will produce an edible root in 21 to

Figure 6-27 An icicle winter radish, also known as daikon. (*Photo courtesy of W. Atlee Burpee and Co., Warminster, PA*)

30 days. These quick-maturing radishes are the red globe-shaped type. The second type, called *Chinese* or *icicle* radishes, takes longer to mature, but produces larger roots which are white. Globe radishes are suitable for planting in the spring and should be picked as soon as they reach edible size, or they become tough and pithy. Icicle cultivars (Figure 6-27) are preferred as a fall crop where the weather is warm.

Both are cool-season crops and will grow best in temperatures below 70°F (21°C). Under warm temperature conditions they tend to be hot and to bolt. Seed should be sown in early spring to avoid the warm temperatures which slow growth and cause bolting. Plants should be thinned to 1 inch (2–3 centimeters) apart soon after germination. Several sowings made one week apart are recommended to assure a steady supply. Radishes can also be seeded along with carrots or leeks or between transplants of other crops since they mature quickly and can be harvested before crowding occurs.

Squash, Summer

Includes yellow straightneck, yellow crookneck, and zucchini. See "Cucurbits."

Squash, Winter

Includes acorn, butternut, spaghetti, and other thick-rind types. See "Cucurbits."

Swiss Chard

See "Chard."

Tomatoes

Tomatoes (*Lycopersicon lycopersicum*) are the most popular of all garden vegetables. Although they are actually herbaceous perennials, they are grown as annuals in the vegetable garden and develop into branched bushes or vines with compound leaves and yellow flowers.

Botanically, tomatoes are related to eggplants and peppers but are easier to grow than either. The fruit will be most flavorful when allowed to ripen fully on the plant. Green tomatoes can be eaten pickled or fried.

Tomatoes will succeed in a wide range of soil types provided the drainage is adequate. However, they should not be planted in the same place year after year due to the increased danger of wilt diseases and nematodes. A pH between 6.0 and 6.8 is best.

Like peppers and eggplants, tomatoes are a warm-season crop. For fruit to set, temperatures must be warm since pollen is usually not shed when night temperatures are below 59°F (15°C) or above 80°F (27°C). In areas with cool summers or short frost-free growing seasons, early cultivars should be selected since they not only mature more quickly but grow better at cool temperatures. Cultivars that produce small fruits such as cherry tomatoes tolerate a wide range of temperatures.

Although tomatoes can be raised easily from seed planted directly in the garden, they are usually raised from transplants for an earlier harvest. Spacing of the plants will depend on whether the cultivar to be grown is "determinate" or "indeterminate." Determinate tomato cultivars grow into bushy plants which can be either staked, caged, or allowed to sprawl along the ground. Indeterminate types grow long and vinelike and definitely require support.

A balanced fertilizer should be applied prior to planting and a nitrogen sidedressing one month later after which nitrogen applications should be restricted. When planting tomatoes from transplants, a starter fertilizer is beneficial.

Although pruning is sometimes advocated for home-grown tomatoes, most experts now agree that pruning only slows fruit production and increases sun scald of the fruit as well.

Turnips

Turnips (*Brassica rapa*) are a cool-season root crop related to cole crops. Though the root is the primary edible portion, the leaves are also edible; in fact some cultivars are grown mainly for their leaves. Turnip leaves do not store well and should be picked immediately prior to eating. Roots, on the other hand, store for long periods if the tops are removed and if kept in a cool, humid environment. They should be harvested when they are 2 to 3 inches (5–8 centimeters) in diameter.

Turnips grow best in a light- to medium-weight soil rich in organic matter but will produce large crops in almost any soil as long as the drainage is adequate. They are planted as a spring or fall crop in most areas or as a winter crop in mild-winter areas. In locales with hot summers, fall plantings are most successful.

Turnips are grown from seed that germinates rapidly in the garden. The young plants grow vigorously and will need to be thinned so that they are about 3 inches (8 centimeters) apart when they are 3 inches (8 centimeters) tall. Thinning will not be required if the turnips are grown for greens only.

During the growth period the plants should be kept well watered and sidedressed once or twice with a light application of balanced fertilizer. This will help the plants to grow rapidly and will assure tender roots.

Watermelon

See "Cucurbits."

Zucchini

See "Cucurbits."

Minor Vegetables

Black-eyed Peas

Black-eyed peas or cowpeas (*Vigna unguiculata* or *V. sinensis*) is a warm-season crop seeded 2 to 4 inches (5–10 cm) apart in rows 6 feet (2 meters) apart. A slightly acid soil is best. The plants are beanlike, erect, and normally do not require training, although some cultivars of *V. s. sesquipedalis*, also called cowpea, are vining, require support, and produce yard-long (one meter) pods. The crop is harvested when the pods, which are normally 3 to 8 inches (8–20 cm) long, are slightly yellow but before the seeds inside are dry.

Artichokes

Two very different plants are called artichokes. The first is the globe artichoke (*Cynara scolymus*), a tender perennial generally grown only in areas where the ground does not freeze. The second is the Jerusalem artichoke (*Helianthus tuberosus*), a hardy perennial that can be grown throughout the United States and Canada.

Globe Artichokes

The globe artichoke is a large, thistlelike plant up to 6 feet (1.8 meters) in diameter. The large flower buds it produces in late spring and early summer are the edible portion of the plant (Figure 6-28).

The globe artichoke grows best in areas where winter temperatures are mild and summers are cool. Areas along the California coast, where summer fogs are common, are ideal. In hot-summer regions artichokes often produce buds which open rapidly and have a leathery texture. In zones 1 through 7, artichokes can be grown in containers in summer and moved indoors in the winter, or they can be dug after cold has killed the leaves and the crowns stored in damp sawdust until spring. A heavy mulch will generally provide sufficient protection in zones 8 and 9 to overwinter the crowns.

Artichokes are usually grown from divisions that consist of a woody stalk, a root, and leaves. This division should be planted vertically, with the base of the leaves slightly below the soil surface. In most areas late winter or early spring is the best time to plant. Allow 3 to 4 feet (1–1.2 meters) between plants. A starter fertilizer is beneficial at planting.

After the plants begin to grow, plenty of water and nitrogen fertilizer should be provided. When buds appear, they should be harvested before they begin to open by cutting the stem ½ to 1 inch (1–3 centimeters) below the base and then refrigerated until ready to eat. An artichoke will keep two to three weeks if it is kept in the coldest part of the refrigerator in a plastic bag.

After the buds are cut, the supporting stem will shrivel and should then be cut back to healthy growth. New shoots will sprout from the base to provide next year's harvest. In cold-weather regions nitrogen fertilizer should be withheld after midsummer to enhance cold tolerance.

Jerusalem Artichokes

The Jerusalem artichoke is a sunflower with coarse, hairy leaves which grows to a height of 6 feet (1.8 meters). The edible portions of the plant are the underground tubers which are produced abundantly in the fall. The tubers are harvested after frost or may be left in the ground all winter and dug as needed. After harvest the tubers can be stored from two to five months under high humidity and at temperatures as close as possible to freezing.

The Jerusalem artichoke can be grown in zones 2 through 9. However, the best quality and highest yields occur in the colder zones. Jerusalem artichokes thrive in any well-drained soil. Highest yields are obtained in

Figure 6-28   A field of globe artichokes. (*USDA photo*)

soils with poor nitrogen content, so Jerusalem artichokes may be grown in the poorest section of the garden.

Propagation is from tubers which may be purchased from nurseries or seed companies or the produce department of a grocery. Tubers are planted 3 to 5 inches (8–13 centimeters) deep (depth is not critical) and 5 to 10 inches (13–25 centimeters) apart in spring. The plants are very vigorous and generally do not require any fertilization or care until harvest. When harvesting, remove all tubers except a few to start the next season's crop. If many tubers are left in the ground, the plant may become a weed problem.

Celery

Celery (*Apium graveolens*) is one of the most difficult vegetables to grow in the home garden because in ad-

dition to its preference for soil rich in organic matter, celery requires a long growing season with cool temperatures (65–70°F, 18–21°C). If temperatures are too warm, it will become tough. If prolonged periods with temperatures below 50°F (10°C) occur, it will bolt before leaf and stalk development take place.

Celery is generally grown from transplants 3 to 4 inches (8–10 centimeters) tall planted 10 inches (25 centimeters) apart. Seeds require three weeks to germinate and another month to reach transplanting size. Regular watering and frequent fertilizing with a high-nitrogen fertilizer will maximize growth. Plan on approximately a five-month period from seeding to harvest.

Blanching excludes light from the stalks and decreases their chlorophyll content. It was a common practice in the past and can be accomplished by mounding soil around the lower stalks or otherwise shading them. Most gardeners today do not blanch celery since it results in a lower nutritional value and many cultivars are "self-blanching."

It is harvested by cutting the plant at ground level or by removing individual stalks as needed. Celery will keep for several weeks if refrigerated and kept in a humid condition.

Chinese Cabbage

Chinese cabbage (*Brassica rapa*, Pekinensis Group), sometimes called *celery cabbage*, is a cool-season annual grown for its leaves (Figure 6-29). Seeds should be sown early in spring or as fall approaches to avoid hot weather and flowering. Seedlings should be thinned to 8 inches (20 centimeters) apart when they begin to crowd; the heads are later harvested individually. This is a very fast-growing leaf vegetable prized for Chinese dishes, and several distinct types with varying leaf shapes are grown.

Curly Endive

Endive (*Cichorium endivia*) (Figure 6-30) is grown for its slightly bitter leaves, which are used fresh as a salad green. Seed should be planted in early spring, and leaves can be harvested whenever they are large enough

Figure 6-29 Chinese cabbage of the Napa type. (*Photo courtesy of W. Atlee Burpee and Co., Warminster, PA*)

Figure 6-30 Curly endive. (*Photo courtesy of W. Atlee Burpee and Co., Warminster, PA*)

Figure 6-31 Kohlrabi 'Grand Duke'. (*Photo courtesy of W. Atlee Burpee and Co., Warminster, PA*)

Figure 6-32 Leeks. (*USDA photo*)

Figure 6-33 'Clemson Spineless' okra. (*Photo courtesy of W. Atlee Burpee and Co., Warminster, PA*)

Figure 6-34 Parsnips. (*Photo courtesy of W. Atlee Burpee and Co., Warminster, PA*)

to eat. For better flavor the mature outer leaves can be tied over the inner leaves to blanch them.

Kohlrabi

Kohlrabi (*Brassica oleracea*, Gongylodes Group) (Figure 6-31) is a relative of cabbage grown for its enlarged stem. Plants are harvested when the enlarged portion is 2 to 4 inches (5–10 centimeters) in diameter. See "Cole Crops" for cultural information.

Leeks

Leeks (*Allium ampeloprasum*) (Figure 6-32) are a non-bulbing relative of the onion with a mild flavor. They are generally grown from seed and require a long, cool growing season. The white underground stem is eaten at any size. Soil can be mounded around the plants to blanch more of the stem, thus producing more edible portion per plant.

Mustard

Mustard (*Brassica juncea*) is a fast-growing, cool-season, vegetable grown for its strong-flavored greens. It is seeded directly in the garden in early spring, and indi-

vidual leaves are picked when they reach 3 to 4 inches (8–10 centimeters). Planting early is important since mustard will flower and die as soon as the weather becomes warm.

Okra

Okra (*Abelmoschus esculentus*) (Figure 6-33) is a large, fast-growing plant that produces edible seed pods. To grow well, okra requires hot weather, well-drained soil, and a relatively long growing season. The pods should be picked before they reach 3 inches (7.6 centimeters) in length.

Parsnips

Parsnips (*Pastinaca sativa*) (Figure 6-34) produce white, carrot-shaped roots with a sweet, nutlike flavor. Cool-season vegetables, they should be seeded in deeply worked soil very early in the spring and grown like carrots. The roots can be harvested at any size and can be allowed to remain in the ground for winter and spring eating.

Parsnip seeds are among the shortest-lived of all vegetable seeds, so storage of old seed is not recommended.

Potatoes

Potatoes (*Solanum tuberosum*) are grown for their underground tubers and will grow in most climate zones. However, quality will be best in a climate with cool nights.

Potatoes are generally grown from "seed potatoes," small pieces of a potato which contain at least one "eye" or bud per section. Only certified seed potatoes that are guaranteed to be disease-free should be grown. A cultivar that can be grown from seed started in flats about eight weeks prior to transplanting is also available.

"New potatoes" are thin-skinned, immature potatoes harvested when the plants begin to flower and the potatoes are still small. They should be eaten immediately after digging in summer. To have potatoes for storing for winter use, delay harvest until the tops of the plants have died in fall.

Unless a large garden space is available, potatoes are not considered worth raising in the home garden. The quality will not differ substantially from that of supermarket potatoes, and monetary savings are usually minimal. In addition, home-grown potatoes require careful attention to control insects and diseases.

Rhubarb

Rhubarb (*Rheum rhabarbarum*) is a perennial grown for its leaf stalks, which have an acid, fruitlike flavor. It will grow in all USDA zones. One- or two-year-old crowns should be planted in spring and light harvesting started the following spring by removing one or two leaves per plant. The leaves should not be eaten, as they contain oxalic acid and are poisonous.

Flower stalks should be removed, as they appear to divert the strength of the plant into vegetative growth. Fertilizing once or twice per year is sufficient, and clumps should be divided about once every four years.

Rutabagas

Rutabagas (*Brassica napus*, Napobrassica Group) are grown for their large turniplike roots, which have a flavor similar to that of a strong turnip. Their culture is the same as for turnips.

Snow Peas, Chinese Pea Pods, Edible-Podded Peas (*Pisum sativum* var. *macrocarpon*)

Culture the same as for peas. See Figure 6-35.

Figure 6-35 Snow peas or Chinese pea pods. (*Photo courtesy of W. Atlee Burpee and Co., Warminster, PA*)

Spinach

Spinach (*Spinacia oleracea*) is a nutritious cool-season vegetable grown for its strongly flavored leaves. It should be seeded in early spring directly in the garden and harvested before the weather becomes hot because the plants flower in warm weather.

Sweet Potatoes

Sweet potatoes (*Impomoea batatus*) can be grown in warm climates with a frost-free growing period of 120 days or longer. They are grown from transplants spaced 24 to 30 inches (0.6–0.7 meter) apart and can be harvested as soon as they reach edible size. If plans include storing the sweet potatoes for winter use, harvest should be delayed until frost kills the tops. The roots are then dug, cured at 85°F (30°C) for three weeks, and stored at 55°F (13°C). Care should be taken that the roots are not bruised and that curing is performed properly, otherwise sweet potatoes will not store well.

Selected References for Additional Information

Abraham, G. *The Green Thumb Garden Handbook.* Englewood Cliffs, NJ: Prentice-Hall, 1977.

Bush-Brown, J., and L. Bush-Brown. *America's Garden Book.* Revised ed. New York: Charles Scribner's Sons, 1980.

Crockett, J. U. *Crockett's Victory Garden.* New York: Little, 1977.

Doscher, Paul, et al. *Efficient Vegetable Gardening.* Globe-Peguot, 1995.

Doty, Walter L. *All About Vegetables.* San Francisco: Ortho Books, 1990.

Fell, Derek. *Vegetables: How to Select, Grow and Enjoy.* Tucson, AZ: HP Books, 1982.

Knott, J. E. *Handbook for Vegetable Growers.* 3rd ed. New York: Wiley, 1988.

Neucomb, Duane and Karen. *The Complete Vegetable Gardener's Sourcebook.* Englewood Cliffs, NJ: Prentice-Hall, 1989.

Ortho Books. *All About Vegetables.* San Francisco: Ortho Books, 1981.

Pierce, Lincoln C. *Vegetables: Characteristics, Production, and Marketing.* New York: Wiley, 1987.

Time-Life Publications. *Summer Vegetables.* Alexandria, VA: Time-Life Books, 1988.

Yamaguchi, Mas. *World Vegetables.* Westport, CT: AVI Publishing, 1983.

7

Growing Tree Fruits and Nuts

Tree crops are raised for their edible fruits or seeds. Most fall into the designations of fruits or nuts, but a few, such as the avocado and olive, grow on trees but fit neither category.

When a gardener decides to grow fruit or nut trees, the species and the number selected will depend on many factors, among them the following:

1 *Climate.* Minimum winter temperature as well as average summer temperature and length will determine which species and cultivars can be grown. A few tree crops such as avocado and citrus are semi-tropical and will tolerate temperatures only to the mid-20°F (−2°C) range. Others such as sour cherries are adapted to climates with widely varying winter and summer temperatures. They will not survive without a cool winter period and will fail to produce fruit and even die in a warm-winter area.

Unexpected spring frosts in temperate areas can also cause problems with some fruits that flower early. Apricots are notably vulnerable because of their early bloom period.

Summer temperature and length can also affect ability to grow some fruit trees successfully. During some years the plant will not receive enough warmth for proper fruit ripening and sugar formation, or the growing season will not last long enough for the fruit to develop completely.

2. *Property size.* The size of the property will also determine how many fruit and nut trees can be raised. Owners of small properties frequently incorporate fruit and nut trees into the landscaped areas in lieu of ornamental trees. Others use dwarf cultivars which grow only about 8 feet (2.4 meters) high. Although this is a solution with most fruit trees, no nut trees are available in dwarf form.

3. *Expected maintenance time.* Although some species have little problem with pests and diseases, others, such as the apple, will require spraying every two weeks in addition to yearly pruning and fertilization. The considerable amount of time that will need to be devoted to maintenance should be taken into consideration.

Site Selection

A property may have several possible sites for planting fruit and nut trees. If so, the following criteria (in order from most to least important) should be taken into account when the spot is selected.

Soil Drainage

Few plants will grow in an area with poor drainage, and trees are no exception. Water must drain away rapidly to allow air to penetrate to the roots. If it does not and the roots are in soggy soil for an appreciable period, the tree will die. *Stone fruits* (those with a single, hard pit such as cherry, plum, peach, and nectarine) are the most susceptible to poor drainage.

Amount of Sunlight

Fruit trees should have as much sun as possible to bear heavily and ripen their crops. Six hours of direct sun per day is considered minimum for average fruit production. Although a sunny spot is also desirable for nut trees and will speed their growth considerably, it is not as essential as for fruit trees. The nut trees, being ultimately larger, will compete for sun, whereas dwarf fruit trees will not.

Land Slope and Exposure

In all but subtropical and tropical climates, situating fruit and nut trees on the upper part of a slope is ideal because it partially protects against frost injury. As discussed in Chapter 1, cool air flows downhill, leaving the higher area of the slope several degrees warmer. The few degrees difference may prevent a frost that kills the season's crop. Unless it is unavoidable, tree crops should not be planted in a low spot or at the bottom of a slope. Flat land is preferable to those areas.

The direction in which the slope should ideally face will depend on the weather conditions in the area. South-facing slopes warm earlier in the spring, and trees will start growth sooner. However, this also means that they may be in a tender state and susceptible to late spring frosts. Consequently, a north-facing slope is suggested in climate areas with spring frost problems, but southern slopes are best in areas that are completely or nearly frost-free. South slopes are also advisable where summer temperatures are likely to be too cool for complete fruit maturation, since a south-facing slope receives direct sun the entire day and will be several degrees warmer.

Incorporating Tree Crops in the Landscape

On small properties, a homeowner may have to work fruit and nut trees into the landscape design in lieu of setting aside a separate orchard area. How the trees can be incorporated will depend on the unique features of the property, but the following suggestions can aid in planning.

1. *Substitute nut trees for deciduous shade trees.* Since nut trees are large, they work well as substitutes for shade trees. Many have a pleasing fall color, although the flowers are inconspicuous. Large fruit trees also make handsome shade trees. However, if spraying is a necessary part of the maintenance program, the job will be harder with a large tree.

2. *Use dwarf trees as a screen.* Three or more dwarf trees planted close together (about 6 to 8 feet or 1.8 to 2.4 meters apart) will eventually grow to form a screen. Staggering the plants in two rows (Figure 7-1) will create the screen effect faster without crowding the trees.

3. *Use dwarf fruit trees as patio trees.* Fruit trees are very ornamental with lovely spring blossoms and heavy sets of ripe fruit. Dwarf types can be conversation pieces, but semidwarfs should be planted if shade is desired.

4. *Espalier fruit trees.* The flat training method called *espalier* (Figure 7-2) can fit several dwarf fruit trees into the smallest yard. The tree will still bear a worthwhile amount of fruit, although probably not as much as it would growing in a normal form.

Tree Selection

In selecting an ornamental plant, the cultivar may be picked for its outstanding flowers or vivid fall color, but neither is crucial to its success in the landscape. But choosing a proper cultivar is very important in fruit and nut growing. Choice of the wrong cultivar can lead to failure, so it must be chosen carefully. Read not only the nursery catalog description but also state Agricultural Extension Service publications for information on the best cultivars for the area. Look for the following information in making a decision.

Figure 7-1 Fruit trees used as a screen in a landscape.

Figure 7-2 Pear trees espaliered diagonally. (*Courtesy of the American Floral Marketing Council*)

Unless you live near a nursery specializing in fruit and nut trees, the best selection is available from mail-order nurseries. Several firms specialize exclusively in fruits and nuts and offer many cultivars of each species. Some are listed under "Selected References" at the end of this chapter.

Winter Hardiness

Some fruit cultivars are bred for growing in the South, whereas others of the same species are bred for the North. A particular southern peach variety may not survive a northern winter, much less produce the expected size and quality of fruit.

Required Fruit Maturity Period

Cultivars of fruits bred for cold climates ripen more quickly than those bred for warm climates where the summer is longer. A cultivar should be chosen that will mature and ripen within the limitations of the frost-free growing season (Chapter 6). Moreover, fruits advertised as maturing extremely early in the season will often be inferior in quality to later-maturing cultivars.

Disease Resistance

Selecting disease-resistant cultivars can eliminate much of the labor of spraying against fungus- and bacterium-caused diseases. Apple varieties, for example, are available that are resistant to scab, fire blight, and cedar rust, three common disease problems. Chinese chestnuts are immune to chestnut blight, whereas the American chestnut is so susceptible that only a few trees are alive in the United States today.

Fruit Use

Some fruit cultivars are primarily for fresh eating, whereas others (originally bred for commercial orchard use) are best for drying or canning. Prune versus fresh-eating plum cultivars are examples. Any fruit bred for processing can, of course, be eaten fresh, but its quality may be poorer than that of a fresh-eating cultivar.

Fruit Appearance

Size, coloring, and shape are included under fruit appearance. Many gardeners have a preference for yellow over red apples, for example, or a cultivar that produces mammoth fruits. But appearance should not be allowed to outweigh the important factors of hardiness, disease resistance, maturity period, and yield.

Pollination Requirements

Many fruits and nuts require cross-pollination with another cultivar or, with nuts, a tree of the opposite sex to bear a crop. Many homeowners have grown healthy trees for years without a crop, due to failure to recognize the pollination requirement of the variety.

Dwarf Fruit Trees

Nut trees are available only in "standard" (normal) sizes. However, most fruit trees come in three sizes: standard, semidwarf, and dwarf. Standard trees grow to the normal size for the species and are large in most cases. A standard apple tree can grow 30 feet (10 meters) across and to an equal height, although 20 to 25 feet (6–8 meters) is more typical.

Semidwarf trees range from 10 to 15 feet (3–5 meters) and dwarfs from 5 to 12 feet (1.5–4 meters). Most fruit trees sold to home gardeners are semidwarf or dwarf for various reasons. First, because they are shorter, picking, pruning, and spraying are easier, and a ladder is not usually required. Dwarfed trees also bear sooner. An average of five to six years from planting is not unusual for a standard tree to bear its fruit crop, but a dwarf will bear the second year after planting and may even produce a few fruits the first year. Dwarf and semidwarf trees occupy less space, and a homeowner with limited land can grow more species and cultivars of fruit in an area than with standard trees.

Dwarf trees are produced in one of two ways. The most common is by grafting onto a *dwarfing rootstock* which prevents the cultivar from growing to full size. The grafting combines the dwarf characteristic of one cultivar with the exceptional fruit of another to produce a plant with the desirable characteristics of each.

Sometimes three cultivars are involved, a rootstock, a scion, and an *interstock* (Figure 7-3). An interstock would be used, for example, when the stock and scion are *incompatible* (will not graft together), but each is compatible with the interstock. Grafting with an interstock might also be used to combine superior fruiting (in the scion), dwarfing (in the interstock), and a vigorous root system (in the rootstock). Many nurseries use this technique commonly to avoid the poor root systems inherent with many dwarfing rootstocks.

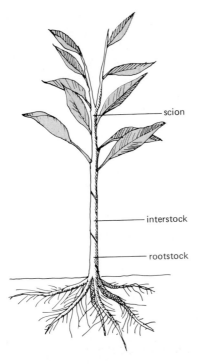

Figure 7-3 An interstock grafted between the rootstock and scion on a fruit tree.

A second type of tree dwarfing is called *genetic dwarfing,* whereby the fruiting portion of the tree remains small due to its genetic makeup rather than because of a dwarfing rootstock. However, genetically dwarf trees are also frequently grafted. The genetically dwarf top is grafted onto a particularly vigorous or disease-resistant root system for the qualities the rootstock can contribute to the tree.

Planting Fruit and Nut Trees

Planting fruit and nut trees is the same as planting an ornamental tree. Deciduous trees are normally sold leafless and bare-rooted for planting in the fall or spring. Evergreens such as olive, citrus, and avocado are sold in plastic sleeves or containers and can be planted at any time of the year.

The hole should be dug large and deep enough to accommodate the root system without crowding. After the plant is settled in the hole, it should be filled around with the removed topsoil. Organic matter such as peat moss or compost can be added but is not necessary. Watering thoroughly afterward is essential.

Most fruit trees and many nut trees are grafted, and the placement of the graft union above ground at planting is important (Figure 7-4). This placement is necessary to retain the special qualities of the rootstock. For example, it may have disease resistance or cause the top to be dwarfed instead of full size. If the scion (top portion) comes in contact with soil, it will send out roots of its own, and the desirable rootstock characteristic will be lost.

Staking for support may be necessary for fruit and nut trees for the first year or two until the root system grows large enough to anchor the plant. Many dwarf cultivars are very shallow and fibrous-rooted, so permanent staking or trellising to prevent blowing over may be advisable. A short stake attached to the truck is usually sufficient. Trees should be tied to the stakes using a wire covered with a section of rubber hose or pieces of rag. Bare wire or string may injure the bark by rubbing or girdling the trunk.

Normally fruit and nut trees are pruned before shipping, but if not, they will have to be pruned after planting. This is discussed under "Pruning."

Maintenance of Tree Crops

Watering

Watering of a fruit or nut tree varies with the species. Some, such as the olive, are very drought-resistant and

Figure 7-4 The graft union of a fruit tree. (*Photo by Rick Smith, Los Osos, CA*)

adapted to arid climates, whereas most others will produce their best crops only with regular rainfall or irrigation.

As a general rule, a newly planted tree should be watered deeply every one to two weeks its first growing season. Decrease the frequency in subsequent years to about every three weeks, depending on the soil water-holding capacity and natural rainfall.

Watering with a drip irrigation system (see Chapter 11) conserves both water and labor. Many commercial orchards use drip irrigation systems, and they can be highly recommended for home use.

Fertilization

Tree crops are normally fertilized yearly on a precautionary basis rather than because the tree has shown any distinct nutrient deficiency. Without seasonal soil tests it is impossible to apply the exact amount of fertilizer required, and yearly fertilizing is used to prevent possible deficiencies.

Methods for applying fertilizer include the drill hole method explained in Chapter 5 and the surface topdressing method. The drill hole method is used when the area under the tree is covered with grass or groundcover plants which would be injured by the fertilizer application.

General recommendations specify applying granular-type fertilizers in spring before or just as new growth starts. About ½ cup (120 milliliters) of nitrogen-rich fertilizer per inch (2.5 centimeters) of trunk diameter will supply an adequate amount of

nutrition. A second application should be given in mid-summer.

Micronutrient fertilizers are frequently used to prevent iron and zinc deficiency on citrus and pecans growing in alkaline soil. Either foliar or soil applications should be made repeatedly until the foliage regains its healthy green color.

Pruning

Some fruit trees, citrus, olive, and avocado, for example, require no pruning to change their shapes. Most nuts also do not require pruning, and indeed many are so tall that it is impractical to do so after the first few years. Most deciduous fruits should be pruned, however.

The three main pruning styles are the central leader, modified leader, and vase shape. Espalier training styles are less commonly used. The pruning style selected depends mostly on the species of tree; some adapt more easily to one style than another. In all cases the object of the training is to produce a tree with well-placed, strong branches capable of supporting a heavy weight of fruit without breaking.

Central Leader Form

Central leader form is used on most nut trees, on sweet cherries, and occasionally for apples and pears. The pruning starts at planting, if necessary, with the selection of the *central leader* and several *scaffold branches* spaced widely around and down the truck. Figure 7-5 shows a young tree pruned to a central leader style. Pruning after the first several years should consist of removing rootstock suckers (Figure 7-6) and competing branches arising from the trunk.

Vase Form

Vase form, called *open-center form*, allows two to five scaffold branches to grow from a short trunk (about 1 to 2 feet, or 0.3 to 0.6 meter). The branches originate close together on the trunk, and the central leader is removed to a side branch.

A vase-shaped tree (Figure 7-5) is pruned to a final form directly after planting by selecting the lowest, strongest branches on the tree. Vase pruning opens the center of the tree to admit light and encourage fruiting and keeps the bearing branches of the tree low for easy picking. Vase pruning is common on peach, apricot, and nectarine.

Modified Central Leader Form

In a modified central leader form (Figure 7-5), the tree is first trained to a central leader form for two to three

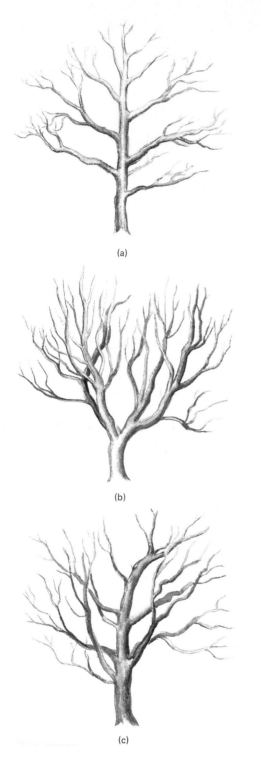

(a)

(b)

(c)

Figure 7-5   Fruit trees pruned to central leader (a), vase (b), and modified central leader (c) forms.

years. The next year the leader is cut back to a strong side branch, leaving three to five scaffold branches on the tree. With the leader gone, the side branches grow stronger and longer, and the height of the tree is kept in

Figure 7-6   Pruning suckers from the base of a tree. (*Photo by George Taloumis*)

check. Modified central leader form is suggested for apples, pears, plums, and sour cherries.

Espalier Forms

Espaliers can take many forms (Figure 7-7). Training must start as soon as the tree is planted; it consists of pruning, bending, and tying the branches as they mature to achieve a symmetrical shape.

Although espaliered fruit trees do not produce as large a crop as conventionally pruned trees, the amount is acceptable considering the small space they use. Apple and pear trees are among the best for espalier training. Peach and plum trees are more difficult but possible. Cherries and nut trees should not be espaliered because their production will be seriously decreased.

Training

Whereas pruning is the primary method for training a tree, several other supplementary techniques help create the desired form. Many fruit trees have branches that grow upward instead of outward, with narrow branch angles called *crotches*. As the branches enlarge, they exert pressure at the crotch, causing it to become weak with a tendency toward splitting under a snow or fruit load. However, when young and flexible, a branch can be trained to develop a wide crotch by two methods.

Tying (Figure 7-8) involves bending a flexible young branch down and securing it to the ground with a stake. The branch should be tied for one or two sea-

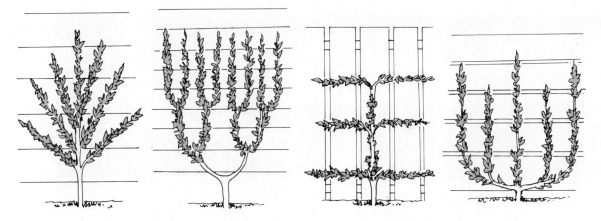

Figure 7-7 Sample espalier forms for fruit trees.

sons and frequently checked to make sure it is not being injured. After the rope is removed, the branch will grow outward instead of upward.

Spreading (Figure 7-9) uses forked or nail-tipped pieces of wood wedged tightly between a young branch and the trunk of a tree to widen the crotch angle. As with tying, the spreader is left in place for one or two seasons to encourage the crotch to remain open.

Fruit Thinning

Fruit thinning is the removal of a portion of the fruits on a tree while they are still small. Many deciduous fruit trees (nut trees are excluded) produce more fruit than can fully mature. If unthinned, the fruit will be undersized and frequently poorly colored. In addition, the excessive fruit load may weaken the tree. Bearing in alternate years may result, with the fruitless year devoted to regaining the vigor of the tree. Since most gardeners want fruit every year, the advantages of thinning are obvious.

Fruits should be thinned when they are ½ to ¾ inch (13–20 millimeters) in diameter, and after the natural drop of young fruit has occurred. Twisting fruit will remove them quickly and easily.

The number of fruits that should be left on a tree depends on the species. Table 7-1 lists approximate distances that should be left between fruits along a branch for moderate fruit load. In practice, fruits are not spaced evenly but are close together on strong branches and far apart on weak ones. The spacing guidelines should be taken only as a guide and not an absolute. The largest, healthiest fruits should always be left on the tree, regardless of their relative spacing on a branch.

If a gardener has a large orchard of, for example, apples, hand thinning may be impractical. In that case the trees may be thinned chemically with a hormone application shortly after bloom. The gardener should consult his county Cooperative Extension Agent for specific recommendations.

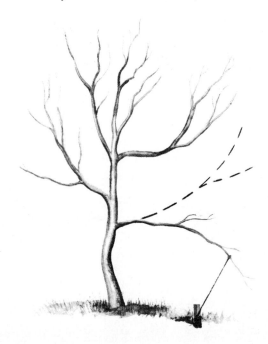

Figure 7-8 Tying a fruit tree branch to the ground to encourage a wide branch crotch to form.

Renovating Neglected Fruit Trees

When buying an older home, it is common to inherit at the same time one or more neglected fruit trees. The tree frequently will be bearing sparsely, and the fruit produced will be small and of poor quality. It may have dead limbs or be diseased.

The question of whether or not to save the tree is a personal choice. It will probably never bear the quality of fruit of a newer cultivar, but fruit quality can be improved to an acceptable grade if the tree is given some

Figure 7-9   A branch spreader in place in a tree. (*Photo by Rick Smith*)

attention. The tree may be valuable for shade and flowers and thus be worth renovating from this standpoint.

Renovation pruning should be spaced over three years to avoid stimulating excess sucker growth. The first pruning should remove dead wood, suckers, and some of the crossing branches. (See Chapter 11 for pruning specifics.) Leaves and fruits that drop in the fall serve as a source of disease reinfection. They should be removed, and a disease and pest control program should be started.

During the next two seasons excess branches can be gradually removed to open up the head, admitting light and improving fruit production. Fertilizer can be withheld until after pruning is complete but should be applied immediately if deficiency symptoms are present.

Failure to Bear

The failure of a tree to bear may be caused by one or a combination of the following factors.

Age

The age at which a tree can be expected to produce its first crop varies with the species. Most dwarf and semidwarf types bear by the third or fourth year after planting.

Pollination Problems

In most fruit and all nut trees, there must be pollination and fertilization to set a crop. Exceptions to this rule are rare but include fruits such as navel oranges, persim-

Table 7-1   *Suggested Fruit Load after Thinning for Common Fruit Species*

Species	Distance between fruits
apples	6–8 in. (15–20 cm)
apricots	1½–2 in. (4–5 cm)
avocados	Thinning not necessary
cherries	Thinning not necessary
citrus, figs, mulberries	Thinning not necessary
nectarines, olives	4–5 in. (10–13 cm)
peaches	6–8 in. (15–20 cm) for early varieties, 4–5 in. (10–13 cm) for later ones
pears	Thinning not necessary but can be thinned lightly to increase size
persimmons	Thinning not necessary
plums, European	Thinning not necessary
plums, Japanese	2–4 in. (5–10 cm)
quinces	Thinning not necessary

mons, and some figs and pears. These fruits are able to set *parthenocarpically,* that is, without fertilization.

The remaining tree crops require pollination, either self or cross. Self-pollination is the rule on most nut trees, the majority of which bear separate male and female flowers on the same tree and are pollinated by wind. Self-pollination is also found in a few fruit trees, but more often fruit trees are incapable of pollinating themselves and are called *self-sterile.* Pollination with another *compatible* cultivar flowering at the same time is necessary if a crop is to be produced.

The causes of self-sterility are primarily the following:

1. *Incompatibility of the pollen with the female flower parts.* This can be compared to an allergic reaction in which the pollen grains do not function correctly when in contact with the female flower parts of the same plant. Pollen may fail to germinate or grow too slowly to the egg, even though it will perform normally with other cultivars.

2. *Nonviable pollen.* The pollen does not germinate.

3. *Failure of the pollen and female flower parts to attain maturity at the same time.* Pollen is shed after the female flower parts have passed their receptive stage or before they have attained it.

4. *Dioecious plants.* In the case of a few tree crops such as the date, male and female flowers are produced on separate plants. There is simply no pollen produced with which the tree can self-pollinate.

In summary, cross-pollination with a compatible cultivar flowering at the same time is a prerequisite for successful growing of many tree crops and should be carefully taken into account when purchasing trees.

Insufficient Winter Chilling

In climates with mild winters the winter temperatures may not be sufficiently low to provide the necessary chilling period. Deciduous fruit and nut trees are most susceptible.

Insufficient Sunlight

A tree receiving insufficient sunlight will produce lush vegetative growth but few flowers and fruits.

Frost or Rain during the Pollination Period

These two weather conditions can damage flowers or restrict bee activity. Thus no crop will set.

Individual Crops and Their Growing Requirements

Almonds (*Prunus amygdalus* var. *dulcis*)

Almond growing is restricted mainly to USDA zones 7 to 9 and to the western states. The primary problem of growing in colder zones is not minimum winter temperature but an early bloom period which hinders pollination and sometimes frosts the blossoms. The spring rains and high humidity of the east are also a hindrance to pollination and nut production in that climate. Lack of winter chilling excludes almonds from zones 10 and 11.

Almond trees are available in only one size, and they grow 20 to 30 feet (6–9.1 meters) high. They should be pruned to a modified leader or vase form and can be expected to bear the third or fourth year after planting. Cross-pollination with another cultivar is required in almost all cases.

The almond is produced inside a shell, which is in turn encased in a peachlike hull. In fall the hulls split, dropping the ripe nuts in their shells to the ground.

Apples (*Malus pumila* descendants)

Apples are grown in zones 4 through 9. Special tropical cultivars must be selected for zone 9 since apples have a chilling requirement. Most apples are self-fertile, and are available in dwarf, semidwarf, and standard sizes.

For greatest strength, apples should be pruned to a modified leader system. Dwarfs can be expected to bear the first or second year after planting.

Apples are highly susceptible to disease and insect problems, and regular spraying will usually be required for good fruit. However, disease problems can be minimized by selection of disease-resistant cultivars.

Apricots (*Prunus armeniaca*)

Like apples, apricots require winter chilling. They can be grown in zones 5 through 8 and some parts of zone 9. Apricots do not have a cross-pollination requirement and are available in standard and semidwarf sizes.

Apricots bear one to two years after planting and are among the most ornamental of fruit trees. However, their early bloom period exposes them to frost damage. Training can be done in either a modified leader or a vase shape, and pruning should be done yearly to restrict height and increase spread.

Avocados (*Persea americana*)

Avocados are strictly subtropical and tropical (zones 9–11) and are hardy only 20–24°F (−7 to −5°C). Their deep green, evergreen foliage and round forms make them valued as ornamentals.

Avocados divide into three races: the Guatemalan, with large fruit; the West Indian, which has large fruit but is more tropical; and the Mexican, which is hardier but has smaller fruit. Most cultivars are available only in standard size though the cultivar 'Little Cado' is dwarf. When mature they range from 30 to 60 feet (10–20 meters) tall (depending on the cultivar) and may be of equal width. Avocados tend to biennial bearing, and although cross-pollination is not essential, it will result in a better fruit set. Avocado cultivars are divided into two groups, A and B, based on their pollination requirements. For best results choose one tree from each group.

Avocados require little pruning. Flowers appear from winter through spring, depending on the cultivar, and the fruit will mature seven to twelve months later. About three years is average from planting until the first set of fruit.

Cherries (*Prunus avium* and *P. cerasus*)

The sweet cherry (*Prunus avium*) is known for fresh eating, whereas the tart cherry (*Prunus cerasus*) is most useful for pies and other cooking uses. Both have a chilling requirement. The sweet cherry can be grown in zones 5 to 7; the tart cherry is more cold-hardy and will survive to zone 4.

Sweet and tart cherries are available in dwarf, semidwarf, and standard sizes. While tart cherries do not require cross-pollination to produce fruit, sweet cherries have very particular pollination requirements. Careful attention must be paid to selecting compatible cultivars from among the many sweet cherry cultivars available.

Pruning of cherries should be modified leader for sweet types and modified leader or vase for sour cultivars.

Chinese Chestnuts (*Castanea mollissima*)

The chestnut that was originally grown for eating in North America was the American sweet chestnut. But in the early 1900s a blight disease killed nearly all and now the Chinese chestnut is the primary source of nuts. This blight-resistant tree will grow in zones 4 through 8 and averages 40 to 50 feet (12–15 meters).

The chestnut will set nuts without cross-pollination, but a heavier set will result if another tree is situated nearby for pollination. Bearing will begin after five to eight years if the plant is from seed, or after two years with a grafted nursery plant. The nuts are perishable and should be stored in the refrigerator in a plastic bag to prevent drying.

Citrus (*Citrus* spp., *Fortunella* spp.)

All citrus are evergreen shrubs or trees adaptable to growing in zones 9 and 10. A few such as the kumquat will survive in zone 8. The major citrus grown in North America are grapefruit (*Citrus* × *paradisi*), kumquat (*Fortunella margarita, F. japonica*), orange (*Citrus sinensis*), lemon (*Citrus limon*), lime (*Citrus aurantifolia*), and tangerine or mandarin orange (*Citrus reticulata*).

Most are available in standard and dwarf sizes and do not require cross-pollination. Citrus require little pruning. Suckers and crossing limbs may have to be removed occasionally.

Because of their year-round growth cycle, citrus should be fertilized twice per year instead of once. They bear when very young and most species retain their fruits in excellent condition on the plant for many months, the exception being tangerine.

Figs (*Ficus carica*)

The fig is a deciduous tree or large shrub grown in zones 7 through 10. It can even be grown as far north as zone 5 with winter protection. Although the top will die back to the ground, it will resprout each spring from the roots and produce a crop.

Depending on the cultivar, the fig can grow from 30 to 80 feet (10–25 meters) at maturity. No dwarf forms are available. The fruit sets parthenocarpically (without pollination) when the plant is about four years old. In zones 9 and 10 two crops per year (one in June and one in August-November) can be expected. In colder zones only one will set and is matured in late summer.

Since fig trees can grow very large, pruning is necessary to keep them low enough to harvest the fruit. Young trees should be trained to a low, flat, vase shape, but after achieving this form, no seasonal pruning is recommended.

European Filberts or Hazelnuts (*Corylus avellana*)

The filbert (Figure 7-10) is a small tree or shrub which grows 10 to 15 feet (3–5 meters) in height and equal that in width. It is hardy in zones 4 through 9.

Filberts bear separate male and female flowers on the same plant and are pollinated by wind. Although

Figure 7-10 Filberts are excellent nuts for cultivation on small properties. (*USDA photo*)

Figure 7-11 Hickory nuts. (*Photo courtesy of Stark Bros. Nurseries, Louisiana, MO*)

they will set some nuts without cross-pollination, a better set can be expected with it.

Little pruning is required for filberts except for removal of suckers and thinning to permit light to enter the center of the plant. Filberts bear when 2 to 3 years old, and a harvest of 5 to 10 pounds (2–5 kilograms) of nuts is average for a mature 10- to 15-year-old plant with cross-pollination. The nuts are harvested after drop.

Hickory Nuts (*Carya ovata* and *C. lacinosa*)

Hard-shell hickories (Figure 7-11) are related to pecans; in fact there are many hybrids between them called *hicans.* They are grown in zones 4 through 8. Hickory trees are rugged looking, with taproots and a strong central leader form. They bear at 3 to 4 years of age if grafted but may take 10 to 15 years from seed.

A mature hickory is quite tall (up to 70 feet; 20 meters), and pruning is practical only when the plant is young. Many cultivars are self-sterile and require cross-pollination. The nuts are harvested from the ground.

Mulberries (*Morus alba, M. rubra, M. nigra*)

Mulberry trees of one of the above species grow in any zone of the United States. Many are used as ornamentals; fruitless and/or grafted weeping types are sold for this purpose.

The species planted will largely determine its ultimate height. The white mulberry (*Morus alba*) will grow to 80 feet (25 meters) and produce white, pinkish, or purple fruits. *Morus nigra* is the shortest at 30 feet (10 meters), and the southern red mulberry (*Morus rubra*) is intermediate at 60 feet (20 meters).

Mulberries are monoecious, with flowers of separate sexes. Pruning should aim to maintain the naturally spreading branches low so that the fruits will be within reach when ripe.

Nectarines (*Prunus persica* var. *nucipersica*)

Nectarines are a mutation of peaches and not a hybrid between a peach and plum, as is popularly believed. They have been cultivated since the days of the Roman empire.

Culture of nectarines is the same as for peaches, except that they are more susceptible to fruit rots and may require more spraying. They do not require cross-pollination and should be trained to a vase shape. Both dwarf and standard sizes are available. Standard trees are moderately small at maturity, 12 to 15 feet (4–5 meters). USDA zones 5 through 10 and sometimes 11 are suitable for growing nectarines.

Olives (*Olea europaea*)

Olives will grow in zones 9 and 10 or wherever winter temperatures do not fall below 12°F (−11°C). They require hot summer temperatures to mature their fruits, but they are grown for their ornamental value alone in coastal climates where excess humidity decreases the fruit production.

Olive trees are small, 25 to 30 feet (8 to 9 meters), single- or multitrunked trees with silver-gray foliage. They transplant easily (even at a very old age) and are moderate to fast growing. Pruning does not seek to create any particular form, but it should develop strong scaffold branches and remove suckers which would eventually cause a bushlike form. Dead wood and twigs are common and should be removed yearly.

Cross-pollination is not a prerequisite for olive production. Biennial bearing is common but can be minimized by fruit thinning to leave two to three olives per foot of branch. Olives are picked green and processed in lye water and brine to make Spanish or black olives. Picked ripe, they are soaked in brine and spices to make Greek-style olives.

Peaches (*Prunus persica*)

Peaches grow in zones 5 through 10, although care must be exercised in selecting a cultivar with a low winter chilling requirement for zones 8 through 10. They are generally trained to a vase shape atop a short 18-inch (0.5 meter) trunk. Pruning should be done every year to thin the head and keep the top low. Bearing begins at about the third year and will peak during the eighth to twelfth years, with no cross-pollination required. Compared to other fruit trees, peaches are short-lived; trees tend to decline rapidly after 10 to 12 years as a result of winter injury and wood rot infection.

Pears (*Pyrus communis*)

By choosing cultivars carefully, pears can be raised in zones 4 through 9, although zones 5 through 8 provide the best climate. Most pear cultivars are vulnerable to the bacterial disease "fire blight," and highly susceptible cultivars such as 'Bartlett' should not be planted east of the Rocky Mountains.

The trees are long-lived, and although some are self-fertile or set fruit parthenocarpically, most need cross-pollination. Pears are available in dwarf, semidwarf, and standard sizes and should be trained to a modified central leader. They naturally grow very upright, and spreading of the branches may be required.

Pecans (*Carya illinoinensis*)

The pecan (Figure 7-12), like many other nut trees, has a taproot and a strong central leader growth habit. It is available in only one size and will easily grow to 60 or more feet (18 meters) tall when mature.

Zones 6 through 8 are best for pecan growing. The tree is hardy even further north, but the growing season is usually not long enough to mature the nuts.

Pecans begin bearing at four to seven years. Trees are monoecious, but more and larger nuts will be borne

Figure 7-12   Pecan nuts. (*Photo courtesy of Stark Bros. Nursery, Louisiana, MO*)

if cross-pollination is provided. A biennial bearing tendency may develop in older trees. Compared to other nut trees, pecans require relatively high maintenance. In alkaline soils, zinc deficiency is common.

Persimmons (*Diospyros kaki*)

The persimmon (Figure 7-13) is a deciduous tree useful as both an ornamental and a fruit tree in zones 6 through 10 and occasionally in zone 5. The fruits, which are apple size and brilliant orange, mature after the leaves drop. Some people question the edibility of persimmon fruits because of their unpleasant astringency when they look and feel ripe. Astringency is partly varietal, but it can be avoided by ripening the fruits until they are extremely soft or by freezing them before eating.

Persimmons are available only in standard or semidwarf size, with standards growing into round-headed trees 20 to 60 feet (12–18 meters) tall at maturity. The species has brittle wood and should be pruned to modified leader form for strength. Annual pruning may be used to maintain a small size. Most bear fruit within 2 to 4 years after planting, and most do not require cross-pollination.

Plums (*Prunus domestica* and *P. salicina*)

Plums are of two types, the European (*Prunus domestica*) and the Japanese (*P. salicina*). The European group includes the blue and prune plums including the well-known 'Damson' cultivar. Japanese plums encompass the larger red and yellow plums popular for fresh eating. They are very early bloomers and should be rejected in favor of Europeans in climates where frosts frequently kill fruit tree flowers.

Plum trees are available in both semidwarf and standard sizes. Their training method and pollination requirements vary by type. Europeans are most often trained to a modified central leader and will usually set fruit without a cross-pollinating cultivar. Japanese usually require crossing with another Japanese cultivar and are normally trained to be vase-shaped.

Zones 5 through 10 are suitable for plum growing. The trees are long-lived, produce outstanding spring flowers, and can be expected to bear the first crop three to four years after planting.

Quinces (*Cydonia oblonga*)

The quince was a much more popular fruit a few years ago than it is now. It is used today primarily for preserves and for baking whole like an apple. The fruit is greenish-yellow when mature and about the size and

Figure 7-13　Persimmon fruits. (*Photo courtesy of W. Atlee Burpee and Co., Warminster, PA*)

shape of an apple. It is covered with a peach-like fuzz which is wiped off before the fruit is used. The fruit is picked while hard, preferably after a light frost, and should be treated gently to avoid bruising.

Quinces grow in zones 5 to 10 on spreading trees which grow to 15 feet (5 meters) in height. They are slow growing, and little pruning is required beyond the initial vase-shape training. The fruit is produced on the current season's growth.

Bearing age for quinces is three to four years, and about one bushel of fruit per tree can be expected at peak bearing. Cross-pollination is not required, but spraying for diseases and insects is usually necessary because fire blight is a common problem.

Walnuts (*Juglans regia* and *J. nigra*)

Walnuts are of two types: English (Carpathian) (*Juglans regia*) (Figure 7-14) and black (*Juglans nigra*) (Figure 7-15). The English type is the smooth-shelled walnut raised commercially and sold in nut baskets. The black walnut is a rough-shelled, native American tree more commonly grown for shade than nuts. Both are hardy in zones 5 through 10 and reach an average height of 50 feet (15 meters) or more when full grown. However, the summer in the colder zones may be insuffi-

cient to develop kernels of English walnuts, resulting in empty nuts.

Normally, walnut trees are not pruned except to remove crossing and dead branches; they naturally grow in a vase or modified central leader shape. Cross-pollination is not necessary, and the nuts are harvested

Figure 7-14　English walnuts. (*USDA photo*)

Figure 7-15   Black walnuts. (*USDA photo*)

after they drop. Small crops can be expected starting two years after planting.

Care should be taken to plant walnut trees away from ornamental plantings and vegetable gardens. This is necessary because the roots secrete a chemical that is toxic to many other plants.

Selected References for Additional Information

American Horticultural Society. *Fruits and Berries.* Alexandria, VA: American Horticultural Society, 1982.

Bountiful Ridge Nurseries. *Retail Catalog.* Princess Anne, MD 21853.

California Nursery Company. *Retail Catalog.* Niles District, Fremont, CA 94536.

Childers, N. F. *Modern Fruit Science: Orchard and Small Fruit Culture.* 9th ed. New Brunswick, NJ: Horticultural Publications, Rutgers University, 1983.

Cumberland Valley Nurseries. *Retail Catalog.* McMinnville, TN 37110.

Hansen's New Plants. *Retail Catalog.* Fremont, NB 68025.

Hilltop Orchard and Nurseries. *Retail Catalog.* Rt. 2, Hartford, MI 49057.

Lundy's Nursery. *Retail Catalog.* Rt. 3, Box 35, Live Oak, FL 32060.

Rom, Roy C., and Robert F. Carlson (eds.). *Rootstocks for Fruit Crops.* New York: Wiley, 1987.

Stark Bros. Nurseries. *Retail Catalog.* Louisiana, MO 63353.

Westwood, Melvin N. *Temperate-Zone Pomology.* 3rd ed. Portland, OR: Timber Press, 1993.

8

Bush and Other Small Fruits

The species classified as "small fruits" are the small perennial plants such as strawberries, grapes, raspberries and blackberries (called "brambles"), and blueberries. Currants, gooseberries, and cranberries are less well-known members.

Because of their size, small fruits can be grown on almost any property. On a large property, enough fruit can be grown for canning and freezing as well as fresh eating. The small fruits are easier to grow than many tree crops. Cross-pollination is seldom required, and spraying for disease and insects is uncommon. Only one or two species are regionally limited, and most bear within one or two years after planting. By carefully choosing cultivars to mature serially throughout the growing season, a harvest of small fruits can be assured from early summer until frost.

Planning the Small Fruit Garden

Site Selection

Rapid soil drainage and full sunlight are the two site prerequisites for successful small fruit growing. An area where water stands after a rain is unsatisfactory and will kill plantings quickly. Without full or nearly full sun, the plants will grow only vegetatively and will fail to bloom or ripen fruit. Beyond these considerations, virtually any site is acceptable. Areas with good garden soil are ideal but even poor-soil areas can be amended to a satisfactory condition with compost or other organic soil additives.

Many gardeners prefer to plant all their fruits in one area. If this is the case, care should be taken to space and arrange the plants to prevent the taller species from shading the smaller species (strawberries). Since the sun in the northern hemisphere is always slightly in the southern sky, this requires placing the taller plants north of the intermediate and short ones. Figure 8-1 gives two sample layouts for fruit garden areas.

Small fruits can also be incorporated in a landscape. Currants, gooseberries, and brambles can be used as hedge plantings which will grow to 4 to 5 feet (1.2–1.5 meters) tall when mature. Highbush and rabbiteye blueberries grow even taller, reaching 10 feet (3 meters). Low-growing strawberries produce lovely white blossoms in spring and make an excellent border for a perennial flower bed. They can also be used as groundcover plants in small sunny areas or in planters around a patio or deck area. Finally, vining grape plants can be trained to an overhead arbor or espaliered against a wall or fence.

Plant Selection

Of the small fruits discussed in this chapter, all but the cranberry are highly climatically adaptable. The choice is therefore a matter of personal preference and, most important, selecting a cultivar adapted to your climate. The county or state Cooperative Extension Service will furnish literature on the small fruit cultivars most successful in your state.

The following sections have basic information on the small fruits and their growing requirements.

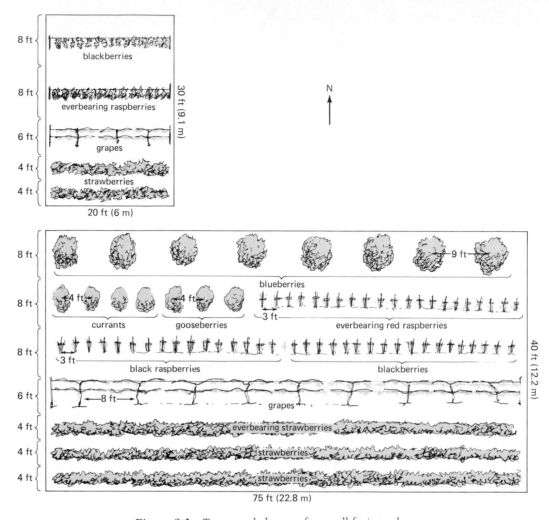

Figure 8-1 Two sample layouts for small fruit gardens.

Blackberries (*Rubus* spp.)

Blackberries are one of the bramble fruits. They look essentially like raspberries except the fruits are larger, and when picked, the center core remains in the berry (with a raspberry it pulls out at picking). They withstand more heat and drought but less cold than raspberries.

Blackberries fall into two categories based on growth habit, height, and climate adaptation. "Erect" blackberries grow 3 to 5 feet (1–1.5 meters) high and grow without supports. The canes of "trailing" blackberries reach 6 to 8 feet (2–2.4 meters), and the plants must be grown on a trellis or staked for support. Erect blackberry types are most often grown in the East, Southwest, and Midwest and are the more cold-hardy of the two types, surviving winter temperatures of down to −20°F (−29°C). Trailing types are the traditional blackberries of the Pacific Coast states and southern states, although they are occasionally grown as far north as Michigan.

There is often confusion in blackberries when the names dewberries, boysenberries, loganberries, and youngberries are used. Dewberries can be one of any species or cultivar of trailing blackberries. Loganberries, actually the cultivar 'Logan,' are a cross between a blackberry and a raspberry. Another blackberry/raspberry cross cultivar is 'Phoenomenal,' which was later crossed with a dewberry to produce the cultivar 'Young,' hence youngberries. The origin of boysenberries is similar to youngberries.

Blueberries (*Vaccinium* spp.)

Blueberries will grow from northern Minnesota and Canada to Florida and to the West Coast; however, they have a soil pH requirement of between 4.2 and 5.5. Since soils that are naturally this acidic are uncommon, pH modification is essential for blueberry growing in most cases. Chapter 5 discusses the use of ammonium sulfate and other compounds for lowering soil pH. Blueberries also grow best in soils high in organic matter. The addition of organic matter such as compost or peat moss will satisfy this need and help to keep the soil pH low.

Table 8-1 *Cultivated Blueberry Species for Home Gardens*

Species	Characteristics	Range
Rabbiteye blueberry (*Vaccinium ashei*)	6–8′ (2–2.4 m) with large fruits; moderately heat- and drought-resistant	Southern states and other areas having a frost-free growing period of at least 160 days
Highbush blueberry (*Vaccinium corymbosum*)	10–15′ (3–4.6 m) tall (unpruned); not drought-resistant; sand and peat soil best; hardy to −20°F (−29°C); requires winter chilling and should only be grown north of zone 8	Zones 5–7; East Coast, Pacific Northwest, Midwest
Evergreen blueberry or evergreen huckleberry (*Vaccinium ovatum*)	Evergreen with ornamental foliage used by florists; grows to 15′ (4.6 m) tall; fruits are small, shiny, black, and strong-flavored	Pacific Coast from central California to British Columbia

The species of blueberry that should be raised depends on the regional climate. Table 8-1 lists the three main cultivated blueberry species and their identifying characteristics and geographic range. Two cultivars of blueberries should be planted for cross-pollination, as some are nearly self-sterile and both more and larger fruits will be produced if a pollenizer is provided.

Besides the cultivated species, additional species such as the dryland blueberry (*V. pallidum*), mountain blueberry (*V. membranaceum*), and lowbush blueberry (*V. angustifolium*) grow in the United States and Canada. However, they are normally wild, and cultivars selected for home growing should be chosen from the three cultivated species.

Cranberries (*Vaccinium macrocarpon*)

Cranberries have exacting growth requirements: moist, peat-bog conditions and a pH of 4.2 to 5.0. Since these conditions are hard to create, cranberries cannot be recommended as a home fruit.

Currants (*Ribes sativum* and *R. nigrum*)

Currants (Figures 8-2 and 8-3) are very resistant to cold temperature and will survive in even the coldest areas. Their growing area spans the northern United States wherever moist, cool conditions exist. In the South they grow less satisfactorily due to prolonged summer heat.

The plants are small attractive shrubs ranging in height from 3 to 5 feet (1–1.5 meters). The plants should be set 4 to 5 feet (1–1.5 meters) apart when grown in the garden or about 3 feet (1 meter) apart when used as an ornamental hedge.

The most popular currants are the red varieties (*Ribes sativum*), which produce large clusters of fruits and are easy to grow. White (*Ribes sativum*) and black (*Ribes nigrum*) currants are relatively rare, but the black varieties are popular in Europe, where the fruits are made into juice.

Figure 8-2 White currants. (*USDA photo*)

Gooseberries (*Ribes hirtellum* [American] and *R. uva-crispa* [European] cultivars)

The growing requirements of gooseberries are similar to those of currants: moist soil and average-to-cool summers such as are found in the Midwest, North, and northern Pacific states. These species are very cold tolerant but will not thrive in a hot summer climate. In warm summer areas planting in partial shade will reduce summer heat and may allow the crop to be raised successfully.

Gooseberries are unknown to most home gardeners. They are shrubs about 3 feet (1 meter) high and densely covered with thorns (with the exception of one or two cultivars). The fruits are the size of large

Figure 8-3 Red currant cultivar "Rondom." *(Photo from* Small Fruit Crop Management *by G. J. Galletta and D. G. Himelrick (eds.), 1990. Reprinted by permission of Prentice Hall, Englewood Cliffs, NJ)*

grapes and are green to red when mature, depending on the cultivar. They can be eaten fresh, as well as made into pies, jam, or jelly.

Grapes (*Vitis* spp. and hybrids)

Grapes can be grown for fresh eating, wine-making, or making into jellies and juice. When a cultivar is selected, the choice should be based on climate and the main use for which the fruit will be raised. Native American grape cultivars (*Vitis labrusca*) are the easiest to raise and very productive. The familiar Concord grape is one American type used commonly for jellies and juice, but others are available which are of fresh-eating or winemaking quality.

The European grape (*Vitus vinifera*) is grown in warmer areas and is the primary species for wine-making. Excellent cultivars for fresh eating such as "Thompson's seedless" are also included in this group.

The French-American hybrid grapes combine the vigor of American cultivars with the winemaking quality of European grapes. The resulting varieties are excellent for wine use and are raised throughout the country for that purpose.

Finally, muscadine grapes (*Vitis rotundifolia*) are the least common, being confined to the southern states where temperatures do not fall below 10°F (−23°C). The fruits are large and are borne in small clusters.

Raspberries (*Rubus* spp.)

Raspberries (Figure 8-4) are the second bramble fruit, growing wild in some areas of the country. Red (*Rubus indaeus*) and black (*Rubus occidentalis*) raspberry cultivars are the most common, with the red having a slightly tarter taste than the black, more cold tolerance,

Figure 8-4 Raspberries in spring, showing the previous season's canes above and the current year's canes growing from below. *(Photo from* Small Fruit Crop Management *by G. J. Galletta and D. G. Himelrick (eds.), 1990. Reprinted by permission of Prentice Hall, Englewood Cliffs, NJ)*

less susceptibility to disease, and greater yield potential. Novelty cultivars in yellow and purple are also available.

Raspberry cultivars (along with strawberries) can be spring-bearing or "everbearing" (also called "primocane"). Everbearing plants produce two harvests per year: one in early summer and a second in fall.

Strawberries (*Fragaria* × *ananassa* cultivars)

Strawberries are one of the most popular small fruits. Although they are perennial, the productive life of a strawberry planting is limited to about two to four years. The first year following planting, a heavy crop will be produced. The second year the plants will bear only two-thirds as much, and the third year only one-third as much. The decline is due to invasion of the plants by viruses which lessen their productivity, hybrid vigor decline, and competition from weeds. The bed should be dug out and replanted.

Like raspberries, strawberries are available in spring-bearing (also called June-bearing) and everbearing (called day-neutral) cultivars. The everbearers produce lightly throughout the summer until fall, but the quality of the berries is sometimes inferior to that of spring-bearing plants.

Preparing and Planting the Area

Soil Improvement

The amount of soil improvement necessary will depend on the present condition of the soil. If the area selected has been a garden, little soil preparation may be necessary. However, if an unworked area is chosen, considerable effort may be needed to create a favorable soil environment. If practical, a soil test should be taken, and fertilizer and pH-altering materials added to the area based on the results of the soil analysis. Organic matter should be incorporated deeply to loosen a clay soil or to improve the water-holding capacity of a sandy one. If a sodded area is selected, the sod should be turned under the fall prior to planting so that it will decay and can be worked into the soil in spring. The success or failure of the crops will depend greatly on the soil conditioning prior to planting.

Amount to Plant

Unlike tree crops, in which one or two plants assure a plentiful harvest, small fruit plants must be ordered in larger quantities. Table 8-2 lists the small fruits, the estimated yield from each plant at maturity, and the suggested number of plants for a family of five.

Planting

Most small fruits are planted in spring and begin producing fruit the following season. The ground is cultivated in early spring, and the plants are transplanted while dormant, using the spacings listed in Table 8-2, and then watered.

Small fruit cultivars are commonly ordered from mail-order nurseries and will arrive without any soil around the roots to decrease shipping cost. Planting is best done immediately, but if it must be delayed, the shipping box should be opened and the plants moistened if they are dry. The plants should then be heeled in, or the roots covered with plastic, and set in a cool (30–60°F; −1 to 15.6°C) area.

When planting it is generally advisable to plant at the same level at which the plant was growing in the nursery, and this is especially important with strawberries. An exception, however, is currants and gooseberries, which should be planted slightly deeper as this encourages a strong root system and abundant new shoots from the bases of the branches.

Pruning and Training Small Fruits

All the small fruits will produce a better crop if they are pruned and trained. On shrub fruits, pruning may be needed only to remove weak and unproductive wood, whereas on the grape, pruning will be needed to control the productivity and shape of the vine.

Table 8-2 *Small-Fruit Cultivation Information*

| Fruit | Minimum planting distance between | | Estimated yield per mature plant | Suggested number of plants per family of 5 |
	plants	rows		
blackberry, erect	4–5' (1.2–1.5 m)	6–8' (2–2.5 m)	1 qt (1 l)	15–20
blackberry, trailing or semierect	6–10' (2–3 m)	6–8' (2–2.5 m)	4–10 qt (4–10 l)	8–10
currant, black	4–5' (1.2–1.5 m)	6–8' (2–2.5 m)	2–5 qt (2–5 l)	4–6
currant, red	3–4' (1–1.2 m)	6–8' (2.–2.5 m)	5–10 qt (5–10 l)	2–3
blueberry	6–8' (2–2.5 m)	8–10' (2.4–3 m)	3–4 qt (3–4 l)	8–10
gooseberry	3–4' (1–1.2 m)	6–8' (2–2.5 m)	4–5 qt (4–5 l)	4–6
grape	3–4' (1–1.2 m)	8–10' (2.5–3 m)	¼–½ bushel (9–18 l)	5–10
raspberry, spring-bearing	3–4' (1–1.2 m)	6–8' (2–2.5 m)	1–1.5 qt (1–1.5 l)	20–25
raspberry, ever-bearing	3' (1 m)	8' (2.5 m)	1 qt (1 l) in spring and ½ qt (0.5 l) in fall	15–20
strawberry, spring-bearing	1–2' (30–60 cm)	4' (1.2 m)	½–1 qt/ft of row* (0.5–3 l/m of row)	100
strawberry, everbearing	1' (30 cm)	1–1.5' (0.3–0.4 m)	½ qt/ft of row (1.5 l/m of row)	100

*Depending on training system used.

Brambles

The pruning of brambles revolves around the knowledge of when the canes produce fruit. On typical brambles, the fruit is borne only on canes produced the previous growing season. These canes grown from the roots the first year, form flower buds that fall, bloom and fruit the following summer, and then die. The fruit is borne on side shoots which grow from buds on the parent cane.

On everbearing raspberries, two crops are produced per cane. The first crop is borne on the tips of that season's canes in fall, followed the next summer by fruiting further down the cane.

Support Systems for Brambles

Most brambles can grow without support, however trailing blackberries and red raspberries have longer canes and require tying to a wire or stake support system to be manageable and relatively easy to pick. If they are grown as individual plants in a row (system called the hill system), support can be provided by either individually staking the plants or by a trellis in which one or two wires pass through all the plants and are attached to posts at either end of the row. If the plants are planted and allowed to grow together to form a hedge, a wire trellis is the only option.

Pruning and Training Raspberries

The pruning method varies greatly by whether the raspberry is red or black, spring-bearing or ever-bearing.

Red and Yellow Raspberries. These are pruned exclusively during the dormant period in a four-step pruning/training process (Figure 8-5).

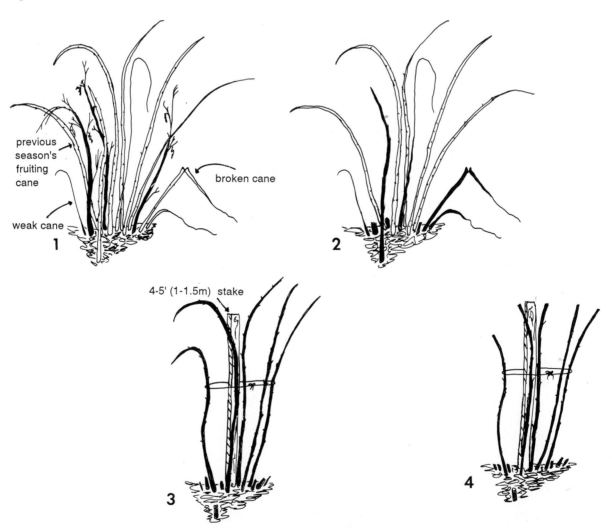

Figure 8-5 Pruning of red and yellow raspberry in four steps, all done during the dormant period. (1) Remove all canes that fruited the previous season; (2) Remove weak, broken, and poorly placed canes; (3) Tie canes to stake or along trellis wire; (4) Cut canes back to 5 feet (1.5 meters).

1. Remove all the canes that fruited the previous summer, cutting them off at ground level.
2. Cut off broken and weak canes and those growing outside of the row or hill.
3. Tie the remaining canes to the wire trellis or a 4 to 5 foot (1.2–1.5 meter) stake.
4. Cut them off at the 5-foot (1.5 meter) height. This will make picking easier the following year.

Everbearing Raspberries (Reds, Yellows, and Purples). Everbearing raspberries fruit first on the tips of the current year's canes in fall and further down those same canes the next spring. They should *not* be pinched and should not be pruned to the ground after fall fruiting.

Instead, only dormant spring pruning to eliminate thin weak canes and tip back the canes that fruited in fall should be done. After the spring fruiting of these canes, they can be removed.

Black and Purple Raspberries. These differ from red raspberries in that although they do not require support, they must be pruned twice per year: in summer and in winter or whenever they are dormant. Pruning encompasses two steps (Figure 8-6):

1. Remove the tips of new canes in summer.
2. In winter, remove canes that fruited and shorten the side branches of canes produced the previous summer.

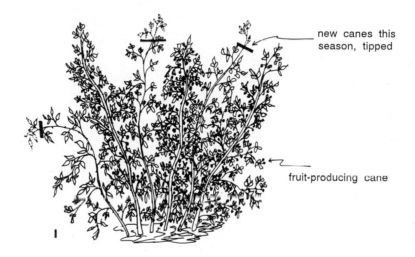

new canes this season, tipped

fruit-producing cane

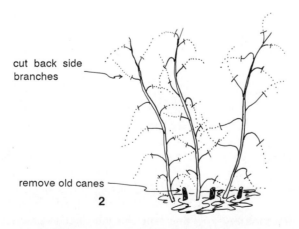

cut back side branches

remove old canes

Figure 8-6 Pruning of black and purple raspberries and erect blackberries. (1) In summer, tip new canes; (2) in winter, cut out old canes and cut back branches on new canes.

The summer pruning is to prevent the canes growing to excessive lengths that would make picking difficult by creating a thicket. To this end, the tops are cut back 2 to 3 inches (5–8 cm) as they reach 14 to 20 inches (35–50 cm) for black raspberries unsupported, 24 to 40 inches (60 cm–1 meter) for black raspberries staked to trellises, and 22 to 32 inches (unsupported) to 32 to 52 inches (80–130 cm) (supported) for purple raspberries, since they are generally more vigorous. Because the canes grow at different rates, summer tipping will be a one-month process with pinching at weekly intervals.

Winter pruning first removes dead canes that fruited, and of course weak, broken, and poorly placed canes as well. Then branches that formed on canes tipped the previous season are shortened, and this improves fruit quality. Weak branches should be left 2 to 4 inches (5–10 cm) long so that they will produce only a light load of fruit, but strong branches can be left 14 to 18 inches (30–45 cm) long.

Erect Blackberries. These are self-supporting and need no stakes or trellisses, but require pruning two times per year: in summer and in winter. Erect blackberries are pruned in a manner identical to that for black and purple raspberries (Figure 8-6) except that the blackberry canes are summer tipped at 3 to 4 feet (1–1.2 meters) instead of lower, and side branches are cut to 12 to 18 inches (30–45 cm) in winter.

Trailing and Semierect Blackberries. These very vigorous plants need support always, but pruning only once per year in the dormant season. Pruning can be accomplished in three steps (Figure 8-7):

1. Cut out the old canes that fruited the previous summer.

2. Select and tie to the support 8 to 10 strong canes and remove the rest.

3. Shorten the tops of these selected canes to 6 to 8 feet (1.8–2.4 meters) and their side branches to 12 to 18 inches (30–45 cm).

Blueberries

The blueberry bears fruit on the growth produced the previous season, and the amount of pruning will vary with the species.

Highbush. Highbush blueberries overbear and, if not pruned annually, will yield small berries. In addition, the energy expended in producing the overabundant crop will inhibit vegetative growth and drastically reduce the next year's yield. Accordingly, highbush blueberries should be pruned "hard" annually in early spring before new growth starts; that is to say, a significant amount of wood should be removed.

Upright-growing cultivars will benefit from thinning out the older center branches (Figure 8-8). This will reduce the berry load, admit more light, and stimulate growth. Drooping lower branches on spreading cultivars should be cut back to avoid breakage during fruiting.

Rabbiteye. Rabbiteye blueberries do not require pruning. The bushes are strong enough to mature large crops of fruit without damaging the plant. An exception to the no-pruning rule should be made for older bushes, which benefit when older stems are thinned out lightly.

No trellising or other training is used on blueberries, which grow naturally into large, attractive shrubs bearing white, bell-shaped flowers in spring.

Currants and Gooseberries

Currants and gooseberries, like brambles, grow into shrubs from canes produced by the roots. The fruit is borne in 1-, 2-, and 3-year-old canes, with older canes becoming progressively less productive.

Pruning both currants and gooseberries should be done once per year before spring growth begins. The aim is to remove wood more than three years old. Excess weak wood should also be removed and the bush shaped into a compact shrub 3 to 5 feet (1–1.5 meters) tall. No training (other than pruning to a shrub form) is practiced.

Grapes

There are many methods of training grape vines. The most commonly used are the following:

Four-Arm Kniffen System. This system uses a two-wire trellis in which the wires are strung between posts in the grape row. It is popular throughout the United States for vigorous native American grapes such as 'Concord.' To train a vine, the strongest new shoot of the vine should be tied to the bottom wire and the rest of the shoots pruned off the first winter after planting (Figure 8-9). The next summer the shoot should reach the top wire, where it is again tied and tipped (Figure 8-9) to encourage side shoots. The side shoots that will then develop are trained and tied out on the wire, with one cane in each direction along each wire. After this is done, annual pruning as described in the following section will be sufficient.

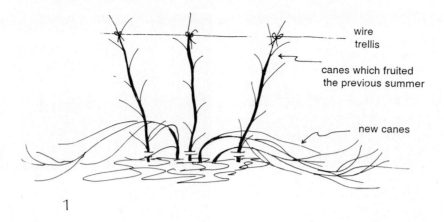

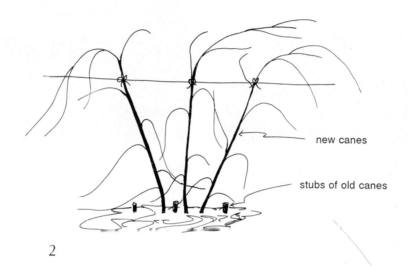

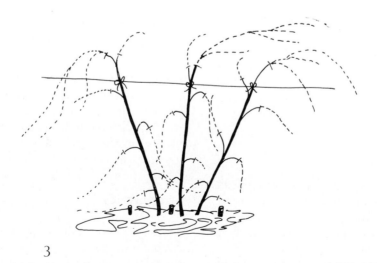

Figure 8-7 Pruning of trailing blackberries in three steps, all done during the dormant period. (1) Cut out old canes that fruited the previous summer; (2) select and tie up 8–10 strong new canes—remove the rest; (3) shorten the tops and side branches on the tied canes.

Figure 8-8   (a) Before dormant pruning of highbush blueberry, (b) after pruning. *(USDA photo)*

Arbor Training. Arbor training (Figure 8-10) produces both fruit and shade in the landscape. Since an arbor-trained vine will need to grow to a large size, only a vigorous grape cultivar should be used.

Arbor training a vine should start at planting. The vine should be trained as described under "Four-Arm Kniffen System." During each growing season that follows the main trunk shoot is continually trained upward. Gradually it will produce side shoots (fruiting canes) to cover the arbor. These fruiting canes should be spaced by pruning and tied to the support and replaced annually as described in the following section.

Annual Pruning

Grape vines bear their fruit only on canes that grew the previous summer. Heavy annual pruning should remove most of the wood growing from the trunk and leave only four shortened canes for fruit bearing, one to grow in each direction along the trellis wires (Figure 8-11).

The canes chosen to remain for fruit bearing should be slightly larger than pencil thickness for best fruit production; they should not necessarily be the largest canes on the vine. Each should be cut to a length

that allows for 15 buds. In addition to these four canes, four canes shortened to two to four buds should be left near the two wires. These short canes (called *renewal-spurs*) will produce the fruiting canes for the following season.

Renovation Pruning

Grape vines that have been neglected for many years require severe pruning to produce quality fruit again. The plants should be cut back severely before growth starts, removing as much of the accumulated older wood as possible and following the specifications for annual pruning. Any old wood that cannot be removed at this pruning may be taken out with regular pruning the following year.

Strawberries

Three systems of training are used for strawberries: the matted row, spaced runner, and hill systems. Except in the hill system, flowers that form on plants during the year they are planted should be removed instead of being allowed to develop fruit. This will increase the harvest the following year.

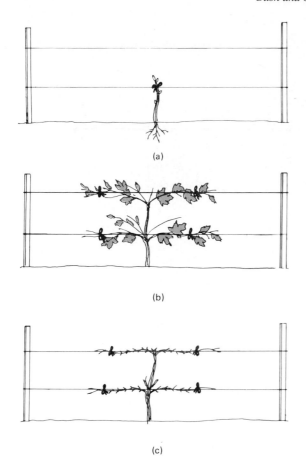

Figure 8-9 Training a grapevine using the four-arm Kniffen system. (a) First winter after planting, (b) the following summer, (c) the following winter after pruning.

Figure 8-10 A grapevine trained to an overhead arbor. *(Photo by Andrew Schweitzer)*

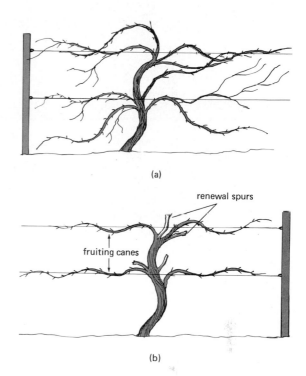

Figure 8-11 (a) Before annual pruning of a grapevine trained to a four-arm Kniffen system, (b) after pruning.

Matted Row (Figure 8-12). This system involves transplanting the young strawberries 2 feet (60 cm) apart in a row, with the rows 3 to 4 feet (1–1.2 meters) apart. The runners produced in late spring and summer are allowed to root at random, forming a dense mat about 2 feet (0.6 meter) wide. This training system is the least productive but the easiest.

Spaced Runner (Figure 8-12). In the spaced runner system, mother plants are spaced at the same distance as in the matted row system, but the runners are trained to root not closer than 4 inches (10 centimeters) apart. After a row 2 feet (60 cm) wide forms, all new runners are removed. This system leaves more room for each individual plant and results in larger and better-quality berries than the matted row. Although there is more work involved, the yield is usually greater.

Hill System (Figure 8-12). The hill system is the best training method for everbearing strawberries and is also useful for spring bearers. The plants are set 1 to 1½ feet (30–45 centimeters) apart in rows 1 to 1½ feet (30–45 centimeters) apart. All runners that appear the planting year are pinched off, resulting in a strong mother plant that will bear profusely the following season. Yields from strawberries trained to a hill system can sometimes double that from a matted row. The

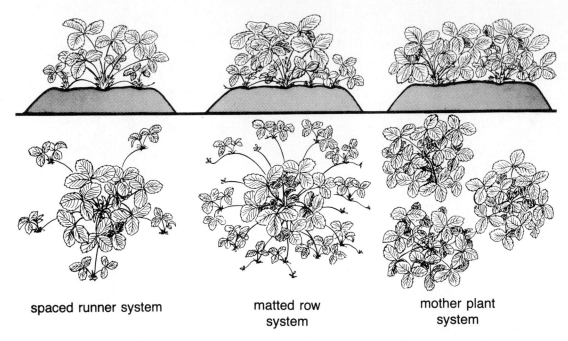

spaced runner system   matted row system   mother plant system

Figure 8-12 Strawberry training systems. [Illustration *from* Small Fruit Crop Management *by G. J. Galletta and D. G. Himelrick (eds.), 1990. Reprinted by permission of Prentice Hall, Englewood Cliffs, NJ*]

system is used primarily in warm areas since plants are more likely to be damaged by cold.

Renewing the Strawberry Patch

Properly renewed, a strawberry planting can last up to four years, although the yield will not approach that picked the first season.

Renewal of the patch should begin as soon as harvesting is complete. The old foliage is mowed off to 1 inch (3 cm) above the crown, and the debris is raked off. A balanced garden fertilizer should be broadcast over the area, all weeds removed, and the rows narrowed to 1 foot (30 cm) wide by hoe or rototilling. The patch should be watered thoroughly. New runners for the next year's crop should appear within one month.

General Maintenance of Perennial Small Fruits

Fertilizing

Most small fruits are shallow-rooted and will benefit from an annual spring application of balanced fertilizer. Sidedressing with granular fertilizer no closer than 6 inches (15 centimeters) from the crown of the plant is advisable for brambles, blueberries, grapes, currants, and gooseberries. Strawberries can be broadcast fertilized

and should be watered thoroughly afterward to keep fertilizer granules from burning the leaves. Blueberries should be fertilized annually with an acid-forming fertilizer (such as ammonium sulfate) to keep the pH of the soil in an acceptably low range. Rotted manure applied in fall will also help achieve a low pH.

Actual fertilizer rates will vary with the natural soil fertility. General recommendations for plants fertilized with a balanced formula of not more than 12 percent nitrogen are listed in Table 8-3.

Watering

The amount of supplemental watering that small fruits require will depend on the natural rainfall. Provisions

Table 8-3 *Fertilizer Recommendations for Small Fruits**

Fruit	Amount of fertilizer annually
Brambles	½ cup (120 ml) per plant
Blueberries	½–1 cup (120–240 ml) per plant
Currants	¼ cup (60 ml) per plant
Gooseberries	¼ cup (60 ml) per plant
Grapes	1 cup (240 ml) per plant
Strawberries	1 cup (240 ml) per 20 ft (6 m) of 2-ft (0.6-m)-wide row

*Recommendations are made for mature plants and are based on a balanced (1–1–1 ratio) fertilizer containing less than 12 percent nitrogen.

for supplemental watering should be made in any climate though, since the shallow-rootedness of these fruits makes them particularly susceptible to drought damage.

It is possible to overwater small fruits and decrease the fruit quality by making the fruits larger but the flavor less intense. This is particularly a problem in strawberries which should not be irrigated more often than is necessary to keep the soil moist.

Mulching

Small fruits should be mulched in summer with a plastic or organic mulch for weed control and soil moisture retention. In strawberry plantings trained to the hill system, plastic mulch is often preferred because it warms the soil and can increase and/or speed up fruit production by increasing root growth. Normally the strip of plastic is laid over the prepared soil and the plants are planted in holes cut through it.

The aim of fall mulching with organic matter is to prevent freezing and thawing of the root systems. Fall mulch should be applied late in the year after temperatures have dropped into the 40°F (5°C) range or lower. If the plants are mulched sooner, the mulch may retain sufficient heat near the roots to continue growth, and resulting tender shoots will be killed.

The shrub-type small fruits (brambles, blueberries, currants, and gooseberries) should be mulched deeply around and covering the crowns and lower stems of the plants.

Strawberries are mulched by completely covering them with straw or a similar lightweight material such as wood shavings. The mulch should be raked from the patch promptly when new growth begins in the spring to admit light for photosynthesis.

Grapes should be mulched thickly in a circle extending out about 4 feet (1.2 meters) from the trunk. This will assure that the entire root zone is protected.

Selected References for Additional Information

American Horticultural Society. *Fruits and Berries.* Alexandria, VA: American Horticultural Society, 1982.

Columbia Basin Nursery. *Retail Catalog.* P.O. Box 458, Quincy, WA 98848.

Farmer's Seed and Nursery Company. *Retail Catalog.* Faribault, MN 55021.

Galletta, Gene J., and David G. Himelrick, eds. *Small Fruit Crop Management.* Englewood Cliffs, NJ: Prentice-Hall, 1990.

Hill, Lewis. *Fruits and Berries for the Home Garden.* Pownal, VT: Garden Way (Storey Communications, Inc.), 1992.

Kelley Brothers Nurseries. *Retail Catalog.* 940 Maple St., Dannville, NY 14437.

New York State Fruit Testing Association. *Retail Catalog.* Geneva, NY 14456.

Ortho Book Staff. *All About Growing Fruits and Berries.* Revised ed. San Ramon: Ortho, 1982.

Otto, Stella. *The Backyard Orchardist.* Maple City, MI: OttoGraphics, 1993.

Savage Farm Nursery. *Retail Catalog.* P.O. Box 125, McMinnville, TN 37110.

R. H. Shumway Company. *Retail Catalog.* Rockford, IL 61101.

Striblings Nursery, Inc. *Retail Catalog.* 1620 W. 16th St., Merced, CA 95340.

Sunset Book Staff. *Berries: How to Grow.* Menlo Park, CA: Sunset Books, 1984.

Waynesboro Nurseries. *Retail Catalog.* Box 987, Waynesboro, VA 22980.

See also Cooperative Extension Service publications from individual states.

9

Flower and Herb Gardening

Flower gardening is a popular type of home gardening, and many gardeners specialize strictly in growing annual and perennial ornamentals. Closely allied to flower gardening are herb, shade, rock, and container gardening. Each is discussed within this chapter, along with roses and flowers for cutting and drying.

Types of Garden Flowers

Annuals

Every spring there is a brisk sale of annual seedlings for transplanting into home gardens. Although they live for only one season, during that brief time they more than justify their expense by flowering continuously and profusely with minimal care (Table 9-1, Figures 9-1, 9-2, 9-3, 9-4, 9-5, and 9-6).

Commercial breeding of annuals for brighter colors, larger flowers, and different sizes is a continual process. There is an annual to fit almost every garden situation: sun or shade, poor or good soil, 3 inches (8 centimeters) or 3 feet (1 meter) tall. New cultivars are evaluated and introduced yearly. Those judged most valuable are given the All-America Selection (AAS) award for their beauty and adaptability to diverse climates.

Most annuals are purchased as greenhouse-grown transplants in packs of four, six or eight plants. By buying seedling plants instead of sowing seed, the gardener saves several weeks of growing time and can transplant nearly mature plants into the garden for immediate bloom. A few annuals that either do not transplant well or mature very rapidly are grown from seed. Among these are nasturtium, California poppy, and balsam.

Most annuals are not frost-tolerant. However, a few, called *cool-season annuals,* are tolerant of light frost and grow better in temperatures in the 50–60°F (10–16°C) range. These annuals (Table 9-1) can be planted in spring before the last frost date. In mild-winter areas they are planted in fall for winter and spring bloom.

Perennials

Many garden flowers are herbaceous perennials (Figures 9-7, 9-8, and 9-9) with foliage that dies yearly but with a crown and roots that renew the plant the next season. Only two common perennials, roses and tree peonies, are woody.

Unlike annuals, which flower continuously, most perennials have a 2- to 3-week bloom period and flower less profusely. However, they increase in size every year and, in established clumps, create a massive display of flowers and foliage. Perennials are more expensive than annuals. They are sold individually in containers or can be ordered from mail-order nursery catalogs and shipped while dormant. Annuals must be replanted yearly, but perennials are permanent and will continue to flower every year with minimal care. For this reason, they are the mainstay of a flower garden (Table 9-2).

Biennials

Only a few popular flowers are biennials. Foxglove and sweet William are two of the better-known examples. Biennials are sold as 1-year-old plants ready for blooming or can be grown from seed.

Cold-hardy Bulbs

Spring-flowering tulips and summer-flowering lilies are two examples of cold-hardy bulbs. This group also includes plants like crocus which are not bulbs in a botanical sense but function similarly.

153

Table 9-1 *Selected Flowering Annuals*

Common name	Botanical name	Cool season/ warm season	Remarks
Ageratum	*Ageratum houstonianum*	Warm or cool	Pink, white, or most commonly blue flowers like small powderpuffs; not frost-tolerant
Snapdragon	*Antirrhinum majus*	Warm or cool	Frost-tolerant
Calendula	*Calendula officinalis*	Warm or cool	Yellow and orange shades of flowers of a daisylike shape
Bachelor button	*Centaurea cyanus*	Warm or cool	Blue, white, pink colors; frost-tolerant
Marguerite daisy	*Chrysanthemum frutescens*	Warm	White or yellow daisies; very fast growing to 3 ft (1 m) and floriferous
Cosmos	*Cosmos bipinnatus*	Warm	Red through white, purple, and orange; many tall and some short cultivars
Cape marigold	*Dimorphotheca sinuata*	Warm	Frost-tolerant; sow in place; daisylike flowers in orange and yellow
California poppy	*Eschscholzia californica*	Warm or cool	Frost-tolerant; does not transplant well
Gaillardia	*Gaillardia pulchella*	Warm	Orange, yellow, red combinations; good in poor soils
Balsam	*Impatiens balsamina*	Warm	Spike flowers in many colors; fast growing from seed
Sweet pea	*Lathyrus odoratus*	Cool	Pink, white, red, or lavender flowers; vining or bush types
Sweet alyssum	*Lobularia maritima*	Warm or cool	Matlike habit; white or purple fragrant flowers; frost-tolerant
Stock	*Matthiola incana*	Cool	Pink, white, red, lavender, or yellow spike flowers; very fragrant
Petunia	*Petunia × hybrida*	Warm	All colors, bushy or vining cultivars
Moss rose	*Portulaca grandiflora*	Warm	Succulent and low-water-requiring; many colors of delicate flowers; creeping habit
Salvia	*Salvia splendens*	Warm	Red and, occasionally, pink or purple shades; many heights available
Marigold	*Tagetes erecta* and *patula*	Warm	White, orange, yellow, red, and combinations; many heights available
Black-eyed susan vine	*Thunbergia alata*	Warm or cool	Reliable vine; flowers orange-yellow with black centers; perennial in zones 9 and 10

Hardy bulbs differ from perennials in several ways. Unlike perennials which can be planted from spring until fall, bulbs should be moved only in late summer and fall when dormant. They then generate roots in fall and winter and flower the following year. Also, unlike other perennials, which retain their foliage throughout the growing season, bulb foliage dies in midsummer, leaving only the underground bulb through the latter part of summer.

The usual source of garden bulbs is a local nursery or mail-order catalog. Although the "major bulbs" like tulip, hyacinth, and daffodil are best known, there are many "minor bulbs" which are easily grown and are

Figure 9-1 'Dolly' marigold. Marigolds are among the most popular and dependable annuals. (*This item available from Park Seed Co., Greenwood, SC*)

Figure 9-2 A planting of petunias will bloom continuously. (*This item available from Park Seed Co., Greenwood, SC*)

Figure 9-3 Impatiens, a favorite annual for shady locations. *(This item available from Park Seed Co., Greenwood, SC)*

Figure 9-5 Gaillardia, a floriferous annual with flowers ranging from yellow through red. *(Photo courtesy of W. Atlee Burpee and Co., Warminster, PA)*

a worthwhile addition to any garden. Table 9-3 lists a number of these lesser-known bulbs.

Summer or Nonhardy Bulbs

Although perennial in mild climates, non-hardy bulbs cannot survive winter temperatures in most parts of the country. Consequently, they are planted in spring for summer flowering and either discarded or stored in fall for replanting the following year. Tuberous begonias (Figure 9-10), dahlias (Figure 9-11), cannas, and

gladiolus are included in this group. They should be dug from the garden immediately after the first killing frost, dried in the shade, and stored in dry peat moss in a cool 50–60°F (10–16°C) area. Table 9-3 lists selected summer bulbs.

Planning the Flower Garden

Site Selection

Good soil drainage is an important factor in selecting a site for a flower garden. Few plants can be grown in wet

Figure 9-4 Bachelor buttons can be blue, white, or pink. *(Photo courtesy of W. Atlee Burpee and Co., Warminster, PA)*

Figure 9-6 Cleome, a tall annual in pink and white. *(Photo courtesy of W. Atlee Burpee and Co., Warminster, PA)*

Figure 9-7   Poppies, a garden perennial. *(Photo by the authors)*

soil, and it is particularly unsuitable for bulbs. When drainage is questionable, a raised bed is a solution that will not only aid drainage but elevate the flowers for better viewing.

    The amount of sunlight an area receives will determine which plants can be successfully grown. A full-sun area is preferable for growing the greatest variety of flowers. Many flowers can be grown in partially or fully shaded locations, but they bloom less profusely and produce more foliage. Others weaken from the lack of light and die. Accordingly, if a shady location is chosen for a flower garden, plant selection must be made carefully and limited to species adapted to shade (Table 9-4).

Figure 9-8   Garden pinks, an old-fashioned favorite. *(Photo courtesy of W. Atlee Burpee and Co., Warminster, PA)*

Figure 9-9   Daylily, a mainstay perennial of almost every garden. *(Photo courtesy of W. Atlee Burpee and Co., Warminster, PA)*

    Finally, the site chosen for a flower bed should be readily viewable. The area around a patio or deck is a good choice.

Flower Bed Styles

The Flower Border

A border of flowers with shrubs or a fence used as a background is a popular traditional bed style (Figure 9-12). The border can be as much as 10 feet (3.2 meters) wide and is designed to be viewed from only one side. Accordingly, tall plants are used in the rear, moderate-sized ones in the center, and short ones along the front.

Figure 9-10   Tuberous begonia, a well-known summer bulb for shade. *(Photo by the authors)*

Figure 9-11   Dahlia. A summer bulb in much of the country, but a perennial in USDA zones 9, 10, and 11. *(Photo courtesy of W. Atlee Burpee and Co., Warminster, PA)*

The Freestanding Bed

A freestanding or island flower bed (Figure 9-13) is well adapted to modern landscape designs and is meant to be viewed from all directions. Tall plants should be placed in the center of the bed and gradually shorter plants stepped down to the edge. If a curv-

ing bed style is to be used, a garden hose can be moved in different ways to create a pleasing outline. The edge of the bed can then be marked by spading along the hose.

The Formal Bed

The formal bed or border (Figure 9-14) originated in Europe and was used for flowers and herbs around palaces and large manor houses. It is a tightly restricted, symmetrical style and not commonly used today. Many gardeners prefer formal beds for herb gardening; compact annuals can be easily included also.

The Cottage Garden

The cottage garden style is an informal massing of flowers of many species and sizes (and sometimes herbs and small vegetables as well) to achieve the effect of profusion of color and variety of form. The effect is a jumble of eye-catching color that arrests the attention of the viewer and causes him or her to wish to stop and look at each of the individual plants in turn. Originally an English style, it is becoming popular in North America because many people now

Figure 9-12   A flower border.

Table 9-2 *Selected Flowering Perennials*

Name	USDA zones	Height	Bloom period	Remarks
Astilbe, false spirea *Astilbe* spp.	3–9	18–36" (0.5–1 m)	Mid- through late summer	White, pink, lilac, or red feathery plumes
Aster, michelmas daisy *Aster* spp.	1–10	6"–4' (15 cm-1.2 m)	Late summer to frost	Many colors
Bellflower *Campanula* spp.	2–10	1–3' (0.3–1 m)	Late spring to summer	White or blue bell-shaped flowers
Chrysanthemum *Chrysanthemum* × *morifolium*	3–10	12–48" (0.3–1.2 m)	Late summer to fall	All colors except blue; pinch off flower buds to delay bloom until fall
Feverfew *Chrysanthemum parthenium*	4–10	2–3' (0.6–1 m)	Early summer to fall	Small, white flowers with yellow centers; produced profusely
Shasta daisy *Chrysanthemum* × *superbum*	2–10	24–36" (0.6–1 m)	Midsummer to fall	White daisy with yellow center; profuse flowers
Delphinium *Delphinium elatum*	1–10	24–48" (0.6–1.2 m)	Mid- to late summer	Blue, white, purple spikes
Pinks *Dianthus* spp.	2–10	8–18" (20–50 cm)	Early through late summer	Pink through white and red carnationlike flowers
Leopard's bane *Doronicum cordatum*	4–10	24" (60 cm)	Early spring	Yellow daisylike flowers; foliage dies in summer
Daylily *Hemerocallis* spp.	2–10	15"–4'(0.4–1.2 m)	Early to late summer, depending on variety	Yellow through wine-red flowers; not suitable for cutting, as flowers close at night
Bearded iris *Iris* spp.	5–10	12–36" (0.3–1 m)	Early summer	Many colors
Siberian iris *Iris* spp.	6–10	24" (60 cm)	Spring to early summer	Flowers differently shaped than bearded iris
Lythrum *Lythrum salicaria* and *virgatum*	2–10	24–48" (0.6–1.2 m)	All summer	Pink through red-purple flowers; tolerant of poor growing conditions including wet soil
Bee balm *Monarda didyma*	3–7	36" (1 m)	Midsummer	Pink, lavender, red, or white flowers; leaves have mint smell
Peony *Paeonia* spp.	4–9	36" (1 m)	Late spring to early summer	Large globular flowers in white through red
Oriental poppy *Papaver orientale*	2–8	24–36" (0.6–1 m)	Early summer	Pink through white and red carnationlike flowers
Hardy phlox *Phlox paniculata*	4–10	24–40" (0.6–1 m)	Midsummer	Large flower heads in many colors
False dragonhead *Phystostegia virginiana*	6–10	36–48" (1–1.2 m)	Midsummer to fall	Pink or lavender spikes; very vigorous

occupy homes such as condominiums that have limited gardening space.

A cottage garden frequently includes other *design elements* such as benches and paths for viewing the flowers. Plants normally chosen are traditional old-fashioned species such as hollyhock, foxglove, and iris. The taller, less compact cultivars are chosen to amplify the informality that is the hallmark of this type of garden.

Plant Selection

Most gardeners prefer a mixed garden of herbaceous and woody perennials, bulbs, annuals, and sometimes herbs. A mixture combines the virtues of each: the dependability of perennials, the floriferousness of annuals, and the earliness of the hardy bulbs. The following criteria should be used to select the species that will be planted.

Table 9-3 *Selected Bulbs and Related Plants*

Name	Cold hardiness (USDA zones)	Time of bloom	Remarks
Ornamental onion, allium *Allium* spp.	4–10	Late spring	Many species in white, blue, and pink; globe-shaped flower heads
Belladona lily *Amaryllis belladonna*	5–10	Summer	Pink lilylike flowers
Anemone *Anemone* × *hybrida*	6–9	Midspring	3–15″ (10–40 cm) tall with pastel flowers
Tuberous begonia *Begonia* × *tuberhybrida*	9–10 perennial, 2–10 summer bulb	All summer	Single and double flower varieties and many colors; good in hanging baskets
Caladium *Caladium* × *hortulanum*	9–10 perennial, all others summer bulb	Foliage only all summer	Grown for foliage only, which is multicolored
Canna *Canna* hybrids	7–10 perennial, all others summer bulb	All summer	Tall (36–48″) (1–1.2 m) background plants in red, yellow; dwarfs available
Glory-of-the-snow *Chinodoxa luciliae*	3–10	Early spring	Blue, pink, or white flowers; less than 1′ (30 cm) tall
Autumn crocus *Colchicum autumnale*	4–10	Fall	Planted in August for fall flowers; leaves appear the following spring
Crocus *Crocus* spp.	3–10	Early spring	Plants 2–4″ (5–10 cm) tall
Dahlia *Dahlia* hybrids	9–10 perennial, summer bulb other zones	Summer	Easy to grow; many colors and sizes
Winter aconite *Eranthis* spp.	4–9	Early spring	Yellow buttercuplike flowers; plants 2–4″ (5–10 cm) tall
Fritillaria *Fritillaria imperialis*	3–10	Midspring	Unusual and attractive, 2½–4′ (0.7–1.2 m) tall
Gladiola *Gladiolus* × *hortulanus*	8–10 perennial, summer bulb other zones	Summer	Tall, to 48″ (1.2 m), with spike flowers
Amaryllis *Hippeastrum* hybrids	9–10 perennial, houseplants all others	Winter	Lilylike flowers on tall stem
Hyacinth *Hyacinthus orientalis*	4–10	Midspring	Fragrant blooms in many colors, 6–12″ (15–30 cm) tall
Bulbous iris *Iris* spp.	5–10	Spring	Purple, blue, yellow, white
Garden lily *Lilium* spp.	3–10	Early summer	Many species and cultivars 12–36″ (0.3–1 m) tall
Grape hyacinth *Muscari* spp.	2–10	Early spring	Blue, white or pink flowers on 3–4″ (7–10 cm) plant
Daffodil and narcissus *Narcissus* spp.	4–10	Spring	Many species available; 12–18″ (30–45 cm) tall
Oxalis *Oxalis* spp.	6–10, depending on species	Winter in 9–10, spring all others	White, pink, yellow flowers on clover-shaped leaves; plants 2–5″ (5–12 cm) tall
Ranunculus *Ranunculus*	8–10 perennial, all others summer bulb	Midspring	Delicate flowers in pastel colors on 12–18″ (30–45 cm) stems
Scilla, squill *Scilla* spp.	4–10	Spring	Blue and other colors; plants about 6″ (15 cm) tall
Tulip *Tulipa* spp.	3–7	Early to late spring	Multitude of cultivars 3–18″ (10–50 cm) tall
Calla lily *Zantedeschia aethiopica*	8–10 perennial, summer bulb all other zones	9–10 winter, other zones summer	Outstanding foliage and white to yellow flowers; 36–48″ (1–1.2 m) tall

Figure 9-13 A freestanding flower bed. *(Photo by George Taloumis)*

Height. Knowing the ultimate size of the flowers will enable a gardener to select plants of a variety of heights and to locate them in the correct place in the bed. Many annuals are available in several heights, and varieties of the same species can be used as either a foreground or a background plant.

Amount of Sunlight Required. A few versatile flowers grow in any light level, but most are restricted to either shade or sun. Choosing plants that are proven performers under the light available will maximize the beauty of the garden.

Figure 9-14 A formal flower border.

Continual Bloom. Planning for continual bloom involves choosing a variety of plants so that flowers will appear in succession all season. Bulbs are the usual choice for early spring and chrysanthemums the choice for fall blossoming. The other perennials have 2- to 3-week blooming periods from late spring through summer, and their flowering sequence relative to each other can be found in reference books on flower gardening. Since annuals flower continuously, they tie together the bloom periods of perennials.

Disease and Pest Susceptibility. Although flower gardens nearly always require more maintenance than landscapes of shrubs and groundcovers, there are degrees. For example, by planting carefully selected species it is possible to avoid all need to spray pesticides under normal circumstances. However the gardener who elects to grow hybrid tea roses should anticipate spraying every 10 days throughout the growing season.

Water Requirement. Although flowers generally require frequent copious watering to bloom well, people living in areas with water shortages can also landscape with flowers provided they select the right plants. *Xeriscaping,* landscaping with drought-resistant plants including drought-resistant flowers, is practical, environmentally conscious and beautiful. It is not, as some people suspect, landscaping with cacti and rocks. Instead, floriferous improved succulents such as moss rose, and naturally drought-tolerant common garden flowers such as nasturtium, verbena, lantana, and gaillardia are mixed with newly improved native species to create a garden that few would suspect has been designed to be water conserving.

Preparing and Planting the Bed

The preparation needed for a flower garden will depend on the present condition of the soil. Recently gardened soils may need little preparation other than the incorporation of fertilizer. Areas that have been uncultivated or in turf will require more work. The area should be spaded to a depth of about 18 inches (60 centimeters), the average root depth for perennials, incorporating a large volume of amendment such as compost or peat moss. About ½ cup (100 milliliters) of balanced fertilizer per 10 square feet (1 square meter) should be spread over the area and mixed in the spaded layer.

If the bed adjoins a lawn area, invasion by grass will be a constant problem. Edging material outlining the perimeter will reduce this problem and give a neater appearance. A row of bricks set flush with the ground level or slightly higher is one attractive edging

Table 9-4   Selected Shade Garden Plants

Common name	Botanical name	Classification	Remarks
Ferns	*Adiantum, Athyrium, Polystichum,* etc.	Perennial	Many cold-hardy species of varying appearance
Ajuga	*Ajuga reptans*	Perennial	Low growing with blue to pink flowers in spring, depending on variety
Columbine	*Aquilegia × hybrida*	Perennial	Unusual-shaped flowers in blues, yellows, pinks, and so on; plants to 30″ (0.7 m) tall
Astilbe	*Astilbe × hybrida*	Perennial	Pink or white plumes atop fernlike foliage; plants up to 3′ (1 m) tall
Tuberous begonia	*Begonia × tuberhybrida*	Summer bulb	Large weeping plants with oustanding flowers in many colors
English daisy	*Bellis perennis*	Perennial	Small plant which blooms in spring to early summer with white to red flowers
Bergenia	*Bergenia × schmidtii*	Perennial	Pale pink flowers on an evergreen groundcover plant
Caladium	*Caladium × hortulanum*	Summer bulb	Grown for its foliage, which can be a mixture of red, pink, white, and green
Periwinkle	*Catharanthus roseus*	Annual in all but the mildest climates	Continual blooms in pink to white; grows 8–18″ (20–45 cm) tall, depending on variety
Coleus	*Coleus × hybridus*	Annual	Grown for colorful foliage in red, green, yellow, and all shades between
Lily-of-the-valley	*Convallaria majalis*	Perennial	Fragrant bell-shaped white flowers; spreads to form a dense carpet
Cyclamen	*Cyclamen persicum* and other species	Cold-tender perennial	Clumping plant producing nodding flowers in pink, red, lavender, or white
Bleeding heart	*Dicentra spectabilis*	Perennial	Distinctive red and white flowers on plants 1–3′ (0.3–1 m) tall
Fuchsia	*Fuchsia × hybrida*	Annual in all but the mildest climates	A reliable shade performer with flowers in shades of white through red; cascading and upright forms are sold
Coral bells	*Heuchera sanguinea*	Perennial	Continual blooms in red to white; rise to 2′ (0.6 m)
Plantain lily	*Hosta fortunei*	Perennial	Solid green or variegated leaves with fragrant lavender flowers
Impatiens	*Impatiens walleriana*	Annual in most climates	Very reliable bloomer producing red, pink, white, or orange flowers
Forget-me-not	*Myosotis sylvatica*	Annual	Bright blue flowers on a plant up to 2′ (0.6 m) tall
Primrose	*Primula* spp.	Perennial	Spring flowers in many colors in a tight head
Violet	*Viola odorata*	Perennial	Tiny white, yellow, or blue flowers on plants to 6″ (15 cm) tall
Pansy	*Viola × wittrockiana*	Perennial	Low-growing favorite; many varieties available
Calla lily	*Zantedeschia aethiopica*	Summer bulb in all but the mildest climates	Large glossy leaves and white funnel-shaped flowers. Pink and yellow spp. also available.

(Figure 9-15). Thin edging boards (bender boards) or plastic edging material is also satisfactory.

Flowers should be transplanted into the soil at the same level at which they were originally growing, and watered deeply. Bulbs should be planted according to Figure 9-16 or by following the directions on the package. Mark bulb areas with small stakes to avoid disturbing them accidentally when working in the bed.

Annuals grown from seed should be sown according to package directions and thinned to final spacing before they become crowded. Only the largest, most vigorous plants should be kept. The others should be pulled out or transplanted.

Flower Garden Maintenance

Watering

When natural rainfall is not sufficient, watering will be a frequent part of flower garden maintenance.

Figure 9-15 Ornamental concrete used as an edging for a flower bed.

Watering should soak the root zones of the plant as often as necessary to prevent wilting. Because annuals are relatively shallow-rooted, annual beds will require more frequent but shallower irrigation than perennial or bulb beds. In mixed beds, watering should be deep enough to soak the most deeply rooted plants.

Mulching

Mulching with compost, grass clippings, or other organic materials will conserve water, reduce weed problems, and give a neat appearance to the bed. Mulch should be applied in spring as soon as annuals have been transplanted and new growth of perennials appears. Over the summer the mulch can be reapplied on top as the old mulch decays underneath.

In cold-winter areas an additional layer of mulch can be laid down over the bed after the cold weather begins. This layer will reduce freezing and thawing of the bed, a common cause of *winterkill,* the death of plants over the winter. The mulch layer should be pulled back from the crowns of perennial plants in spring as new growth starts.

Fertilizing

Frequency of fertilization of perennials will depend on climate and soil type. An annual fertilization in spring will be sufficient for perennial beds in cold-winter climates with loam to clay soils. On sandy soils, which lose nutrients rapidly, and in warm climates where growth continues all year, fertilization should be more frequent. Fertilizing every two months may be necessary on perennial beds, top-dressing with a granular fertilizer.

Fast-growing annuals will benefit from frequent fertilizer applications. Topdressing every four to six weeks starting at transplanting will assure that sufficient nutrients are available for rapid growth.

Bulb fertilizing can be done at planting or when the bulb is actively growing. Some gardeners mix about a teaspoon (15 milliliters) of phosphorus-rich granular fertilizer or bone meal into the soil beneath each bulb as they plant it. Others wait until foliage appears and fertilize by topdressing. Either method is satisfactory.

Pruning

Pruning of annuals, perennials, bulbs, and herbs encourages maximum bloom and keeps the garden neat.

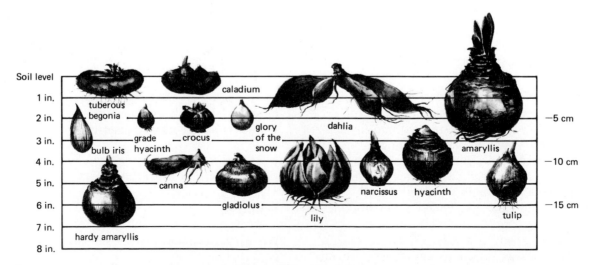

Figure 9-16 Suggested planting depths for common bulbs.

Pinching or clipping off dead flowers should be a routine, particularly with annuals. This prevents energy from being diverted into seed development and increases bloom production.

Dead leaves should be removed from all plants as necessary to improve the appearance of the garden and to eliminate a source of disease. The leaves of hardy bulbs will yellow and die naturally in midsummer, and they can be cut off at ground level at that time. Leaves should never be cut back until they are almost fully yellow, since to do so would remove the source of energy for the next year's blooms.

An alternative to waiting and cutting the leaves when they are yellow is bunching them. In this technique the gardener waits until flowering is finished then collects the bulb leaves together in a bunch while they are still green and loops them over on themselves, securing the bunch with a rubber band. This neatens the garden but still allows the bulbs to continue photosynthesizing to a limited extent. When the leaves turn yellow, they are cut off.

Division

The crowns of herbaceous perennials increase in size every year and, after several years, may become overcrowded. Bloom size and number may decrease due to this overcrowding, and therefore the clump should be divided into several smaller pieces and replanted. Chapter 4 gives instructions for division of perennials and bulbs.

Staking and Supporting

Tall flowers and those with profuse or heavy flowers may require staking to keep them from breaking or falling over. Staking will be especially necessary where frequent rains weight the blooms or where wind is common.

One or several stakes may be required to support a plant, depending on its size. Figure 9-17 shows ways of supporting plants by staking. Regardless of which method is used, the stakes should be placed behind or close to the plant so they are hidden by the foliage. The plant should be secured to the stake with twine, strips of cloth, or covered-wire twist tabs.

Figure 9-17 Two methods of staking to support a flower. (*Photos by George Taloumis*)

Specialty Gardens

Rose Gardens

Roses are the most popular garden flower, so gardens devoted exclusively to rose growing are very common.

Many types of roses are grown; the most well-known types are listed:

Hybrid Teas. Hybrid tea roses (Figure 9-18) produce large blossoms singly on moderate-sized, bushy plants and are favorites for cut-flower use. They are the most

Figure 9-18   A tea rose. *(Photo by Andrew Schweitzer)*

prone to disease problems and are not always reliably cold-hardy.

Floribundas. Floribundas (Figure 9-19) bear their blooms in clusters through summer and have blooms generally smaller than those of teas.

Grandiflora. Grandiflora is the name given to the roses that are crosses between floribundas and hybrid teas. They produce fewer blooms per cluster than floribundas, but each blossom is nearly as large as a hybrid tea bloom.

Climbing and Rambling Roses. These have long canes which grow from the base of the plant. They grow very large and are trained along fences and garden arches, called pergolas (Figure 9-20).

Tree Roses. Tree roses can be hybrid tea, grandiflora, or floribunda roses. They are grafted onto long-cane hardy species to achieve a tree form. Tree roses are frequently grown in pots for patio decoration in summer. In freezing areas they are then buried, still potted, in a sheltered place for overwintering.

Figure 9-19   A floribunda rose. *(Photo courtesy of Marge Coon)*

Figure 9-20 A climbing rose. *(Photo courtesy of Marge Coon)*

Miniature Roses. Despite their frail appearance, miniature roses are very cold-hardy. They produce thumbnail-sized blooms singly or in clusters and can be raised indoors in winter on a sunny windowsill.

Rose Culture

Roses are often high-maintenance plants, requiring regular spraying for diseases and insects and annual pruning. Their cultivation is the subject of many books which deal in depth with the subject. The following gives only a brief overview.

Soil. Quick- and thorough-draining soil is essential for roses. Large quantities of amendment should be added if the soil contains a large percentage of clay. Soil should be worked and amended to a depth of 20 to 24 inches (50–60 centimeters), and pH should be slightly acid, in the range of 6.5 to 7.0.

Fertilizing. Granular balance fertilizer should be applied in spring and after each bloom flush.

Watering. Soil should be kept moist to a 2-foot (60-centimeter) depth. Rose foliage should not be wet since this encourages the spread of fungus diseases such as blackspot, rust, and mildew.

Overwintering. Being woody, rose canes live over the winter and renew the foliage in spring. The canes of some types are not reliably cold-hardy though. Bush roses should be mounded with soil to 1 foot (30 cm) deep in areas where winter temperatures drop below 10°F (−12°C). Rambler and climber canes may be pinned to the ground and covered with soil in areas where they are not reliably winter hardy.

Pruning. Pruning should remove only dead wood, weak canes, and crossing branches. Death of the top portions of canes during winter is a very common problem and will necessitate shortening the canes in spring to living wood. Unless the rose is very vigorous, pruning should be minimal. The removal of large quantities of wood will slow growth and reduce the number of flowers produced.

Rock Gardens

Rock gardening (Figure 9-21) combines rocks and plants to create a natural-appearing area for the cultivation of species native to rocky or alpine regions. In some cases a gardener can make use of naturally rocky land with some reorganization of the available rocks, but more often large rocks must be brought to the site and arranged artistically to give the appearance of a natural formation. Herbaceous perennials, succulents, bulbs, annuals, and even shrubs and trees (see Table 9-5) are then planted in the small crevices and soil pockets formed between the rocks. Most rock garden plants are small or even minute compared to other garden plants. When shrubs or trees are used, they are generally dwarf cultivars.

Planting of a newly created site should be postponed until after the soil has settled among the imported rocks. Little soil amending beyond the addition of small amounts of organic amendment will be needed, since most rock garden plants are native to soils of poor natural fertility. However, if acid-loving plants are to be grown, pockets between rocks can be amended with peat moss to create the proper soil environment.

Care of a rock garden consists of weeding, dividing large plants, and watering as needed when the soil becomes dry. In general, rock garden plants will need little moisture. Fall care in very cold-winter areas consists of covering the area (with the exception of evergreens) with a mulch layer of hay or other material. The

Figure 9-21 A rock garden. *(Photo courtesy of Zurn Industries Incorporated)*

mulch minimizes heaving of the ground and uprooting of the plants.

Container Gardens

Container gardens (Figure 9-22) are a popular way of growing flowers and vegetables without land. Almost any plant can be grown in a pot, including trees, vines, vegetables, flowers, and shrubs.

Because of the limited volume of soil available for root growth in containers, the soil should be of the best possible quality. Artificial media (discussed in Chapter 15) are the best choices, or a mixture of natural soil and a large amount of organic soil amendment (See Chapter 5). Pure garden soil is seldom satisfactory because of its tendency to pack and drain poorly when confined to a pot.

Supplying water to a container garden is the most time-consuming chore. Water is used up very quickly from the soil and must be replenished often. Daily watering will be necessary during hot weather or when clay or small pots are used. However automatic irrigation systems designed for pots are available (see Chapter 20).

Frequent fertilization is essential for container-grown plants. A soluble fertilizer (see Chapter 15) applied in the irrigation water is one solution. Top-dressing with slow-release or granular fertilizer is also satisfactory.

Overwintering perennials, trees, and shrubs in the North in containers may need winter protection. Being aboveground, the roots will be subjected to lower temperatures than roots growing in the earth, and death from cold is likely. If possible, the pots should be planted in the ground in fall and mulched heavily. If this is not practical, they can be overwintered in a garage or placed in a sheltered location next to a building.

Cutting and Dried-Flower Gardens

Although many gardeners harvest flowers directly from their display beds, others prefer cultivating a separate area for flowers to be cut and brought indoors. A cut-flower garden is usually located in the service area of the landscape out of view and includes varied tall annual and perennial cultivars that lend themselves to arrangement and last well when cut.

Flower preservatives to lengthen the life of cut flowers are an effective and worthwhile investment. However even more economical is a mixture of ⅓ tonic water and ⅔ water. This solution supplies sugar for respiration, a preservative to slow growth of bacteria in the

Table 9-5   *Selected Rock Garden Plants*

Common name	Botanical name	Type of plant	Remarks
Bugleweed	*Ajuga genevensis* and *reptans*	Perennial groundcover	Grows 2–6″ (5–15 cm) high, depending on variety, and thrives in sun or shade; produces blue flowers
Sea pink or thrift	*Armeria maritima*	Evergreen perennial	Small grasslike clumps of deep green leaves produce pink ball-like flower heads
Basket-of-gold	*Aurinia saxatilis*	Perennial groundcover	Low growing and covered with bright yellow flowers
Crocus	*Crocus* spp.	Hardy bulb	Bell-shaped flowers on tiny plants in early spring
Broom	*Cytisus* spp.	Small shrubs, deciduous or evergreen	White to yellow pea-shaped flowers on bright green plants
Garden pink	*Dianthus* spp.	Herbaceous perennial	Flowers pink, white, red, purple on short plants, to 12″ (30 cm) tall
Heaths and heathers	*Erica* spp., *Calluna* spp.	Evergreens	Bell-shaped flowers on plants 1–2′ (30–60 cm) tall; need acid soil
Candytuft	*Iberis sempervirens*	Evergreen groundcover	Deep green leaves and white flower heads
Iris (bulb type)	*Iris reticulata* and other small species	Hardy bulb	Yellow, white, or purple flowers in very early spring on plants 4″ (10 cm) tall
Juniper	*Juniperus* spp.	Needle evergreen	Many types available with blue-green to deep green foliage; choose low or ground-cover varieties
Alyssum	*Lobularia maritima*	Annual in all except the mildest climates	White, pink, or purple flowers cover the creeping matlike plant; reseeds naturally
Grape hyacinth	*Muscari* spp.	Hardy bulb	Purple, blue or white flowers on tiny plants, up to 4″ (10 cm) tall
Creeping phlox or moss pink	*Phlox subulata*	Herbaceous perennial	White or pink flowers produced profusely over a matlike plant
Dwarf alberta spruce	*Picea abies* (dwarf cultivars)	Dwarf evergreen tree	Very compact, slow growing, with deep green foliage
Mugo pine	*Pinus mugo* cv. *mugo*	Needle evergreen shrub	Shrublike pine growing to 18″ (45 cm) high but spreading widely
Lavender cotton	*Santolina chamaecyparissus*	Herbaceous perennial	Finely cut silver-gray foliage is its outstanding trait; yellow to white flowers in summer
Stonecrop	*Sedum* spp.	Perennial succulent	Species vary from 1″ (2 cm) to 2′ (60 cm) high and have bright flowers
Hen-and-chickens	*Sempervivum* spp.	Perennial succulent	Small rosette plants in green or with tinges of purple or red
Thyme	*Thymus* spp.	Herbaceous perennial or annual	Dark green to gray leaves make up a creeping plant
Species tulip	*Tulipa kaufmanniana*	Hardy bulb	Small plants, about 6″ (15 cm) high, with the traditional tulip flowers

water (bacteria block xylem and cause flower wilting), and acid to lower the pH of the water (a condition that also prolongs flower life).*

*This formula given to the author in conversation with Dr. Michael Reid, Professor of Horticulture with specialization in post-harvest physiology of flowers at the University of California at Riverside.

Flowers in the same area can be grown for drying and use in permanent arrangements (Figure 9-23). These should be harvested at their peak of bloom, bunched loosely with the stems together, and hung upsidedown in a dry area with good air circulation. Hanging out of direct sunlight will help preserve the natural color. Table 9-6 is a list of flowers suitable for drying and cutting.

Figure 9-22   Containers of flowers abound on this patio area and keep it from appearing stark. *(Photo courtesy of Landscape Images, a Division of British Indoor Gardens, El Toro, CA)*

Figure 9-23   Bells of Ireland, a pale green flower excellent for both cutting and drying. *(Photo courtesy of W. Atlee Burpee and Co., Warminster, PA)*

Table 9-6 *Selected Garden Flowers for Cutting and Drying*

Common name	Botanical name	Annual/perennial	Remarks
Yarrow	*Achillea filipendulina* and *millefolium*	Perennial	Yellow, white, or pink umbrella-shaped flowers; use fresh or dry
Snapdragon	*Antirrhinum majus*	Annual	Spike flowers in many colors; select tall varieties for fresh use
Aster	*Callistephus chinensis*	Annual	For fresh use, with large heads in pink through red and purple; pick disease-resistant cultivars
Cockscomb	*Celosia cristata cultivars*	Annual	Red or yellow blooms in plume or cockscomb shape; use fresh or dried
Bachelor button	*Centaurea cyanus*	Annual	Blue, pink, or white button-sized flowers for fresh use
Marguerite daisy	*Chrysanthemum frutescens*	Annual in all but the mildest climates	Very fast growing and produces enormous amounts of daisy flowers in white and yellow; for fresh use
Chrysanthemum	*Chrysanthemum* × *morifolium*	Annual or perennial	Many colors and forms for fresh use
Larkspur	*Consolida orientalis*	Annual	Spike-edged white, pink, or purple flowers for fresh use
Cosmos	*Cosmos bipinnatus*	Annual	Tall-growing plant which may need staking; pink through red colors for fresh use
Dahlia	*Dahlia* hybrid	Summer bulb	Both dwarfs and full-size cultivars are useful; long lasting for fresh use and available in most colors
Gladiola	*Gladiolus* hybrid	Summer bulb	Tall spike flowers in many pastel shades; fresh use only
Globe amaranth	*Gomphrena globosa*	Annual	Cut fresh or for dried use; large cloverlike blossoms in white, pink, red, or purple
Baby's breath	*Gypsophila elegans* and *paniculata*	Annual or perennial	Fresh or dried use, in white or pink, grows into a shrub 3′ (1 m) tall
Strawflower	*Helichrysum bracteatum*	Annual	Use fresh or dried; 2″ (5 cm) blooms in yellow, red, white, purple, and shades between
Helipterum	*Helipterum manglesii*	Annual	Daisylike flowers in white or pink are used for drying
Statice	*Limonum sinuatum*	Annual except in very mild climates	Papery flower clusters in many colors; useful fresh or for drying
Bells of Ireland	*Molucella laevis*	Annual	Green bell-shaped flowers on a 3′ (1 m) plant; use fresh or dry
Narcissus	*Narcissus* spp.	Hardy bulb	Many types available; useful for fresh cutting in early spring
False dragonhead	*Physostegia virginiana*	Perennial	White, pink, or purple flowers in spikes; a rampant grower for fresh-cut use
Blue salvia	*Salvia farinacea* and *Salvia patens*	Annual in all but mild climates, but re-seeds readily	Deep blue spike flowers for use fresh or dried
Pincushion flower	*Scabiosa fischeri*	Annual	White, pink, or blue flower heads on slender stems; for fresh use
African marigold	*Tagetes erecta*	Annual	Yellow, orange, white flowers for fresh use; choose tall cultivars; easy to grow
Immortelle	*Xeranthemum annuum*	Annual	Fluffy, papery flowers in pink through red; use fresh or dried
Zinnia	*Zinnia elegans*	Annual	Flower heads 1–4″ in diameter (2–10 cm), depending on cultivar; available in any color; for fresh use

Table 9-7 *Herb Garden Plants*

Common name	Botanical name	Annual/perennial hardiness	Height	Uses
Garlic	*Allium sativum*	Tender perennial zones 9–11 (treat as annual)	To 2' (0.6 m)	Bulbs used for seasoning
Chives	*Allium schoenoprasum*	Perennial zones 2–11	To 2' (0.6 m)	Leaves used for seasoning, very ornamental flowers
Dill	*Anethum graveolens*	Annual all zones	To 4' (1.2 m)	Leaves used fresh for seasoning and seeds dried for pickling
Tarragon	*Arthemisia dracunculus*	Perennial	To 2' (0.6 m)	Leaves used for seasoning
Caraway	*Carum carvi*	Biennial zones 3–11	To 2' (0.6 m)	Seeds used for seasoning
Coriander or cilantro	*Coriandrum sativum*	Annual	To 3' (1 m)	Leaves used fresh for seasoning, and seeds also
Cumin	*Cuminum cyminum*	Annual	To 6' (15 cm)	Seeds used as seasoning
Bay	*Laurus nobilis*	Woody shrub zones 7–11	To 2' (0.6 m)	Leaves used for seasoning
Lavender	*Lavandula angustifolia*	Perennial zones 5–11	To 3' (1 m)	Leaves and buds used for sachets
Mint	*Mentha* spp.	Perennial zones 3–11	To 2' (0.6 m)	Leaves used for seasoning and tea
Sweet basil	*Ocimum basilicum*	Annual all zones	To 2' (0.6 m)	Leaves used for seasoning
Sweet marjoram	*Origanum marjorana*	Perennial zones 9–11 (treat as annual in other zones)	To 2' (0.6 m)	Leaves used for seasoning
Oregano	*Origanum vulgare*	Perennial zones 3–11	To 2½' (0.8 m)	Leaves used for seasoning
Parsley	*Petroselinum crispum*	Biennial treated as annual	To 15" (0.4 cm)	Leaves used for seasoning and as garnish
Anise	*Pimpinella anisum*	Annual	To 2' (0.6 m)	Seeds used for seasoning
Rosemary	*Rosemarinus officinalis*	Perennial zones 6–11	To 5' (1.5 m) (prostrate forms common)	Leaves used for seasoning
Sage	*Salvia officinalis*	Perennial zones 3–11	To 3' (1 m)	Leaves used for seasoning
Thyme	*Thymus vulgaris*	Perennial zones 3–11	To 1' (0.3 m)	Leaves used for seasoning
Ginger root	*Zingiber officinale*	Perennial zone 10[†]; (annual or houseplant in other zones)	To 3' (1 m)	Root used for seasoning and Chinese cooking

Herb Gardens

Herb gardens are among the oldest types of gardens and have recently come back in favor. They are both ornamental and utilitarian, supplying fragrant leaves and seeds for decorating, cooking, teas, and perfuming.

The growing requirements of herbs are less stringent than those of other garden flowers. Full sun is required for strong growth and full flavor, but herbs will perform well in relatively poor, rocky soils and in drought. A few, including the mints, will even grow in wet soils. Table 9-7 lists the more common herbs, their hardiness, size, and uses.

Selected References for Additional Information

American Rose Society. *The American Rose Annual* (issued yearly by the American Rose society, P.O. Box 30,000, Shreveport, LA). Kingsport, TN: Kingsport Press, 1977.

*Ball, Jeff. *How to Design a Flower Garden*. Irwindale, CA: KVC Entertainment.

*Ball, Jeff. *How to Grow and Cook Fresh Herbs*. Irwindale, CA: KVC Entertainment.

*Indicates a video.

Burke, Ken, ed. *All About Roses.* San Ramon, CA: Ortho Books, a Division of Chevron Chemical Co., 1983.

Cresson, Charles O. *Rock Gardening.* Burpee American Gardening Series. New York: Prentice Hall, 1994.

Crockett, James A. *Crockett's Flower Garden.* New York: Little, Brown and Co., 1981.

Edinger, Philip. *Flower Garden Plans.* San Ramon, CA: Ortho Books, 1991.

Fell, Derek. *The Easiest Flowers to Grow.* San Ramon, CA: Ortho Books, 1990.

Hay, R., and P. M. Synge. *The Color Dictionary of Flowers and Plants for Home and Garden.* New York: Crown, 1982.

Hill, Lewis, and Nancy Hill. *Successful Perennial Gardening: A Practical Guide.* Pownal, VT: Garden Way, 1988.

McKinley, Michael D. and A. Cort Sinnes. *Color with Annuals.* San Ramon, CA: Ortho Books, 1993.

Parks Seed Company. *Annual Retail Catalog.* Greenwood, SC.

Sinnes, A. *Shade Gardening.* San Ramon, CA: Ortho Books, 1990.

Sinnes, A. Cort, and Ortho Book Staff. *Shade Gardening.* San Ramon, CA: Ortho Books, 1990.

Sinnes, A. Cort, and Larry Hodgson. *All About Perennials.* San Ramon, CA: Ortho Books, 1992.

Sunset Magazine and Book Editors. *Annuals and Perennials.* Menlo Park, CA: Sunset-Lane, 1993.

Sunset Magazine and Book Editors. *Herbs.* Menlo Park, CA: Sunset-Lane, 1993.

Sunset Magazine and Book Editors. *Bulbs for All Seasons.* 4th ed. Menlo Park, CA: Sunset-Lane, 1985.

Wayside Gardens. *Retail Catalog.* Greenwood, SC.

IO

Home Landscape Planning and Installation

Landscaping one's home is not usually thought of as an act of ecological awareness, but it is. Global warming, or the greenhouse effect, has recently come to the foreground of our society's consciousness as an environmental change that could radically affect all our lives. The use of fossil fuels such as gasoline, oil, and coal; accumulation of propellant and refrigerant chlorofluorocarbons; and rapid deforestation, particularly in tropical areas such as the Amazon basin, are all causes of increased temperatures in the world. The warming is significant; the five warmest years of the century were in the 1980s. Although global warming was anticipated, the rate is three times higher than was originally predicted, and there is justification to say that North Americans are largely responsible because we produce nearly one-fourth of the world's carbon dioxide from burning fossil fuels.

All this may sound daunting and not a matter that can be remedied by individual action, but such is not the case. Global Releaf is a public education campaign whose aim is to increase people's awareness of individual actions they can take to help combat this environmental problem. Three suggestions for steps to take start with *planting trees!* Although trees absorb carbon dioxide and thus directly reverse its accumulation in the atmosphere, their main effectiveness is through the shade they give in summer. By shading buildings and paved areas (which generate reflected heat onto buildings) and by their transpiration, trees decrease the need for air conditioning. This indirect cooling reduces carbon dioxide emissions by decreasing electrical use for air conditioning by 10–50 percent.

Trees form the backbone of most landscapes, but other landscape plants can also be ecologically benefi-

cial. Groundcovers prevent erosion of topsoil. In winter, evergreens can shield a house from heat-robbing winds and lessen heating bills. Where smog is a problem, plantings of shrubs and trees absorb pollutants.

Finally, trees and shrubs shelter squirrels, birds, and other wildlife, and their berries and nuts serve as food. By adding a source of water such as a pool or fountain, all the requirements of small wildlife can be met and many species will inhabit the yard.

The importance of planning your landscape cannot be overemphasized. As land becomes more expensive and the lots for houses smaller, haphazard planting of shrubs and trees not only creates a cluttered effect but becomes impractical. In the end it is nearly as expensive and labor-consuming as a landscape based on a plan would have been.

A thoughtfully planned landscape offers many rewards. Most obviously it enhances the design and beauty of the home, thus increasing its value. But a landscape has other functions which are not as immediately apparent. For example, it can extend the living area of the home by a patio or deck screened for privacy and protected from wind and sun. An outdoor play area can provide an outlet for overflowing childhood energy and the noise that normally accompanies it.

The Professional Landscape

Landscaping a home is often expensive because, unless one is blessed with a home in a naturally beautiful setting with enough land to assure privacy, extensive planting and construction will frequently be necessary.

If cost is not the main consideration, a professionally designed and executed landscape is easiest. The plan will be artistic, the shrubs and trees transplanted correctly, and the entire package will come with a guarantee.

But for many homeowners the cost of a professional landscape is prohibitive. For these people, two alternatives should be considered: (1) paying for a professional plan but installing the landscape personally and (2) planning and installing the landscape yourself. Each has advantages and drawbacks.

A professional design may cost many hundred dollars, but it is usually money well spent since most people lack the experience and imagination to design a professional-looking landscape. It is, however, possible for a homeowner to design an attractive landscape, provided considerable time and effort are devoted to the project. The following section gives directions and suggestions for the do-it-yourself landscape designer.

The Homeowner-Designed Landscape

Needs Analysis

The first step is to compile a needs analysis, that is, an inventory of what should be accomplished by means of the landscape and what the landscape will include, based on the preferences and life style of the owners.

To facilitate the needs analysis, the property should be divided into three areas (Figure 10-1). The *public area* includes all property viewable from the street: the drive, walk, front door, parking area, and sometimes the sides of the house. The *private living area* includes the patio or deck, play areas for children, barbecue and outdoor eating areas, and any game areas. The *service* or *utility area* is the area reserved for necessary but unattractive features such as recreational vehicle storage, dog kennel, and clothesline. A needs analysis for the public area should answer such questions as: "Is privacy needed from the street?" and "Does the entryway need lighting for night entry?" To design the private living area, the life style of the owners should be taken into consideration, asking such questions as: "Does the family entertain frequently outdoors?" and "Is a play area needed for children?"

Figure 10-2 is a checklist covering most of the options one would consider including in a typical landscape. Desired options should be checked on the list, and the completed form compared to the final design to make sure all requirements have been met.

Site Analysis

Site analysis, the second stage of planning, is an inventory of the property in terms of house architecture, views, soil, land slope, lot size, existing landscape, and the like. It lists both good and bad points of the property, so that they will be dealt with effectively in the design. For example, are there established trees which should be incorporated into the design or an air conditioning unit which should be screened from view? Are there poorly draining areas where nothing grows or a steep slope where erosion is a problem? Figure 10-3 is a checklist much like the needs analysis sheet.

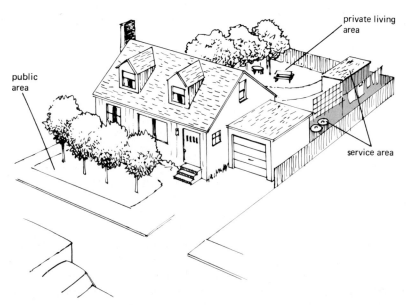

Figure 10-1 The public, private living, and service areas of a typical landscape.

Public area

night lighting
privacy from street
extra parking
improved walkway/entryway
entry court
private living area
irrigation system
shade
lawn
terracing for changes in level

Private living area

patio/deck (specify size and normal use times)
roof for patio/deck
shade
screening from neighbors
irrigation system
night lighting
plants attractive to birds
settings for ornaments (statuary, birdbath, and the like)
ornamental or swimming pool

barbecue grill/fire pit
children's play area (specify included equipment)
game area (specify games and space needed)
view emphasis
wind control
terracing for changes in level
auxiliary patio/deck (specify location)
lawn

Service/utility area

recreational vehicle storage
vegetable garden
fruit/nut trees
pet area
tool storage shed
clothesline
trash can storage
irrigation system
firewood storage
cut-flower garden
greenhouse
night lighting

Figure 10-2 Needs analysis checklist.

The Preliminary Drawing

The first step in designing the landscape is to create a preliminary drawing which shows the basic layout and permanent features of the property and includes the information from the site analysis. A scale drawing of the lot, house, and valued plants should be made on a large sheet of graph paper. This drawing is the skeleton of the design. Sheets of tracing paper can then be laid on top for experimenting with different ideas.

The drawing need not look professional to be serviceable; it is more important that it be accurate and include all the necessary information. Unless there are trees or particularly valued shrubs that will definitely remain, it is best not to include plants in the preliminary drawing. They limit creativity and may influence the designer to see the property only in terms of the present design. However, permanent features such as retainer walls, power lines, good and bad views, and the windows and doors of the house should be included, since modifying them would involve major changes.

Area Layout

After the basic drawing is complete, the parts of the property to be included in the public, private, and service areas should be chosen. In most established yards these areas will already have been defined through use, but unless there are absolutely no alternatives, it is worthwhile to experiment with shifting the locations of

these areas. A service area could be moved from the back to the side of the house, or a private living area from the back to an enclosed courtyard in front. Figure 10-4 shows how a property could be designed with the service area in three different locations.

Public Area Design

The public area is the first area a visitor sees. This area conveys an initial impression about the residents of the house. If poorly designed and ill-kept, it will give an impression of sloppiness; if sparsely planted or rigidly pruned, it will convey formality. If walled as a courtyard, it will project a sense of seclusion.

The main functions of the public area landscape are to blend the house with its surroundings and to provide a pleasant and readily accessible entry to the house. To perform these functions, most landscapes include several basic elements: the entryway and walkway, the driveway, and the auxiliary plantings.

Entryway and Walkway Areas

In most homes the builder includes a standard narrow walkway leading from the drive and an undersized concrete slab at the door to serve as a waiting area for guests.

Many homeowners consider these permanent features and are reluctant to use a design that involves removing and replacing them. While it is true that

Overall lot

1. Is the overall soil condition good, average, or poor?

2. Are there any areas with poor drainage where water stands after a rain?

3. Are there steep or gradual slopes to incorporate into the design? rocks or rock outcrops?

4. Are there good or objectional views, and from where can they be seen?

5. Is the property higher than, lower than, or equal in elevation to surrounding properties (for screening purposes)?

6. Could grading improve drainage or add interest to the lot?

The house

1. Is the home one story, two story, or split level?

2. Is the design modern or old? Are there historic considerations in designing the landscape?

3. Could changes in the house (painting, adding natural wood siding, extending windows) improve the effectiveness of the landscape?

4. Where are the windows which will look out on the landscape?

5. Are there unattractive permanent architectural features which should be concealed?

Weather and microclimates

1. In what USDA hardiness zone is the house located?

2. Are there frequent strong winds, and from what direction?

3. Where are the shady and sunny areas of the property?

4. Are there microclimate areas, and where? (See Chapter 1 for a discussion of microclimates.)

5. What is the pattern of sun and shade in the private living area from dawn to dusk?

Existing plantings

1. Are there large existing trees; where are they located?

2. Do any trees need to be removed for shade, disease, or design reasons?

3. Does the existing lawn area need improvement?

4. Is there any native vegetation to be worked into the design?

5. Are there any plants sufficiently desirable to justify being saved (other than trees)?

6. What are the species of any existing plants which will be saved?

Existing construction

1. Are existing fences and plantings sufficient to provide privacy?

2. Should the drive be replaced or rerouted?

3. Is the entryway to the house sufficient, or should it be redesigned on a larger scale or in a more convenient location?

4. Are there utility meters, natural gas tanks, or air conditioners which require screening?

5. Is any existing construction in need of repair or replacement (fences, walks, retainer walls, and the like)?

6. Does the patio or deck area need enlargement?

Figure 10-3   Site analysis checklist.

removing concrete is a major undertaking, if the walk is too narrow and the waiting area at the door uncomfortably tiny, the labor of removing these features and replacing them with a gracious entry is justified. They can also be covered and extended with a new paving material.

At a minimum, any walk should be 4 feet (1.2 meters) wide, the least width that allows two people to walk side by side. Five or six feet (1.5 to 1.8 meters) wide is even better. At this width a person can walk without watching his feet to avoid stepping off the paving. A walk of this width may at first sound exces-

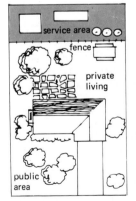
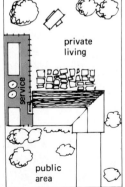
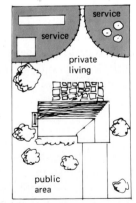

Figure 10-4   Three drawings of the same house, showing how the service area could be positioned in several locations.

sively wide and unattractive, but the monotony can be broken by small trees framing the walk, groundcover along the edge, or an interesting paving material.

Steps that need to be included should be wide and rise slowly, with treads at least 18 inches (45 centimeters) deep. Monotonous flights of evenly spaced steps should be avoided. Instead, several steps followed by a section of walkway and then more steps should be used to complete the level change. Changes in direction to create a curving walk should be considered.

One interesting and little-used public area design is the entry patio. It combines the walk and entryway into one paved area interspersed with beds of landscape plants (Figure 10-5). Such patios require little maintenance and create a feeling of spaciousness.

A variation of the entry patio is the closed entry courtyard (Figure 10-6). A wall or fence is used to delineate an area leading to the front door. The area is then planted with shrubs and small trees or paved and designed to include planting beds. An entry courtyard is useful when the front of the house is not overly attractive, when the house is situated close to the street, or when the home has large front windows that would otherwise have an unattractive street view. An entry court can also double as a private living area.

Before constructing an entry court, it is wise to check the zoning ordinances in the community regarding the construction of walls or fences in the public domain. Occasionally, walls of any type will be prohibited, but more often there are height restrictions.

The Driveway

Unless the driveway is a sweeping circular drive or leads through a large property, it will not contribute to

Figure 10-5 An entry patio requires little maintenance and is visually pleasing.

Figure 10-6 An entry courtyard in Spanish style. *(Photo by Andrew Schweitzer)*

the appearance of the public area and should be as inconspicuous as possible. Unnecessary curves should be eliminated, and the edges should never be outlined with flowers. Such treatment serves only to emphasize its presence.

However the designer should beware of attempting to minimize the driveway by limiting its size. This is not practical as it limits its functionality and the ease with which it can be used. Figure 10-7 gives sample driveway and parking area designs along with the minimum measurements acceptable for each type.

In many homes the driveway doubles as a walkway to the point where the regular walk begins. This is completely acceptable provided the driveway is not blocked by cars; in fact it is preferable to bisecting the yard with a walk to the street. Another effective driveway treatment is to extend the driveway/walkway with a strip of paving (Figure 10-8) merging into a patio attached to the house. Visitors parking in the drive can then step onto a paved area instead of the yard.

The selection of an attractive paving material can do much to improve the appearance of this otherwise mundane landscape element. Blacktop may be cheapest in terms of labor and materials, but if a homeowner can do the installation, more decorative paving materials may be affordable. Table 10-1 lists paving materials.

Auxiliary Plantings

Auxiliary plantings are the third element of the public area design. They encompass the trees, shrubs, and groundcovers which are the living part of the landscape.

Trees form the backbone of the landscape, framing and forming a backdrop for the house. If a house has large, well-placed trees, it will always be inviting,

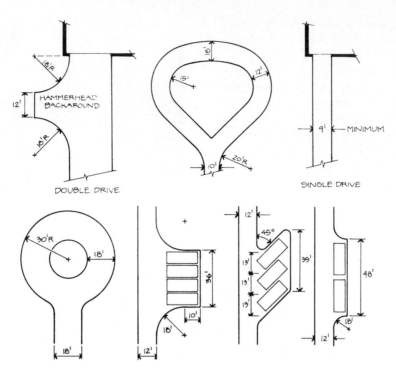

Figure 10-7 Specifications for driveways and small parking areas of various shapes. (*Taken from* Landscape Design, A Practical Approach, *2nd ed. by Leroy Hannebaum, 1990. Reprinted by permission of Prentice Hall, Englewood Cliffs, NJ*)

regardless of what else is planted in the yard. The cool, restful feeling of a tree-shaded front yard is unequaled.

If large trees are already present, their location must be worked around when designing the landscape. But in a treeless yard the selection and placement are of utmost importance.

Evergreen trees such as pines are useful because they retain their foliage throughout the year and are a permanent screen from wind, sun, and street traffic. Deciduous trees, on the other hand, offer shade during the summer months but do not screen out welcome winter sunlight.

For easy maintenance and protection from lawnmower nicks, it is helpful to locate trees in beds of shrubs or groundcover or to surround them with mulch. This also eliminates the need for hand trimming grass.

Trees of moderate to large size should not be planted closer than 20 feet (6 meters) from the house, since their spread and height will increase as they mature. Small trees can be planted closer since their ultimate size is not as great, but they should never be planted under the eaves of the house. In a similar manner, trees should never be planted where they will grow into overhead lines.

Auxiliary plantings of shrubs and groundcovers are used in most landscapes. Beds can be used to screen the house from the street (Figure 10-9), to balance other large plantings in the yard, or as accents. They can be used to incorporate shade trees since they are not usually close to the house.

Planting beds connected to the house can be either corner or foundation plantings. Corner plantings extend out from the front corners of a house, blending it into the site and eliminating the harsh right angle where the house meets the ground. They are among the most common landscape plantings, often including a small tree, several shrubs, and an attractive groundcover (Figure 10-10).

Foundation plantings were a necessity in the 1930s when homes were sometimes built with about 18 inches (45 centimeters) of unattractive foundation showing above ground level. The foundation had to be covered, and the foundation planting concept was formed. The idea persists today, and most homeowners still plant a line of shrubs along the front of their houses. A foundation planting should no longer be considered essential, and planting beds can be designed to cover none, all, or part of the front of the house.

Figure 10-8 Extending a driveway with a strip of walkway creates a graceful entrance and avoids bisecting the lawn. *(Photo by Andrew Schweitzer)*

Regardless of whether the bed is freestanding or connects to the house, most beds follow several simple rules. First, the shrubs are used in masses of three or more of the same species and are planted to grow together to form a mass of foliage. Second, the planting bed is always delineated by edging or paving, and all the included area is covered with groundcover or mulch. The clean lines and mulch- or groundcovered area create a finished, professional appearance. Third, sweeping lines are used to form large, curved, or geo-

Table 10-1 *Paving Materials*

asphalt or blacktop
brick or brick pavers
concrete (of any color)
    exposed aggregate concrete
    precast colored concrete slabs
    interlocking precast concrete forms
gravel
railroad ties
stone
tile
combination of brick and concrete

metric beds. The homeowner is frequently timid when designing planting beds, making them small and close to the house. A better effect can be achieved by large beds with bold shapes that jut out into the yard.

Guidelines for Public Area Design

There are many clues that separate the professionally designed from the obviously homeowner-designed landscape. By using a few techniques from professional landscapers, the finished design will be more pleasing.

First, do not try to save every plant in the yard and all the presently paved areas. If the design was poor initially, adding on will not improve it.

Second, if the lot is flat, consider construction such as terracing, mounding, or building fences, walls, or raised beds. The changes in level that construction creates add interest to the yard even before the plants grow and will eliminate the flat, monotonous feeling.

Third, use only a few species of plants, but many of each of those species. For an average front yard, two to three species of shrubs, one species of shade tree, one or two species of small trees, and one groundcover should be sufficient. Using individual plants of dozens of species creates an unharmonious and unattractive landscape.

Fourth, use groundcovers and mulches. Often the finishing touch can be put onto a landscape by underplanting trees and shrubs with groundcover or mulching with bark or stone (see Chapter 11 for mulch suggestions).

Fifth, avoid most ornaments. Birdbaths, gazing balls, wrought-iron chairs, and the like attract attention to themselves, deemphasizing the house and the remainder of the landscape. Ornaments should be restricted to the private living area or an enclosed courtyard and incorporated within decorative plantings.

Finally, do not be reluctant to use appealing design ideas from other people's landscapes or from books. Many do-it-yourself landscaping books give sample designs which, with slight altering, could be adapted to any house. The references at the end of the chapter include several idea books which are helpful in this regard.

Private Living Area Design

The private living area of your landscape should be just that: PRIVATE. Screening neighbors from both view and hearing is very important if the living area

Figure 10-9   A planting that screens the house from the street.

is to be used to full potential. Screening from wind and shading from sun become likewise important to overall comfort and will affect the usefulness of the area. A well-designed private living area can be pleasant for a Saturday morning breakfast, an afternoon nap in the shade, a dinner buffet at 6, and a party until 2. The elements of the private living area vary enormously with the family and climate, but almost everyone wants a deck or patio, screening for privacy, and shade.

Figure 10-10   A corner planting softens the edge of the house. (*Photo from* Landscape Design, A Practical Approach, *2nd ed. by Leroy Hannebaum, 1990. Reprinted by permission of Prentice Hall, Englewood Cliffs, NJ)*

Patios and Decks

A patio or deck is an outside room with a floor, walls, and a ceiling. The floor is the paving or decking. The walls can be fences, plants, or walls around it, and the ceiling can be the sky, a canopy of tree branches, or a constructed roof.

Most builders locate the patio area behind the house and adjacent to the living and dining areas. Here it is conveniently accessible from the house, an important consideration. However, other factors such as weather and privacy should also be considered when deciding on the location of this important area.

The orientation of the patio on the north, south, east, or west side of the house affects its livability and determines the amount of construction and shading needed. A patio situated to the east of the house is best in most cases, since it is warmed early in the morning by the sun but shaded in the afternoon by the house. A west patio, by comparison, will remain cool through the morning and require shading in the afternoon for comfort.

South-facing patios receive direct sun all day. In warm climates shade will be needed almost all day. However, in cold climates the continual sunniness warms the patio and extends its usefulness into early spring and late fall. Conversely, a northern patio is shaded through most of the year by the house and remains relatively cool. This orientation could be ideal in the Southwest.

The size of a patio can greatly determine its usefulness. Homes are usually built with 12 × 12-foot (4 × 4-meter) patios, barely adequate for a table and two chairs and room to walk around them. Preferable minimum size is 15 × 15 feet (5 × 5 meters). The extra 3 feet (1 meter) per side will accommodate a table and four chairs or a picnic table easily, with room left for a lounge chair and/or two regular chairs. Although 15 × 15 feet (5 × 5 meters) is suggested as a minimum, building an even larger patio is desirable. A spacious patio enhanced by incorporated planting beds (Figure 10-11) will be well used and will become a focal point for gatherings of family and friends.

While a main patio should be large and spacious, auxiliary patios or decks off a bedroom or bath need not be as large. Often they are designed for viewing from inside the house with only occasional usage. A 6 × 10-foot (2 × 3-meter) area will be adequate to provide room for a small table and two chairs.

Visual and Environmental Screening

Most homes located in an urban or suburban area need screening, not only for privacy but also to keep animals

Figure 10-11 A raised berm used to screen. *(Photo by Andrew Schweitzer)*

and children in or out and sometimes for wind protection. The traditional fence along the property line does make maximum use of the available land but can create a boxed in feeling if not lavishly planted. In addition, local zoning laws may prohibit fences higher than 5 or 6 feet (1.5–1.8 meters) along the property line; if neighboring properties are higher, this may not be sufficient to ensure privacy. Alternatives to property line fencing include berms, raised beds, interior fences, and visual screening by plants.

Berms are mounds or ridges of soil 1 to several feet (0.3–1 meter) high (Figure 10-11). Although some properties have natural berms, usually they are created by grading. They are useful because they add interest to the landscape with elevation changes and, when planted with shrubs and trees or topped by a fence, can stay within zoning restrictions and still provide privacy.

Raised planters (Figure 10-12) of wood, stone, brick, or other materials holding tall shrubs and groundcover can also screen a patio. Almost any plant that will grow in the ground will grow in a raised planter. Moreover, a raised planter is ideal for growing plants with special soil requirements because the bed

Figure 10-12 A raised brick planter such as this one can delineate a patio.

can be filled with the soil mixture of the owner's choice. In areas with alkaline soil, for example, a raised planter can be filled with an acidic mixture of soil and peat moss for growing azaleas and rhododendrons.

Interior fences are constructed within rather than along the boundary of the property. Where existing property line fences are too short, an interior fence can create a small private area and eliminate the necessity of replacing existing fencing. Interior fencing is also useful for screening the service/utility area from the view of the private living area, and it provides an attractive background for plants and garden ornaments (Figure 10-13).

Plants alone can be used to provide screening, but the effect is not immediate as it is with a fence. For the homeowner with patience, plants are among the best visual screening materials.

Plants are also better than most fences for controlling wind. If a fence is used for wind control, the wind frequently goes over the top and comes immediately back down to the ground within 2 to 3 yards (2–3 meters) of the fence. Plants, on the other hand, break up wind into small eddying currents of less velocity.

When planning for wind control, the seasonality of the wind should be considered; in other words, from which direction does it come during each season? Most people would not want to restrict a cool summer breeze in late afternoon, for example, but would like to decrease or eliminate a cold northern wind in fall with the hope of extending the usable season of the patio by a few weeks.

Many species of plants can be used for screening. Pines provide year-round evergreen screening; they can grow up to 100 feet (30 meters) high. A dense hedge of barberry plants will keep animals and trespassers out because of the sharp spines. When choosing plants for screening, ultimate height, denseness, growth rate, and whether the plants are evergreen or deciduous should be considered. For noise problems, plants are not the best choice. The main factors that influence noise level are distance from the source and the mass (weight) and height of the material used as a noise barrier. In view of these facts, significant noise control will really only be achieved by erecting a wall of brick, stone, or concrete of sufficient height so that the source of the noise is 3 to 6 feet (1–2 meters) from the top. For noise from traffic, this means a wall of 6 feet (2 meters) if the source is primarily automobiles, or even higher if trucks are the problem.

Shade

Afternoon shade is essential to outdoor comfort in most climates. Trees are one way to provide shade, and their location should be chosen carefully to ensure they will cast their shade at the right time of day and over the desired area. The disadvantage of using trees for shade is that the effect is not immediate. The homeowner must wait several years until his trees reach an appreciable size.

Roofs offer immediate shade, although their cost is relatively high. They can create a rain-proof area for protecting patio furniture which trees cannot. Aluminum and fiberglass are two popular roofing materials. Wood roofs made of 2 × 6's or lath spaced in varying patterns give a feeling of protection without complete shade, and they are architecturally pleasing in both public and private areas.

One relatively new product for use in shading a patio is woven or knitted sunscreen fabric (Figure 10-14). Such fabrics were originally used in horticulture for lowering the light intensity for shade-loving tropical plants. However now they are available in a va-

Figure 10-13   An interior fence used to delineate and screen a service area.

Figure 10-14   Knitted shade fabric allows light through but filters the sun and makes it less intense. *(Photo courtesy of Weathashade, a Division of the Gale Group, Apopka, FL)*

riety of colors (formerly they were always black), which makes them much more suitable in a landscape. A sunscreen fabric has the advantage of cooling a patio while still letting air and a significant amount of light through from above.

Optional Elements of the Private Living Area

For a family with youngsters, a play area is an essential feature of a landscape plan and can be included as part of the private living or utility area. The size of the area depends on the number of children who will use it and the total size of the property. For example, 12 feet by 12 feet (4 × 4 meters) is adequate for one or two children, although more room could be included if the yard was large. Surfacing can be grass, although daily use makes it difficult to maintain in good condition. Useful and safe grass substitutes include smooth pea gravel, bark chips, and sand, all applied thickly to cushion against falls. A paved ring for tricycle riding might also be useful. When locating the play area in the landscape, consider it as a temporary feature used only for a few years when the children are small. Accordingly, it should be placed so it can later be included into the landscape gracefully.

Game areas such as for croquet and badminton require large grassy areas, 24 × 54 feet (7.3 × 16.5 meters) for the latter and 30 × 60 feet (9.1 × 18.3 me-

ters) for the former. This size area may be difficult to include in a small lot, however. A tetherball pole requires only a 20-foot (6-meter)-diameter circle and could be included in most landscapes. Paving material such as concrete or gravel is recommended for tetherball circle areas.

Emphasis of a view may be part of the landscape plan. With mountain or lake properties, plantings can be situated to frame the view (Figure 10-15), with taller plants to the sides and short ones in front.

The Service/Utility Area

A utility area is the part of a landscape that provides room for necessities such as clotheslines, garbage cans, and pet runs. The design inside the area and the space allocated for each element should be functional, and the area should have convenient access. The narrow strips along the sides of most houses are convenient service areas, since they tend to be difficult to landscape. If a vegetable garden is included, care should be taken that the site chosen will provide maximum sunshine.

A main consideration for the utility area is that it be screened from the private living area. Fences, beds of dense shrubs, and wire supports with climbing vines all perform this function. A baffle effect with overlapping

Figure 10-15 Landscaping to frame a view. Note how attention is focused on the area between the two plantings and how the shapes of the trees mirror the form of the view.

sections of fence can substitute for a gate and give easy access while providing complete visual screening (Figure 10-16).

Choosing Plant Species for the Landscape

Selecting plants is the final step in designing a landscape. Establishing the basic design, with patios, walkways, and the like drawn in the plan, is preliminary; the plants to be included at this stage should be represented only by circles or similar symbols and labeled

Figure 10-16 Sections of fence used as baffles to hide a service area.

tree, shrub, groundcover, or whatever. The actual choice of plant species is made later to avoid the confusion of coordinating plant species and design elements at the same time. Regardless of whether the plant is a tree, shrub, or vine, the following factors should be taken into account before specifying plants in the design.

First, how adaptable is the plant to your climate? Winter hardiness is the most limiting climatic factor and the one over which there is least control. Choosing a plant that is not completely cold-hardy condemns the homeowner to extensive pruning of winter-killed wood every spring and may result in the complete death of the plant during a particularly cold year.

Rainfall is a second climatic characteristic to be considered. Natural rainfall in the eastern and southern parts of the country provides sufficient moisture for most plants. Irrigation is often necessary in the West, where water use for landscaping is fast becoming a luxury. Using drought-resistant plants to eliminate the majority of landscape water use works with, rather than against, the natural climate.

Smog tolerance can also be a limiting factor in the choice of plant species. Although smog will rarely kill a plant, some species, especially pines, can be severely damaged. Most often leaf margins or areas between the veins brown, leaving the plants burnt-looking and unattractive. Since there is no protection against smog damage, choosing smog-tolerant plant is recommended where there is significant air pollution.

Native soil conditions are an environmental factor that can affect the growth of many plants. pH, the acidity or alkalinity of the soil, can be a serious restriction. Plants native to acid soils (azalea and rhododendron, for example) planted in alkaline soils will turn yellow from inability to absorb nutrients at that pH range (see Chapter 13). These plants should be avoided in favor of more adaptable species or placed in raised planters in which the soil can be modified to the correct pH.

The relative amount of sun and shade the property receives should be considered when selecting plants. The yard of a house built in a wooded area will be largely shady, so shade-tolerant plants should be selected.

Resistance to pests and diseases is also important. Some species have a natural susceptibility to insect or disease problems, whereas others remain unaffected. To a large extent, disease and insect susceptibility is determined by the area in which the plant is being grown; species grown out of their adapted habitat are frequently prone to problems. A local nurseryman can offer advice regarding the adaptability of specific ornamentals to your area.

Not all landscape plants require the same amount of maintenance. Selecting the correct species can minimize the amount of time spent on maintenance chores like pruning and raking. Intensive-maintenance plants are those that drop large quantities of dead flowers and fruits or require frequent pruning of dead wood and unruly growth.

The mature height and spread of plants and their growth rate are often overlooked when homeowners choose plants. Only when the mature sizes of plants are known can they be intelligently placed in the design to blend without overgrowing their allotted spaces. If plants are placed too close together, frequent pruning will be necessary to keep them within bounds; if they are placed too far apart, they will never grow large enough to create the intended effect.

Growth rate is a function of both the species and the amount of water and fertilizer received. If the plant has a specific function to fulfill, such as screening, fast-growing species will be desirable. Similarly, fast-maturing trees are frequently desirable because they provide shade quickly. However, ordinarily, moderate- or slow-growing plants are preferred over fast-growing ones because the latter tend to be structurally weaker plants with short life spans.

A personal preference for the plant is, of course, important when selecting plant species to be included in the landscape. The attractiveness of the foliage, size of individual leaves (called fine or coarse texture), and color should be noted, along with outstanding flowers, fruits, or fall color. An assortment of plants with varying foliage textures and colors is preferred to give interest to the landscape throughout the year.

Choosing Tree Species

Trees can be divided into three sizes: large (45+ feet; 14+ meters), medium (30–45 feet; 9–14 meters), and small (less than 30 feet; 9 meters). Large and moderate-sized trees are used primarily for shade, although they can be highly ornamental and are then called *specimens*. Trees in the small category can be up to 30 feet (9 meters) tall but are frequently in the 10- to 15-foot (3- to 4.5-meter) range. Small trees yield some shade when mature and are often used by patios and decks. "Patio trees" frequently have ornamental flowers or fruits and are pruned when young to be multitrunked (Figure 10-17).

Shade trees should be selected so that they will complement the scale of the house when fully grown. One-story houses are complemented by trees in the small and medium categories. Owners of two-story, split-level, and larger houses should choose shade trees of moderate or larger size.

Trees vary in the density of the shade they cast. Those with round heads of large leaves cast deep shade, whereas trees with an open branch structure and small leaves cast "filtered" shade. Trees that will provide the proper amount of shade for the climate should be selected; the denser the shade, the more difficult it will be to grow a lawn or landscape plants underneath.

Root systems are another consideration in tree selection. Some trees have *invasive roots* which grow vigorously and can clog a septic tank or sewer pipe. Planting such trees should be avoided within 30 feet (9 meters) of wells or sewer systems.

Other trees produce surface roots radiating from the trunk. The roots can crack driveways or walkways (Figure 10-18). One solution is not to plant such trees next to a paved area; however, this restriction may mean that parking areas and patios are deprived of shade. A better solution is to install a thin sheet of metal or herbicide-treated plastic between the tree and the paved area at planting (Figure 10-19). This will deflect the young growing roots downward and prevent the problem.

Choosing Shrub Species

Whereas trees provide height and shade, shrubs provide mass and bulk. They are divided into three sizes: small (less than 3 feet; 1 meter), medium (3 to 6 feet; 1–2 meters), and large (6 to 12 feet; 2–3.6 meters). Small and medium-sized shrubs are most frequently used in home landscapes. Those in the large category can be up to twice as wide as they are tall and are suited to commercial buildings and large properties. They

Figure 10-17   A multitrunk tree. *(Photo by Andrew Schweitzer)*

can, however, be pruned to remove their bottom branches and used as small trees.

Along with small trees, shrubs are the main flowering plants in landscapes, and a design should include at least a few flowering species. Evergreen species are essential as well. Their foliage persists throughout the winter, making them a part of the backbone of the landscape.

Figure 10-18   Tree roots that have cracked a paved area. *(Photo by the authors)*

Selecting Groundcover Species

Groundcover plants are low-growing species which add variation in texture and unify groupings of larger plants. They give a finished appearance to a landscape, prevent weed establishment, and act as a living mulch. Although most groundcovers are less than 6 inches (15 centimeters) tall, they can be considerably larger: up to about 2 feet (0.6 meter). Shorter groundcovers usually are well suited to use on small properties; taller species are better suited to large properties because they are more in scale.

A groundcover should be selected for disease and insect resistance and suitability to climate; it should also blend well visually and in terms of its growing requirements with the other landscape plants. A groundcover should be unobtrusive—visually harmonizing rather than competing with other plants in the landscape. The choice of an evergreen versus a deciduous groundcover is a matter of personal preference. In an area with persistent snow cover, the groundcover will be hidden throughout winter anyway, but in other areas the evergreen foliage will be pleasing in the winter.

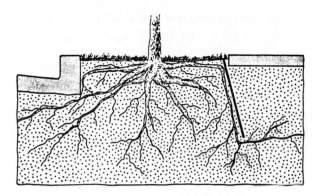

Figure 10-19 Tree roots will not disturb paving if the tree is initially planted with a root-deflecting barrier of a thin, impervious material such as metal or fiberglass. *(Taken from Arboriculture: Integrated Management of Landscape Trees, Shrubs, and Vines, 2nd ed. by Richard W. Harris. Reprinted by permission of Prentice Hall, Englewood Cliffs, NJ)*

Choosing Vines

Vines are not often included in landscapes, although they can be very attractive features trailing along the ground, climbing up a wall, or supported by a trellis or arbor. Vines can form the floor, walls, or ceiling of the garden. They can provide screening and shade, supply colorful flowers or fruits, and soften a wall, among other uses.

The methods with which a vine climbs will determine whether it will require a supporting framework. The stems of some vines grow small "holdfasts" which look like small suction cups and anchor them to any flat surface. Others have adventitious roots along the stems. These vines adhere to a wall or climb a fence without additional support.

Other vines climb by twining. The entire vine may grow spirally or may send out tendrils that twine to support the rest of the vine. Vines in this category need a trellis or other means of support.

Obtaining Landscape Plants

Most people purchase the plants they need for their landscape, but plants can also be propagated at home or transplanted from the wild or from friends' yards.

Home Propagation

Propagating woody plants by cuttings can be difficult for a home gardener. However, cuttings of some shrubs and most herbaceous groundcovers are relatively easy to root. If a source of cuttings can be found, it is prac-tical to propagate the groundcover needed for the landscape at home.

Transplanting from the Wild

Transplanting established trees and shrubs can be successful if the plants are small. Trees should be less than 10 feet (3 meters) tall and preferably growing in full sun, since a move from a shady woodland to a sunny yard can cause serious injury. Cool, rainy weather is best for transplanting (early spring in most cases, winter in warm climates where the ground does not freeze). If the tree is deciduous, it should be moved while leafless. Evergreens should be moved before starting active growth.

The first step is to cut through the roots with a spade in a circle about 15 inches (40 centimeters) from the trunk. The tree can then be uprooted, keeping as much soil around the roots as possible. The soil ball should then be wrapped in plastic or wet burlap for transport to the new location. Replanting should be done immediately, following the suggestions given for balled and burlapped plants in this chapter. Spraying with an antidesiccant (available at nurseries) will slow transpiration in evergreens and improve chances of survival.

Purchasing Plants from Local and Mail-Order Nurseries

Both local and mail-order nurseries are good sources of plants, but each has advantages and disadvantages. Local nurseries may not have the selection of mail-order ones, but the buyer can see the plant and be assured of its quality. Often plants from a local nursery have been grown in containers and can be transplanted with very little chance of loss. Mail-order nurseries dig the plants from the field and ship them with the roots wrapped in damp wood shavings. The plants may be subjected to extremes of heat or cold during shipping and are less likely to survive transplanting.

As a compromise, many gardeners patronize local nurseries for more common plants but use mail-order nurseries for hard-to-find species. However, you should not be reluctant to ask for a species or cultivar that a local nursery does not normally carry. The salesperson will usually be willing to order it, particularly if you are willing to buy a sizable number (10 or more) of the species.

The nursery section of a discount store is another source of plants, and bargains are available there with careful shopping. These stores order rather than grow their own plants. You should ask the manager when the nursery shipment is due and inspect the plants a day or two after they arrive. Plants frequently receive poor care in discount stores, and what were originally healthy plants may deteriorate to poor quality within a couple of weeks after arrival.

Planting the Landscape

Landscapes are planted in spring in most parts of North America. At this time the plants will be actively growing and will transplant with minimal shock. The ground will be thawed (in the North), and the temperature will be relatively cool, thereby slowing transpiration and preventing wilting. Fall is the second-best time for transplanting, again because of the cool temperatures. This prime planting period extends through winter in any climate where the ground does not freeze. Summer is the least preferable time for planting a landscape, since sun and hot temperatures make the problems of wilting more acute. However, if container-grown plants are used exclusively, the roots will be practically undisturbed and will grow even in summer.

Transplanting Trees and Shrubs

Shrubs and trees are sold *bare root, balled and burlapped,* or *container grown.* The selling method determines the price, the chances of surviving transplanting, and the seasons for transplanting.

Bare Root

Only deciduous plants are sold bare root, and they are always dug and sold while dormant. Most bare-root plants are sold only in spring; however, occasionally they are available in fall or throughout winter in areas where the ground does not freeze. Bare root is an inexpensive way to buy plants, since they are grown in fields with minimal maintenance and simply dug up when the selling season arrives. Most mail-order nurseries utilize this selling method because it minimizes shipping weight.

Because bare-root plants are dug from the ground, many roots are lost, so a bare-root plant is more likely to die during transplanting than a container-grown one. It must be given particularly good care to assure its survival and establishment in the landscape. When selecting a bare-root plant from a nursery, you should pick the sturdiest plant with the largest root system. The roots should be covered with or bagged in wood

shavings and be damp to the touch. If they have dried, it is likely that the plant may be dead. You should also scratch a tiny area of bark with your fingernail. If the plant is alive, a green layer will be visible beneath the surface.

A bare-root plant should be planted as soon as possible. If planting cannot take place for several days, the plant should be left in a shady area with the roots covered with moist soil. This "heeling in" prevents drying and the death of the roots.

At planting time the roots should be placed in a bucket of water and the tree taken to the planting site. A hole should be dug large enough to accommodate the roots easily without crowding (Figure 10-20). The topsoil should be piled on a piece of canvas or plastic next to the hole and not mixed with the subsoil. Organic amendment can be mixed with the soil to make a mixture to refill the hole. However, research has shown that in many cases this does not improve growth, and therefore it is probably not necessary unless the soil is of very poor quality.

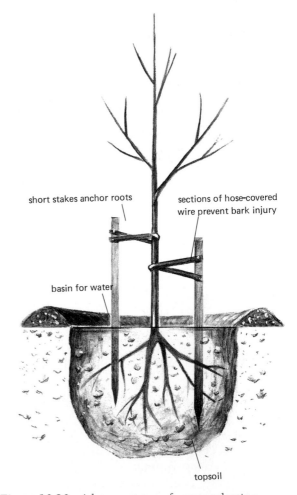

Figure 10-20   A bare-root tree after transplanting.

Next part of the mixture should then be shoveled back into the hole so that when the plant is placed in the hole, it will be sitting at or slightly above its original growing level.

Root pruning should take place next, with any badly broken roots clipped off with hand pruners. The remaining roots should be spread outward in the hole and covered with soil to ground level. A small ridge of soil should be mounded around the planting hole as a basin for water. Finally, the plant should be watered in well by filling the basin three or four times.

Balled and Burlapped Plants

Balled and burlapped (Figure 10-21) refers to trees and shrubs grown in a field but dug to keep a ball of soil around the roots. Both evergreens and deciduous plants are balled and burlapped; since the process involves considerable labor, plants sold this way are expensive. Because the digging disturbs only part of the roots, balled and burlapped plants usually transplant successfully provided the ball remains moist and proper planting procedures are used.

Hole and soil preparation is basically the same for balled and burlapped plants as for bare-root plants, with the stipulation that the hole should be twice as wide as the ball and one to one and one-half times as deep. Balled and burlapped plants are heavy, and care must be taken when transplanting to prevent the soil ball from breaking. Once the plant is positioned in the hole, the strings can be cut and removed and the burlap peeled back from the trunk. If plastic burlap has been used, it should be removed completely.

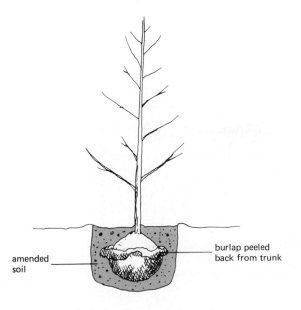

amended soil

burlap peeled back from trunk

Figure 10-21   A balled and burlapped tree after transplanting.

Container-grown Plants

Many plants are grown and sold in containers. Because the root system develops within the container, the plant is only slightly disturbed at transplanting and can be moved whenever the soil is workable.

As with balled and burlapped plants, a planting hole at least twice as big as the container should be dug. Removal of the root ball from the container without breakage is the only difficult part of transplanting. With a plastic container, the plant can be turned upside down and will slide out easily with little shaking. For plants in metal cans, it is easiest to cut the container down opposite sides, peel it back, and remove the plant. Many nurseries cut the can at the time the plant is purchased, saving the inconvenience of cutting it at home.

Boxed Trees

Occasionally, you will see large trees for sale with their roots in boxes. These plants may have been container grown or field grown. With the latter, the boxing is in lieu of balling and burlapping.

Planting hole preparation should be the same as for balled and burlapped plants. Removal of the box must be done carefully to avoid breaking the soil ball. First, the tree should be turned on its side and the bottom of the box removed. Using a board as an inclined plane, the box should then be slid onto the planting hole. Once in place, the sides of the box can be removed and the hole filled with soil.

Pruning Newly Transplanted Trees and Shrubs

Pruning newly transplanted trees and shrubs encourages them to develop proper form (follow the instructions for pruning in Chapter 11). In the case of bare-root plants, it reduces the amount of top growth to a level that the sparse root system can support. Many times bare-root plants are sold prepruned and need no further pruning until the following season. However, if not, about one-third of the top should be removed at transplanting by removing the weaker branches and clipping the remaining branches back to one-half to two-thirds their original length. Branches low on the trunk should *not* be removed as they strengthen it.

Staking and Wrapping Trees

Staking may be needed either to anchor the root system of the tree until it becomes established or to support the trunk in an upright position. Unless it can be

Figure 10-22   Wrapping a tree trunk with paper tape.

determined that staking is fulfilling one of these functions, it is not necessary and will cause weakening of the trunk.

For root anchorage of bare-root trees, two short stakes (1–1½ feet; 30–45 centimeters) should be inserted on opposite sides of the trunk (Figure 10-20). The tree should then be secured to each with strips of rubber, cloth, or wire covered with a section of garden hose. Uncovered wire, rope, or twine is not recommended since friction will cause it to rub off the tender bark.

Trees that require trunk support should be staked as low as will allow them to stand upright under calm conditions. Again, stakes should be positioned on opposite sides of the tree and the trunk tied to the stakes at a single point near the top.

Wrapping (Figure 10-22) involves twining tree-wrapping paper or burlap secured with twine around a tree trunk from the base to the lowest branches. The wrapping protects the trunk from sun damage, a problem on trees grown in rows with their trunks shaded by neighboring trees. Wrapping should be left in place about a year, after which it will begin to decay and can be removed.

Transplanting Groundcovers

Groundcovers are herbaceous perennials or woody shrubs. The following recommendations apply to herbaceous groundcovers. Shrubs should be planted as recommended in the previous section.

Because many herbaceous groundcovers spread by trailing along the surface of the soil, it is important that all soil in a groundcover area be conditioned. The area should be spaded deeply to at least 12 inches (30 centimeters), and amendment should be worked throughout. The groundcover is then transplanted with a trowel throughout the prepared bed and watered thoroughly.

Spacing of herbaceous groundcover plants is determined by economics and how quickly coverage is de-

Figure 10-23   Planting a groundcover through a landscape fabric will make subsequent maintenance much easier. *(Photo courtesy of the DeWitt Co., Sikeston, MO)*

sired. Spacing plants 2 feet (60 centimeters) apart will cost less than spacing them 8 inches (20 centimeters) apart but will take up to a year longer to cover the area. As a general rule, spacing plants 1 foot (30 centimeters) apart gives coverage at a moderate rate and expense.

When landscaping with groundcover, buying *landscape fabric* to cover the groundcover area is a wise investment (Figure 10-23). It is a water-permeable, woven or felted synthetic material which is laid over the area to be planted. Holes are then cut for each groundcover plant; the plants are inserted through the fabric for planting. The advantages of landscape fabric are permanent, total weed control while still allowing water to penetrate. In this regard, it is different from the black plastic mulch described in Chapter 11.

Postplanting Care of the Landscape

The most important part of postplanting care of a landscape is regular watering. During the first growing season plant roots occupy only a small volume of soil and need constant moisture to become established in the new location. Slow soaking with a hose is preferred for shrubs and trees, which should be watered every week to 10 days. Slow watering allows all the moisture to be absorbed into the soil and provides sufficient water to moisten the entire root area.

Groundcover watering is done by hand or sprinkler as frequently as required to keep the soil moist. The generous supply of moisture will enable the groundcover to fill the area within a short period. It is advisable to check the depth of water penetration after watering with a trowel, digging to the depth of the roots and examining the soil moisture. In this way there is no possibility of moistening the surface while the roots remain dry.

Mulching should be a part of postplanting care. It not only keeps in soil moisture but also prevents weed growth, a problem in newly planted groundcover areas. Mulching of landscape plants is discussed in Chapter 11.

Selected References for Additional Information

American Horticultural Society. *Shrubs and Hedges.* Alexandria, VA: American Horticultural Society, 1982.

Brimer, J. B. *Homeowner's Complete Outdoor Building Book.* New York: Sterling, 1989.

*Clokey, Joe. *Xeriscape—Appropriate Landscaping to Conserve Water.* San Luis Obispo, CA: San Luis Video Productions.

*Garrett, Howard. *The Art of Landscaping: Design.* San Luis Obispo, CA: Vocational Education Productions of California Polytechnic State University.

*Garrett, Howard. *The Art of Landscaping: Shopping and Planting.* San Luis Obispo, CA: Vocational Education Productions of California Polytechnic State University.

*Harrigan, Jim. *Landscaping with Container Plants.* San Luis Obispo, CA: Vocational Education Productions of California Polytechnic State University.

Hightshoe, Gary L. *Native Trees, Shrubs, and Vines for Urban and Rural America.* New York: Van Nostrand Reinhold, 1988.

*Labor, Elizabeth. *Plant Selection.* San Luis Obispo, CA: San Luis Video Productions.

Ortho Staff. *Creative Home Landscaping Book.* San Ramon, CA: Ortho Books, a Division of Chevron Chemical Co., 1987.

Phillips, Roger, and Martyn Rix. *The Random House Book of Shrubs.* New York: Random House, 1989.

Reader's Digest Staff. *Practical Guide to Home Landscaping,* Pleasanton, NY: Reader's Digest Books, 1977.

Smith, Michael. D. *Affordable Landscaping.* San Ramon, CA: Ortho Books, 1988.

Sunset Magazine and Book Editors. *Fences and Gates.* Menlo Park, CA: Sunset-Lane, 1981.

Sunset Magazine and Book Editors. *Garden and Patio Building Book.* Menlo Park, CA: Sunset-Lane, 1983.

Sunset Magazine and Book Editors. *Landscaping and Garden Remodeling.* Menlo Park, CA: Sunset-Lane, 1978.

Time-Life (eds.). *Evergreen Shrubs.* Alexandria, VA: Time-Life Books, 1989.

*Vocational Education Productions. *Landscaping Design Series, Vols. I and II.* San Luis Obispo, CA: Vocational Education Productions of California Polytechnic State University.

Wilson-McClune, Nancy Patton. *Ortho's Plant Selector.* San Ramon, CA: Ortho Books, 1991.

Wyman, D. *Shrubs and Vines for American Gardens.* (rev. ed.) Toronto: MacMillan, 1969.

Wyman, D., and Pam Hoenig. *Trees for American Gardens.* New York: MacMillan, 1990.

Zucker, Isabel, and Derek Fell. *Flowering Shrubs and Small Trees.* New York: Friedman, 1990.

*Indicates a video.

11

Landscape Maintenance

Regular maintenance by pruning, fertilizing, watering, and weed control is a necessity if home grounds are to remain attractive and plants healthy. But maintenance does not need to be time-consuming. A landscape planned without a lawn and with a goal of infrequent maintenance can be kept in condition with only about two to three hours of labor per month.

Features of a Low-Maintenance Landscape

Landscapes designed to remain attractive with minimal maintenance include some or all of the following features:

1. Large paved or decked areas that require only occasional sweeping

2. Raised planters that require less bending for maintenance

3. Groundcover areas in lieu of lawns to eliminate mowing

4. Mulches or groundcovers around all trees and shrubs to eliminate hand trimming (which grass requires) and suppress weed growth

5. An automatic irrigation system where natural rainfall is not sufficient

6. A limited number of carefully placed plants rather than large numbers of plants scattered throughout the landscape

7. Plants that do not require frequent pruning and do not drop leaves, fruits, or dead flowers

8. Plants which, when mature, will not require extensive pruning to keep them the proper size

9. Flowering shrubs rather than maintenance-requiring annual and perennial flowers

10. Plants that are naturally disease- and insect-resistant and have been proven to grow well in the locale

Tools for Garden Maintenance

Every gardener needs a selection of tools such as trowels, rakes, and pruners to maintain the landscape. The following sections describe the most frequently used and valuable gardening tools and equipment.

Soil-Working Tools

Trowel. A trowel (Figure 11-1) is used for digging small holes for planting bulbs, setting out transplants, and other similar tasks. When purchasing a trowel, check to make sure the handle attaches firmly to the blade and does not bend.

Round-Point Shovel. A round-point shovel (Figure 11-1) is used for digging larger holes such as for planting trees and shrubs and other general garden digging. The tip is pointed to make it easy to push into the soil.

Hoe. A hoe (Figure 11-1) is used for shallow cultivation of a previously prepared area and will penetrate the soil up to 4 inches (10 centimeters) deep. It is frequently used in vegetable gardening to prepare rows for

193

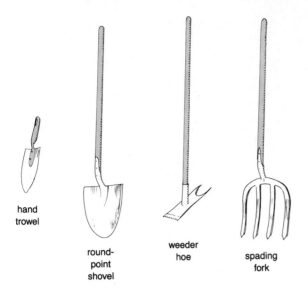

Figure 11-1   Most commonly used soil-working tools for the garden.

seeding. Hoes are also used for weeding. For this purpose, they should be used with a scraping motion that severs the weeds at the soil line. If used in a cultivating manner, additional weed seeds will be exposed, germinate, and intensify the weed problem.

Spading Fork. A spading fork (Figure 11-1) is used to move piles of leaves or refuse, turn compost, and break up large clods in the soil. Although a shovel can be used for these purposes, a spading fork is more efficient and will move a greater volume of leaves or compost with each bite. A spading fork is distinguished from a pitchfork in that it has stronger flat tines.

Watering Equipment

Hose. Many grades of garden hoses are available. Inexpensive plastic ones are stiff and develop permanent kinks which restrict water flow. Reinforced plastic hoses remain pliable, do not kink, and are lightweight. Rubber hoses are the most durable, but they are heavy and expensive.

Pistol Nozzle. A pistol nozzle (Figure 11-2) provides on-off control and can be adjusted to give from a strong forceful stream to a mist. Unfortunately, when the stream is directed, it is too forceful for most watering, and when emitted at a lower volume, it sprays in a circle and cannot be accurately directed. However, some gardeners find the forceful stream useful for washing off insects and the mist ideal for watering seedlings.

Flaring Rose Nozzle (Fan Nozzle). This nozzle (Figure 11-2) often does not provide on-off control. However,

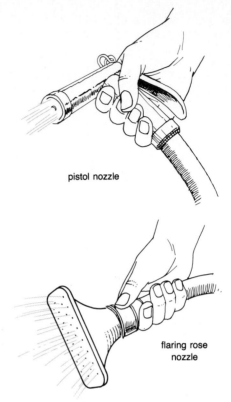

Figure 11-2   Types of hose nozzles.

it applies a large volume of water without force and is good for general overhead watering.

Oscillating Sprinkler. This sprinkler (Figure 11-3) is the most adaptable to general yard use and can be adjusted to cover any size area by altering the water pressure and the oscillation mechanism. Because the head moves back and forth, water is applied over the area at

Figure 11-3   An oscillating sprinkler.

a slow rate, and runoff is less likely than with a non-oscillating type.

Watering Can. Either plastic or galvanized metal watering cans are acceptable. The can should contain a removable sprinkler head so that either a fast volume or a slow sprinkling of water can be applied. Although metal cans last longer, they are heavier and more expensive than plastic cans.

Rakes

Leaf Rake. A leaf rake (Figure 11-4) can be made of lightweight metal or bamboo and has long flexible teeth. It is used for raking leaves, grass, and other lightweight materials.

Garden Rake (Bow Rake). A garden rake (Figure 11-4) is generally metal with short, rigid, widely spaced teeth. It is used for smoothing the soil surface in preparation for seeding a lawn or vegetable garden, removing thatch from the lawn, and other purposes.

Fertilizing Equipment

Hose-End Sprayer. This apparatus (Figure 11-5) attaches to a hose and automatically mixes any liquid concen-

Figure 11-5   Hose-end sprayer. (*Photo courtesy of Vocational Education Productions, San Luis Obispo, CA*)

trate into the water at dilute strength. It is used for fertilizing lawns, flowers, and other shallow-rooted plants and for applying certain insecticides. However, it is not accurate enough for most pesticide applications.

Root Feeder. A root feeder works like a hose-end sprayer, proportioning liquid fertilizer into the irrigation water. It is equipped with a long, needle-like tube for injecting the fertilizer into the root zone of shrubs and trees.

Pruning Tools

Hand Pruners. Hand pruners (Figure 11-6) will cut branches up to $\frac{3}{4}$ inch (20 millimeters) in diameter. They are the most frequently used pruning tool. When using hand pruners it is important never to twist while cutting in an attempt to cut a large branch. This will pull the cutting blades out of alignment and ruin the pruners.

Lopping Shears. This tool (Figure 11-6) will cleanly remove branches up to about 2 inches (5 centimeters) across in a single cut. Loppers are used on larger shrubs and trees.

Pruning Saw. A small, curved pruning saw (Figure 11-6) can be used for removing larger branches and is specially designed to fit in tight places. It can substitute for lopping shears but will require more labor. Both lopping shears and pruning saws are available in pole-mounted models for pruning high in trees.

Hedge Shears. These are needed only if formal hedges are to be maintained (Figure 11-6). Electric types are also available.

Care of Garden Tools

If properly cared for, garden tools will last many years. Regular removal of rust on rakes, trowels, shovels, and

leaf rake

garden or
bow rake

Figure 11-4   Types of rakes.

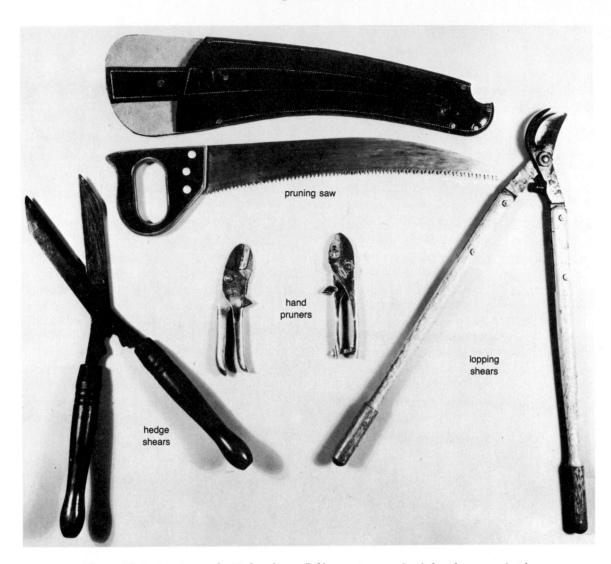

Figure 11-6 Pruning tools. Hedge shears (left), pruning saw (top), hand pruners (middle), and lopping shears (right).

other tools is important. Wire brushing followed by rubbing with steel wool removes most corrosion. A light coating of oil will slow its return and is also useful as a prestorage treatment in areas where garden tools are not used in winter.

Wood handles should be inspected occasionally for cracks and splintering. Small cracks can be fixed with wood cement. Sanding followed by a coating of boiled linseed oil will cut down on splinters. As an alternative to oiling, some gardeners paint tool handles red. This seals the wood and makes misplaced tools easier to find in the garden.

Sharpening of pruning tools at least once per season is recommended so they will make clean cuts. A sharpening steel or whetstone can be used, and professional tool-sharpening services area also available. Spades, hoes, and shovels also require sharpening and

should be honed lightly with a hand file (but not filed razor sharp).

Hoses should be coiled after each use and stored in the shade. Sunlight will deteriorate a hose rapidly. In cold areas the hose should be stored over winter in a garage or basement.

Pruning Landscape Plants

Pruning is probably the most misunderstood maintenance task. Considerable time can be spent on seasonal clipping of trees and shrubs, only to find that in a few months the plants are again in need of pruning. However, this waste of time can be avoided by understanding how to prune to encourage controlled growth, and time invested in pruning can be held to a minimum.

Purposes of Pruning

Size Control

Most people use pruning for size control—to keep plants from growing over windows or a walkway, for example. However, pruning should not be necessary if the plants are selected with their mature size in mind and positioned with forethought as to obstructions they might later create.

Health Improvement

Plant health is maintained or improved by such pruning practices as removing dead branches and diseased limbs. It also includes pruning trees when young to encourage strong branch structure and thus decrease the chances of branches breaking in storms.

Appearance Improvement

Pruning to remove scraggly branches, clip off dead flowers, and encourage bushy compact growth all improve plant appearance.

Pruning Trees

Young Trees

Pruning trees at transplanting and in the first few years that follow will greatly affect their strength and health in later life. During this early period the tree develops the main branches which will form the support structure for the foliage. Pruning directs the growth of those main branches in order to create a strong tree.

Removal of Double Leaders. Most trees have a "central leader," that is, they grow with one main upright shoot (leader) and side branches called *scaffold branches* (Figure 11-7). Occasionally two leaders develop, causing a fork. The angle between the two leaders is a potential weak point of the tree, because as each leader increases in diameter, it exerts pressure against the other at the base. Eventually, conditions conducive to splitting of the trunk develop, and in a wind storm the weight of the leaders can pull the trunk apart. Accordingly, the weaker of the two leaders on a double-leader tree is removed while the tree is young. The remaining one will then develop normally.

Selection of Scaffold Branches. Removal of a double leader is not usually necessary in tree pruning because it will have been done at the nursery, but removal of excess scaffold branches is usually necessary. Ideally a

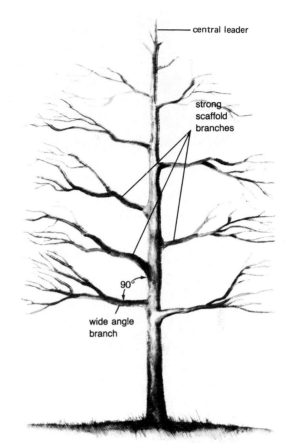

Figure 11-7 A well-shaped tree with a central leader form. Note the even spacing of the scaffold branches down the trunk and the wide branch angles.

young tree should have only a few branches spaced widely and evenly along the trunk. Numerous small branches are not desirable since they will enlarge and crowd each other as the tree grows. Scaffold branches should be selected so that they are evenly spaced up the trunk and not growing directly above one another. Branches growing from the trunk at a wide angle approaching 90° (Figure 11-7) are preferred since they are stronger, although many trees have narrow-angled branches when young.

When a double leader or extra branch is removed, the branch should always be cut just beyond the *bark ridge*, also called the *shoulder ring* (Figure 11-8). This ridge is an area of raised bark or a ring that surrounds the point at which the branch grows from the trunk. It contains cells that are a barrier to infection of the cut and proliferate rapidly to close the wound. Therefore even though a small stub may be left in order to leave the bark ridge intact, it is preferable to do so.

Topping. Topping is a practice used by some gardeners in the hope of forming a tree that is full-headed and

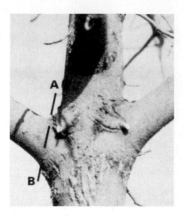

Figure 11-8 Bark ridges on two branches. (*Photo from Arboriculture: Care of Trees, Shrubs and Vines in the Landscape by Richard W. Harris. Reprinted by permission of Prentice Hall, Englewood Cliffs, NJ*)

dense. The central leader is removed; the side branches below the cut then grow rapidly. The ultimate effect is a flat-headed tree with an unnatural appearance and weak branch crotches. Topping is not recommended except for modified-central-leader-trained fruit trees.

Limbing Up. Limbing up is the removal of lower branches of a tree that are positioned too low on the trunk and thus create a hazard or inconvenience. Removal is necessary because the height at which a branch originates is the height at which it permanently remains. The trunk does not lengthen and elevate the branch, as is popularly believed. A young tree 4 feet (1.2 meters) tall with branches 2 feet (0.6 meter) off the ground would have large branches still 2 feet (0.6 meter) off the ground 20 years later if left unpruned.

Limbing up should be delayed for three to four years after planting, if possible, since removal of foliage from a young tree will slow its growth and weaken the trunk. Once the tree is established and growing rapidly, one or two branches can be removed each year until the desired trunk height is reached.

Desuckering. Desuckering is the removal of shoots that grow from the base or lower trunk of trees. If the tree is grafted, these branches often grow from the vigorous root-stock. In other cases, buds on the trunk are located far enough from auxin-manufacturing buds that they are not reached by the dormancy hormone. Such branches are called *suckers,* or *water-sprouts,* and should be removed as they appear with hand pruners. If left in place, they give the tree an unkempt appearance and use up energy that should be diverted to top growth.

Although most waterspouts grow from the trunk at ground level or slightly above, they are sometimes found on scaffold branches. They are easy to locate be-

cause of their vigorous vertical growth and should be removed. If left, they will grow until they cross and rub against the scaffold branches.

Mature Trees

A tree that has been properly pruned when young should need little, if any, pruning when mature. Desuckering may be necessary, along with the removal of branches broken by wind or growing toward the center of the tree (Figure 11-9). Small branches growing close together may rub one another (Figure 11-10); the smaller or more poorly positioned ones should be cut back to their parent branches to avoid injury to the bark and grafting together of the limbs.

Removal of Large Branches. Occasionally, large branches may need to be removed because of disease, incorrect pruning when young, or some other reason. Such branches are frequently heavy and, while being sawed off, can break and rip down the trunk. To avoid the chance of injury to the tree, the three-cut pruning method (Figure 11-11) is recommended for removing branches more than 2 inches (5 centimeters) in diameter. The first cut is made under the branch about 1 foot (30 centimeters) from the trunk and goes halfway through the branch. This cut provides insurance against tearing in case the branch should fall. The second cut is made farther out on the branch and cuts completely

Figure 11-9   Branches that begin to grow inward toward the center of the tree should be cut out as shown.

Figure 11-10 The weaker and more poorly positioned of these two rubbing branches should be removed, even though they are both major branches. (*Photo courtesy of Vocational Education Productions, California Polytechnic State University, San Luis Obispo, CA*)

through and removes the branch. The third cut removes the stub down to the bark ridge, leaving it intact.

If it is necessary to remove a stub that was left because the tree was pruned improperly years ago, the branch collar should be left and only the part of the stub extending beyond it removed (Figure 11-12) even though this may leave a considerable amount of stub. The collar is like the bark ridge in that it prevents entrance of infection.

Tree wound paint, an asphaltlike sealer, is commonly used on wounds larger than 1 inch (2.5 centimeters) in diameter to discourage the entry of

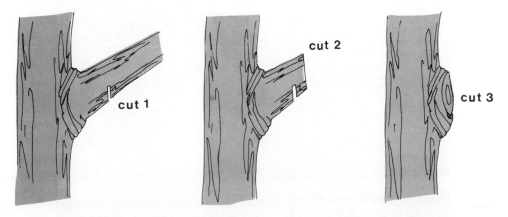

Figure 11-11 The three-cut pruning method for removing large branches. This method prevents the branch from tearing down the trunk.

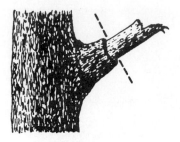

Figure 11-12 This branch stub has a collar of live wood and should be cut back only as far as shown. (*Taken from Arboriculture: Integrated Management of Trees, Shrubs and Vines, 2nd ed. by Richard W. Harris, 1992. Reprinted by permission of Prentice Hall, Englewood Cliffs, NJ*)

microorganisms and insects. However there is no clear scientific evidence proving that it has any actual effect. In most cases the paint cracks and admits water within a year after it is applied, and water can then be trapped in pockets created by the cracks. This is actually a worse situation for disease invasion than leaving the wound open. If wound paint is used it should always be applied after the wound has dried and renewed when it cracks.

Removal of Girdling Roots. A girdling root (Figure 11-13) usually results from the incorrect initial planting of a tree. If the roots are not spread out in the hole, they can enlarge and press against the trunk at or just below ground level. The resulting pressure halts the flow of water and nutrients within the trunk and can eventually kill that side of the tree.

Signs of damage from a girdling root include flattening of the trunk on one side and, on deciduous species, dropping of the leaves on the affected side earlier in fall. The only cure for a girdling root is to cut it away with a chisel or hatchet.

Pruning Vase-shaped Trees

Many patio trees, weeping trees, and some larger shade trees like elm do not naturally have a central leader, and it is a mistake to prune them to have one. Such trees may split into two leaders, forming a vase shape, or have all their branches arising at one point on the trunk (Figure 11-14). Pruning for these trees should include only desuckering and removal of crossing and inward-growing branches.

Pollarding

Pollarding (Figure 11-15) is the formal training of deciduous trees and a common pruning technique in certain areas of the country. It restricts the height of the tree and creates a denser head of foliage than the species would normally have.

Pollarding is usually begun after the tree has attained the maximum height desired. When the tree is dormant it is cut back severely, the main branches sawed off deeply into the old wood. When the tree breaks dormancy, buds will sprout below each cut, producing vigorous watersprouts and creating the dense foliage mass. The next winter and each winter following, the recurring watersprouts are removed. Eventu-

Figure 11-13 A girdling root on a young tree. (*Photo by Rick Smith*)

Figure 11-14   A vase-shaped tree with all main branches arising from one point on the trunk. (*Photo by Rick Smith*)

ally the ends of the branches become gnarled, giving the plant a distinct appearance.

Although pollarding does give a tree a full, pleasing appearance during the growing season, the tree will be very unattractive when dormant. In addition, pollarding generally requires the yearly services of a tree maintenance person who has the equipment needed to reach high in the tree and remove the watersprouts. Consequently, pollarding cannot be recommended for normal home landscapes.

Pruning Needle Evergreen Trees

Pines are the most familiar needle evergreens, all of which are characterized by a pyramidal, or "Christmas tree," form with a central leader. Evergreens generally do not require pruning except to remove broken or dead branches and occasionally to eliminate a double leader. They may be limbed up above eye level if necessary.

When pruning needle evergreens, remember that unlike most other trees, they do not branch if cut back to old wood, so pruning into the woody portions has a permanent effect. The only pruning that can be used to limit size or make the foliage thicker is the removal of some of the new growth, or *candles* (Figure 11-16),

while the tree is still immature. Breaking out a portion of the center candle and/or side candles will cause the growth to become shorter and fuller.

Pruning Shrubs

To preserve their natural appearance, most shrubs are pruned by either of two pruning methods: *thinning* or *heading back.*

Thinning

Many shrubs are composed of a clump of woody stems or one or more short trunks coming out of the ground. They grow by the increase in height and number of these stems or trunks. As the oldest stems age and lengthen, they will have a tendency to shade out the bottom foliage, and the shrub will become bare at the base. Thinning consists of removing the oldest stems at ground level. This shortens the shrub and encourages new shoots to grow and refoliate the bottom. Thinning can be done with hand pruners, if the stems are small, or with loppers, if the shrub is heavily overgrown with large stems (Figure 11-17).

Figure 11-15 A pollarded tree in winter after pruning. (*Taken from* Arboriculture: Integrated Management of Trees, Shrubs and Vines, 2nd ed. *by Richard W. Harris. Reprinted by permission of Prentice Hall, Englewood Cliffs, NJ*)

Heading Back

Heading back (Figure 11-18) is done with hand pruners to remove straggly growth or to limit size. It consists of cutting twigs or small branches back to a point where the cut is hidden by the remaining foliage. If possible, the stem should be cut back to just above an outward-pointing bud or branch (Figure 11-19). This will encourage new growth to develop outward and eliminate crossing branches.

Renewal Pruning

Renewal pruning (Figure 11-20) is used to rejuvenate and shorten overgrown shrubs. It is always done in

Figure 11-16 Pruning a needle evergreen by breaking out the unexpanded candles. (*Photo by Rick Smith*)

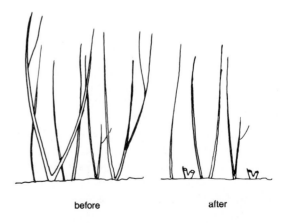

before   after

Figure 11-17 Pruning a shrub by thinning.

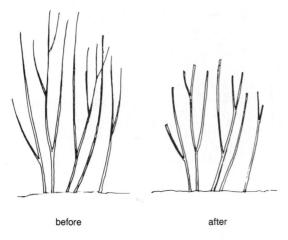

before   after

Figure 11-18 Pruning a shrub to increase its density by heading back.

early spring before new growth starts and consists of cutting back all the branches to stubs 2 to 3 inches (5–8 centimeters) long. This forces new growth from dormant buds on the stubs, which can then be pruned as needed by heading back and thinning to maintain the proper size.

Renewal pruning is a drastic measure and will make the plant unsightly for at least one season. As an alternative, half the branches can be removed the first year and the remainder the second year. In this way the plant will always keep some foliage and have a somewhat better appearance.

Pruning Hedges

Natural hedges are lines of shrubs planted close together to create a thick mass of unpruned foliage. Pruning natural hedges is basically the same as for other shrubs, that is, thinning and heading back once or twice a year.

Formal hedges are pruned to a specific shape (Figure 11-21) and require frequent cutting to maintain a neat appearance; shearing every two weeks during the growing season is typical.

Selection of the correct hedge shape is very important. Too often a hedge is pruned to be narrow at the base and wide at the top. This is incorrect because the wide top will shade the lower portions, causing them to drop their leaves. Pruning the top narrower than, or at the same width as, the bottom (Figure 11-21) will avoid shading of the bottom, and the plants will remain full and leafy to the bases.

Pruning Groundcovers

Most groundcovers do not require regular pruning to improve their appearance or health, although shrub types such as junipers may require heading back to control spread (Figure 11-22). However, occasionally an older bed of groundcover will develop long leafless stems and can be improved by renewal pruning. If the area is small, it can be clipped back severely with hand pruners. For large areas, a lawnmower set to cut as far from the ground as possible (and with a catch bag) can be used during the period of active growth. Hedge shears can be used on woody ground covers. Although the area will be unsightly at first, it will usually refoliate in two to three months.

Timing Pruning

The time of year at which maintenance pruning takes place will not affect the health of a plant and need not be restricted to a particular season. However, early spring is a traditional pruning time for several reasons. First, it is easier to determine which parts of deciduous plans should be removed if the branch structure is not obscured by leaves. Second, by pruning before new growth starts, the growing direction of new shoots can be controlled. By contrast, summer or fall pruning removes poorly placed shoots produced that season.

There are exceptions to the early spring recommendation for pruning. If spring-blooming shrubs are pruned early in the year, some of the flower buds that have already formed will be cut off. Consequently it is

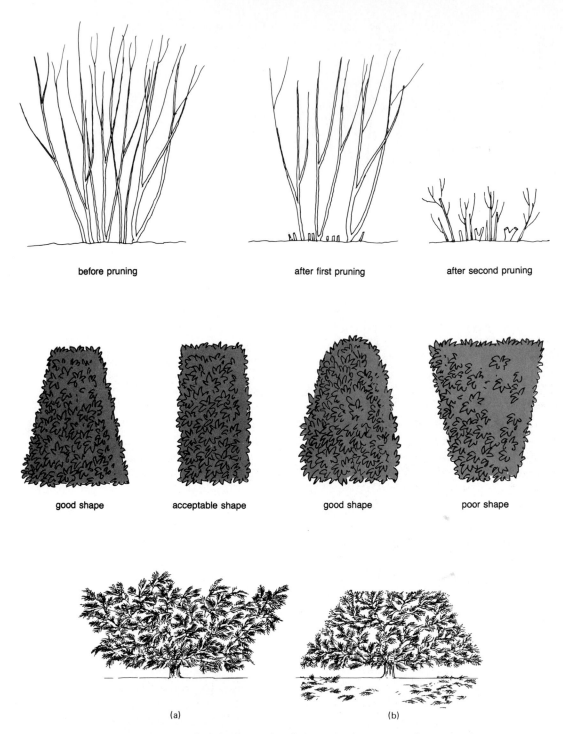

before pruning   after first pruning   after second pruning

good shape   acceptable shape   good shape   poor shape

(a)   (b)

Figure 11-22 Heading back a groundcover juniper to control its spread (a) before pruning and (b) after pruning. Note how the branches are layered with longer branches on the bottom to prevent shading out and death of the base of the plant.

best to prune these shrubs immediately after flowering and before they develop buds for the next year's bloom.

Deciduous trees that experience sap flow in early spring are the second exception. Pruning wounds will exude sap liberally in early spring. This loss of fluid will not seriously injure the tree, but it is unsightly and can be avoided by pruning later in the year.

In conclusion, light pruning twice or three times annually is preferable to severe pruning once a year to maintain an attractive landscape.

Special Shrub-Pruning Techniques

Espalier

Espalier (Figure 11-23) is the stylized training of a shrub or tree flat against a wall or trellis, usually in a symmetrical branching pattern. Since not all plants can be adapted to this type of pruning, consult a knowledgeable nurseryman or one of the references listed at the end of the chapter before espaliering a plant.

Espalier pruning must be started when a plant is small. The trunk is tied to the trellis, and the sites where future branches are desired are selected. The buds or branches closest to these points are then tied to the trellis, and all others are removed. This permits only these preselected branches to grow and achieves the desired shape in several years. Annual pruning will still be required to remove stray growth and keep the plant from outgrowing the trellis.

Topiary

Shrubs and small trees trained into pompoms or animal shapes are examples of topiary pruning (Figure 11-23), a novelty pruning style that originated in

Figure 11-23 Espalier (left) and topiary (right) pruning styles. (*Photo courtesy of Monrovia Nursery Company, Azusa, CA*)

formal European gardens. Topiary training can be thought of as a combination of a formal hedge and an espalier. Branches are left at selected locations on the trunk and then sheared repeatedly to encourage fullness. As with espalier pruning, a topiary should be started while a plant is young, shaping it gradually to its final form over several years.

Topiary plants in the landscape should be considered specimens; one or two are sufficient to enhance any landscape. They can be used effectively in Oriental or modern landscapes and can be purchased pretrained at garden centers.

Watering

The goal of watering is to replenish the moisture in the soil surrounding plant roots. A considerable volume of water is needed to accomplish this, and underestimating this volume is a common error. An inexperienced gardener may sprinkle a plant with a hose for a few minutes and consider it watered. In actuality, the sprinkling has only wet the top 2 to 3 inches (5–8 centimeters) of the soil; the majority of the roots remains dry.

Correct watering of shrubs should give a slow flow of water at the base of the plant for 20 to 30 minutes. The water will penetrate evenly throughout the root area and wet the soil deeply. This will encourage the roots to grow downward into the moisture-retaining soil layers instead of along the surface. Trees should be watered deeply over the entire area covered by the branches, since roots will be growing throughout the area. Exception to the rule of deep soaking must be made for established plants accustomed to sprinkler irrigation. The root system of these plants will be shallow, and the plants will need to be maintained with sprinkler irrigation.

Watering Frequency

The frequency with which a plant should be watered depends on its age, its drought tolerance, and the ability of the soil to retain moisture.

Plant Age. More newly transplanted trees and shrubs die from lack of water than from any other cause. When first transplanted, the root system of a plant occupies a limited volume of soil and must extract all its water and nutrients from that soil. Until it is established and is able to absorb moisture through an extensive root system encompassing a considerable volume of soil, watering should be frequent. Watering every seven to ten days for the first year and every two weeks the second

year is advisable. After that, irrigation will be needed only during periods of drought lasting more than three or four weeks.

Drought Tolerance. Tolerance to drought varies immensely among plant species. Some are able to extract minute quantities of water present in the soil and remain healthy, whereas others under identical conditions would wilt to the point of death.

Cacti and succulents are the best-known examples of drought-tolerant plants, but they are not the only such plants. Trees with tap-roots and shrubs native to arid climates are similarly drought-tolerant and can be used to create a low-maintenance and ecologically sound landscape where natural rainfall is scarce.

Soil Moisture-Holding Capacity. The proportions of sand and clay in a soil greatly affect its ability to retain moisture and consequently affect the watering frequency. Soils containing a large percentage of clay will be the most water-retaining, with loams intermediate and sandy soils least able to hold water. A sandy soil may need watering twice as frequently as a clay.

If a soil is sandy, its moisture-holding capacity can be greatly improved by adding organic matter, as described under "Water Conservation" and in Chapter 5.

Watering Techniques

Sprinklers

Sprinkler irrigation, whether by permanent or portable sprinkler heads, was the main method of irrigation used for many years in areas lacking regular rainfall. It watered large landscaped areas at one time and, if left on long enough, would eventually saturate the ground to the desired depth. The main disadvantage of sprinklers is that they waste water. They wet all the soil instead of only that which contains plant roots, and also waste water by evaporation from the droplets as they pass through the air.

Today, watering by sprinklers can be recommended only for groundcover, lawns, and densely planted areas such as flower beds. However, sprinkler irrigation should not be discontinued on landscapes established by this watering method, since the plant root systems will have spread throughout the sprinkler pattern area.

Hand Watering

In landscapes where supplemental watering is seldom required, hand watering with a hose is practical. The hose can be adjusted to low pressure and moved from plant to plant every half-hour, or the watering basin made at planting can be filled several times.

Leaching

Leaching is the application of a large volume of water to the soil, much more than would be necessary to wet the root zone, for the purpose of flushing accumulated *minerals* or *salts* from the soil. It is sometimes necessary because an excess concentration of minerals accumulates in the soil, whether from constant application of mineral-rich irrigation water, excessive fertilization, or a naturally rich concentration of minerals in the local soil. Under these conditions plant roots are unable to extract water from the soil even though an adequate amount is available, because the natural osmotic process for absorption through the root hairs is disrupted due to excess minerals. To alleviate the problem, it is necessary to dissolve these salts and move them to the deeper soil layers by profuse and prolonged application of water.

One way of leaching is to water a plant deeply, wait a few hours, then repeat the process several more times. During the waiting the minerals dissolve in the water held in the soil, and when more water is applied they are flushed away. Another method is to leave a hose trickling at the base of the plant for half a day or so. This will accomplish the same goal but probably use more water.

The necessity of leaching depends on several factors and is normally most necessary in areas with little natural rainfall, such as the Southwest. In these areas lack of natural leaching by mineral-free rainwater, combined with dry air (which necessitates frequent irrigation with normally saline irrigation water) creates a soil mineral buildup which should be leached every few months.

General symptoms calling for leaching include slowed growth, yellowing of foliage, and sometimes wilting, even when the soil is moist. Some species are more resistant to excessive soil salts than others, and this is another factor to consider in deciding if leaching is necessary. There are attractive plants native to salty soils, and a homeowner should consider planting them if soil salts are a recurring problem.

Drip or Trickle Irrigation

Drip or trickle irrigation (Figure 11-24) is an irrigation technique the advantages of which include low cost, ease of installation, and conservation of water. An indirect advantage of drip irrigation is that because it applies water only at one spot (instead of throughout an area, as a sprinkler), the germination and growth of weeds are inhibited.

A typical drip system would consist of five major parts: a filter, a pressure regulator, ½- to ¾-inch (12- to 20-millimeter)-diameter flexible tubing, and ⅜-inch (3-millimeter)-diameter microtubing and emitters. Water from an outside tap is first filtered to remove particulates and then passed through the regulator to lower the pressure to about 10 percent of normal. Next it passes through the ½-inch (12-millimeter) tubing located near the plants to be irrigated. Short lengths of the microtubing inserted into the main lines carry water to emitters at each plant, with all plants watered at once at a slow, even rate. The "drip" and "trickle" names come from the slow flow of water from each microtube, a total of about 1 gallon (4 liters) per hour.

Trickle irrigation is very easy to install, even for a person with no knowledge of plumbing. Because it operates at low pressure, no glue or threaded connections are needed. The splices between the main lines are held together with self-sealing plastic connectors. Likewise, the microtubes can be inserted into the main line by simply making a hole and pushing in the microtube until it fits snugly. Unlike a sprinkler, no digging is needed to install a drip irrigation. The main lines are laid on top of the ground and covered with mulch.

The main disadvantage of drip watering is the tendency of the microtubes to clog with mineral deposits from the water, soil, or algae. These problems can usually be remedied by flushing the system periodically at high pressure.

Drip systems are most often used on tree and shrub plantings, since each individual plant can have its own microtube. However, the system can be adapted to water groundcover and grass areas with microsprinkler heads (Figure 11-25) which spray over a limited radius.

Water Conservation

Even where water is abundant, conserving water used for irrigating landscape plants should be considered. There is never enough of this valuable natural resource to justify thoughtless use. The following techniques will reduce the amount of water used in a landscape planting.

Improvement of the Water-Holding Capacity

Though the natural soil in an area may have a poor water-holding capacity, it can be improved with the addition of organic matter. Compost, peat moss, shredded bark, coarse sawdust, and similar materials can be incorporated at planting. Herbaceous groundcover beds

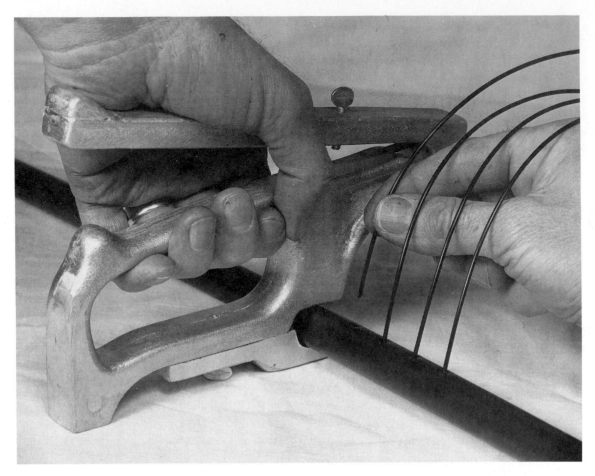

Figure 11-24　Putting together a trickle irrigation system. Note the small microtubes which will carry water to each plant. (*Photo courtesy of Chapin Watermatics, Watertown, NY*)

can be noticeably improved by mixing in organic matter to a 1-foot (0.3-meter) depth, since the root systems of the plants are naturally shallow and will remain within the improved layer.

Prevention of Evaporation

Much of the water lost from landscape plantings leaves by evaporation from the soil. This evaporation rate can be slowed by the use of a mulch and by watering in the early morning when the temperature is lowest.

Use of Proper Irrigation Techniques

The proper irrigation equipment should be used: trickle for trees and shrubs, plus sprinkler for lawns and groundcover. The system should then be used wisely. Be cautious when using automatic timing devices, since they do not take weather conditions into account and may apply water in a rainstorm. Frequent checks for water runoff will prevent one of the most common means of wasting water.

Fertilizing Landscape Plants

Many homeowners never fertilize their trees and shrubs, assuming that fertilizers are needed mainly for vegetables and lawns. But just as a fertilizer promotes the strong rapid growth of these plants, it will also improve the growth of landscape plants. A newly installed landscape that is fertilized regularly will grow much faster than an unfertilized one.

Timing of Fertilizer Applications

The time of year when fertilizer should be applied to landscape plants varies with the climate, but generally it should be restricted to just before or while a plant is actively growing.

In most parts of the country, fertilizer is applied annually in spring so the nutrients will be available for the new growth. However, for faster growth fertilizer could be applied two or three times during the growing season at one- to two-month intervals, for example, on April 1, May 5, and July 1.

Figure 11-25 A microsprinkler head used with a drip irrigation system. (*Photo courtesy of Chapin Watermatics, Watertown, NY*)

In mild climates such as those found in Florida and parts of California, plants grow continuously. Fertilizer will need to be applied every two to three months through the entire year to replenish the nutrients absorbed by the plants.

Rates of Fertilizer Application

Trees

Generally speaking, established trees do not need to be fertilized with phosphorous or potassium unless it has been determined that the soil is naturally deficient in these nutrients. Their main requirement is for nitrogen, which can be applied on a yearly basis by drill hole or surface application (see Chapter 5). Small and newly planted trees are normally fertilized only by surface application.

The yearly amount of fertilizer recommended for trees is about 6 pounds (3 kilograms) of actual nitrogen per 1000 square feet (100 square meters) of area under the branches. Actually, the amount of fertilizer applied is not critical, provided it falls within a reasonable range. In some cases it will not be practical to compute a fertilizer weight for a small tree with, for example, a 4-foot (1.2-meter) branch span. Estimating is completely satisfactory in most cases. About ½ cup (120 milliliters) of ammonium sulfate or sodium nitrate or ¼ cup (60 ml) of urea or ammonium nitrate is adequate for small trees less than 12 feet (4 meters) high. This allotment can be increased to 1 cup (240 milliliters) for 15-foot (5-meter) trees, 3 cups (700 milliliters) for 20-foot (6-meter) trees, and 4 cups (950 milliliters) annually for trees 30 feet (9 meters) or greater in height.

Shrubs

Shrubs, whether evergreen or deciduous, are fertilized in most circumstances at the rate of 2 to 4 pounds (1–2 kilograms) of nitrogen annually per 100 square feet (10 square meters) of bed area per year, using the surface application method. The fertilizer can be the same as is used for trees and can be split into two applications if desired: one in early spring and the second immediately after flowering. If acid-loving plants are being grown, an acid-forming fertilizer can be used (see Chapter 5).

If fertilizer calculations are not going to be made, a moderate amount of fertilizer can be estimated to equal ½ cup (120 milliliters) of 15-5-5 for shrubs less than 2 by 2 feet (0.6 × 0.6 meter), 1 cup (240 milliliters) for shrubs up to 4 feet (1.2 meters) tall and 3 feet (1 meter) across, and 1½ cups (350 milliliters) for larger shrubs. The fertilizer should be watered in after it is applied.

Groundcovers

Groundcover plants will benefit from yearly fertilization in spring using a surface application. Two to four pounds (1–2 kilograms) per 1000 square feet (100 square meters) watered thoroughly afterward is advisable, or ¾ cup (200 milliliters) can be estimated to be sufficient for a 10 × 10-foot (3 × 3-meter) area. Either a balanced or a high-nitrogen fertilizer is acceptable.

Weed Control

Weeds growing in a landscape planting can seriously detract from its appearance. They also compete for the available light, water, and nutrition; in an establishing groundcover, weeds will thrive at the expense of the desirable plants. For these reasons weed control is essential, whether chemical or cultural.

Chemical Weed Control

With few exceptions, chemical weed control is impractical in landscape plantings. Many species are used in a typical landscape, and each will react to weed-controlling chemicals (herbicides) in a different way. An improperly applied weed killer may ruin not only your own plants but also those in neighboring yards.

Exceptions can be made. Herbicides of the "pre-emergence" type kill only germinating seedlings and not established plants. Such chemicals could be applied, for example, to a newly planted groundcover area to alleviate the inevitable sprouting of weed seeds that occurs when the soil is disturbed. "Post-emergence" herbicides may also be used to kill existing weeds prior to planting.

Table 11-1 lists some of the herbicides used in ornamental plantings.

Cultural Weed Control

Cultural weed control consists of hand pulling and cultivating with a hoe, using ornamental plants competitively, and mulching.

Hand pulling and cultivation are effective for controlling annual weeds. But they are only temporarily effective in controlling perennial weeds, since the weeds often break off at the soil line and grow back from the roots. Cultivation exposes fresh soil and additional weed seeds to favorable germination conditions

Table 11-1 *Herbicides for Ornamental Plantings*

Chemical	Weeds controlled	Remarks
EPTC (Eptam)	Germinating annual grasses and some broadleaf weeds; will not control established weeds	For weed control in flower plantings; irrigate after application
Trifluralin (Treflan)	Same as above	For flower and shrub plantings; irrigate after application
DCPA (Dacthal)	Same as above	For use in herbaceous plants; most effective in sandy soils
Dichlobenil (Casoron)	Germinating annual broadleaf and some grassy weeds including quack grass; will not control already established weeds	For use around established deciduous trees and shrubs and around conifers
Amitrol-T	Nonselective on all actively growing weeds	For spot treatment of localized weed problems
Oryzalin (Surflan)	Germinating grasses and broadleaves	For use in woody ornamentals and herbaceous plants
Dichlofop-butyl (Fusilade)	Actively growing annual and perennial grasses except annual bluegrass	Does not harm most broadleaf plants except 'Bar Harbor' juniper. Two applications may be necessary
Oxydiazon (Ronstar)	Germinating broadleaf weeds and some grasses	For use in a wide variety of ornamentals and to prevent annual bluegrass in lawns
Glyphosate (Kleenup®, Roundup®)	Actively growing annuals and perennials	Post-emergence spray or wiping on for weeds around woody ornamentals
Triclopyr	Annual and perennial broadleafed weeds, especially oxalis.	Post-emergence spray

and necessitates repeating the cultivation several times each season. Although pulling and cultivation may be the only remedies for large weeds, mulching is more practical and labor-saving over a long period.

Using ornamental plants competitively can also reduce weed problems. By planting ornamentals thickly and using groundcovers on bare soil areas, weeds will not have an opportunity to become established. They will be crowded out by the desirable plants.

Mulching is the third cultural weed control. A layer of mulch excludes light from the soil, checking the growth of both seeds and existing small weeds. A mulch can be any material used to cover the soil surface.

Mulching plants in the landscape serves other purposes besides weed control. It provides a decorative background for plants, conserves soil moisture, and maintains the soil at an even temperature.

Organic Mulches

The most popular organic (plant-derived) mulches used in landscape plantings are shredded bark and bark chunks (Figures 11-26 and 11-27). They are ap-

Figure 11-27 A chunk bark mulch. (*Photo courtesy of Mountain West Bark Products, Inc. and Development Workshop, Inc. of Idaho Falls, ID*)

plied 2 to 3 inches (5–8 centimeters) thick throughout shrub beds and under trees and give these areas a neat, rustic appearance. Because they are organic, they decompose slowly over a number of years and will need to be renewed occasionally by adding a fresh layer to the existing mulch. Organic mulches useful for the landscape are not limited to bark. Rice hulls, cocoa bean hulls, pine needles, and oak leaves are attractive mulches, but are available in only a few areas of the country. Likewise, corncobs, compost, and grass clippings are also effective but are generally considered somewhat unattractive for landscape use. They are discussed in Chapter 6 as vegetable garden mulches.

Rock Mulches

Rock mulches can be considered permanent since they do not decompose. Included in the inorganic mulches are stone materials such as lava rock, marble chips, smooth stones, and coarse or smooth pebbles (Figures 11-28, 11-29, and 11-30). Their only disadvantage is that, if replanting of the bed is required, they must be removed and replaced afterward. Organic mulches, on the other hand, could be simply worked into the soil.

Synthetic Mulches

Plastic and landscape fabric (discussed in Chapter 10) are some of the most effective mulches because they are completely impervious to weed growth. Although unattractive by themselves for use in landscape plantings, they can be covered with a decorative layer of bark or stone. They will also be concealed when used with a groundcover shrub, which will spread and hide it soon after planting.

Figure 11-26 Shredded bark mulch. (*Photo courtesy of Mountain West Bark Products, Inc. and Development Workshop, Inc. of Idaho Falls, ID*)

Figure 11-28   Lava rock mulch. (*Photo courtesy of Mountain West Bark Products, Inc. and Development Workshop, Inc. of Idaho Falls, ID*)

Figure 11-29   A marble chip mulch. (*Photo courtesy of Mountain West Bark Products, Inc. and Development Workshop, Inc. of Idaho Falls, ID*)

Figure 11-30 A flat stone river rock mulch. (*Photo by the authors*)

Plastic and landscape fabric generally come in rolls of varying widths and are rolled onto the area to be planted with edges overlapping. Plants are then transplanted into slits cut through the material. Finally, the edges are covered with soil to hold them firmly in place in case of wind. For decorative areas such as landscapes, a layer of bark mulch can be applied over the entire area to cover it.

Water will run off plastic during rainfall and be diverted through the holes to the plants. However if plastic wider than 3 feet (1 meter) is used, additional holes should be made for water penetration into the soil.

There is no such problem of water penetration with a landscape fabric, which is an advantage. The fabric is also much less likely to tear, which would permit weed growth through the cut. For these reasons landscape fabrics are becoming more popular than plastic in spite of their higher cost.

Selected References for Additional Information

American Horticultural Society. *Pruning*. Alexandria, VA: American Horticultural Society, 1980.

*Clokey, Joe, and Jim Harrigan. *Efficient Water Management in the Landscape*. San Luis Obispo, CA: San Luis Video Productions.

Edinger, Philip. *Pruning Handbook*. Menlo Park, CA: Sunset Books, 1983.

Haas, Cathy, and Michael MacCaskey. *All About Lawns*. San Ramon, CA: Ortho Books, 1994.

*Harrigan, Jim. *Advanced Pruning*. San Luis Obispo, CA: Vocational Education Productions of California Polytechnic State University.

*Harrigan, Jim. *Elements of Pruning*. San Luis Obispo, CA: Vocational Education Productions of California Polytechnic State University.

*Harris, Richard W. *Arboriculture: Integrated Management of Landscape Trees, Shrubs and Vines*. 2nd ed. Englewood Cliffs, NJ: Prentice-Hall 1992.

National Arborists' Association. *Pruning Tools and Techniques*. Amherst, NH: National Arborists' Association.

Pirone, P. P. *Tree Maintenance*, 6th ed. New York: Oxford University Press, 1988.

Rice, R. P. *Nursery and Landscape Weed Control Manual*. 2nd ed. Fresno, CA: Thomson Publications, 1992.

Sinnes, A. C. *Easy Maintenance Gardening*. San Ramon, CA: Ortho Books, 1982.

Sunset Book Editors. *Basic Gardening*. 3rd ed. Menlo Park, CA: Sunset-Lane, 1981.

*Indicates a video.

12

Lawn Establishment and Care

Lawns have long been a traditional part of the home landscape. Many landscape designers minimize or eliminate them altogether; others assert that a well-kept lawn is the most attractive setting for a home. Although it is true that care of the lawn is one of the most time-consuming maintenance activities, lawns are among the best surfaces for game and play areas.

Establishing a New Lawn

For owners of a new house, establishing a lawn is often the first landscape activity. Usually a lawn is seeded or sodded, but in some regions techniques called *plugging* and *stolonizing* are used. Each of these establishment methods is discussed in the following sections.

Establishment by Seeding

Establishing a lawn by seeding has advantages and disadvantages. First, seeding can require several months to make an attractive turf, whereas a sodded lawn is practically instant. More effort is required to establish the seedlings, and there is greater chance of failure. However, seeding is inexpensive, and when there is a large area involved, cost can be important. Some lawn grasses such as hybrid Bermuda grass and zoysia cannot be established from seed so vegetative means must be used.

Time of Year

The best time of year to seed a new lawn is dependent on the species of grass and the region in which it is be-

ing planted (Figure 12-1). In warm areas like the South and Southwest, warm-season turfgrasses (Table 12-1) are popular because they grow best in the temperature range of 80 to 95°F (27–35°C). Spring is best for establishing these grasses, since the hot summer temperatures to follow will supply ideal growing conditions.

Cool-season grasses such as bluegrass (Table 12-1) are the primary lawn grasses in areas with cold winters. These grasses grow best between 60 and 75°F (15 and 24°C). There, late summer is the preferred seeding period. Sown at this time, the grass will receive cool fall temperatures for establishment, and there will be less competition from annual weeds. Spring planting is also acceptable if fall is not practical.

Site Preparation

Site preparation is the same whether seeding, sodding, plugging, or stolonizing is used, and it is a critical factor in assuring a high-quality lawn. Depending on the condition of the area, some or all of the following operations will be a part of site preparation.

Debris Removal. Debris removal is usually required around a newly constructed house. Pieces of wood, chunks of concrete, and other building materials will doubtless have been left at the construction site. Worse yet, they may have been buried and will later cause sunken areas or interfere with root growth. Debris removal should eliminate any foreign materials in the top 4 inches (10 centimeters) of the soil surface.

Elimination of Existing Vegetation and Preventative Weed Control. The best way to have a weed-free lawn is to

215

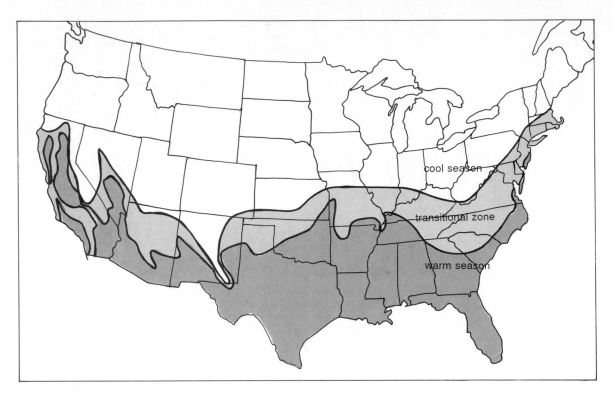

Figure 12-1 Areas for warm- and cool-season grasses.

Table 12-1 *Selected Turfgrass Species*

Name	Cool/warm season	Characteristics
Tall fescue *Festuca arundinacea*	cool	tough, drought tolerant, coarse textured (new cultivars like 'Bonsai' and 'Twilight' are finer)
Creeping red fescue *Festuca rubra*	cool	fine texture, shade tolerant, good bank cover, drought tolerant
Chewings fescue *Festuca rubra 'commutata'*	cool	same as red fescue but lacks rhizomes
Kentucky bluegrass *Poa pratensis*	cool	medium texture, cold tolerant, thick, attractive turf
Perennial ryegrass *Lolium perenne*	cool	fast establishment but short-lived, use for overseeding
Annual ryegrass *Lolium multiflorum*	cool	same as perennial ryegrass but shorter-lived
Creeping bentgrass *Agrostis palustris*	cool	finest texture, luxuriant turf but requires high maintenance, not usually used for home lawns
Bermudagrass *Cynodon dactylon*	warm	tough, medium to fine texture, hybrids best but must be vegetatively propagated, full sun needed
Buffalograss *Buchloe dactyloides*	warm	drought and cold tolerant, low maintenance
Zoysia *Zoysia japonica*	warm	fine texture, slow to establish, long dormant period, prickly blade, low maintenance
St. Augustinegrass *Stenotaphrum secundatum*	warm	coarse texture, not cold tolerant, thick and tough
Centipedegrass *Eremochloa ophiuroides*	warm	low maintenance, medium texture

prevent initial weed establishment. To accomplish this, the area should be evaluated in terms of the weeds already present and the potential for invasion from outside the lawn area. Noting whether the species are annual or perennial will pay dividends after the grass is planted. Perennial weeds can often be recognized by thick, fleshy roots or creeping underground stems. Generally speaking, you can assume that existing annual weeds will be killed during soil preparation, but that seedlings of the same species will soon reappear. Perennial weeds, on the other hand, will reestablish quickly from roots, tubers, and rhizomes after soil preparation is completed. Spading these weeds under or rototilling will spread them, resulting in a severe problem later. It is therefore important that perennial weeds be controlled prior to grading.

Perennial weeds may be controlled in several ways. If a small number of perennials is present or if time is short, they can be removed by digging. All portions of roots and stems must be removed, since even a small section will sprout a new plant. If the infestation is too widespread for hand digging, herbicides can be utilized. In selecting an herbicide, avoid those that kill only the tops, because the roots will survive and resprout. An ideal herbicide translocates through and kills the entire plant without leaving a harmful residue in the soil. The herbicide glyphosate known commercially as Kleenup®, Roundup®, or several other names will kill both annual and perennial weeds and can be recommended for this purpose.

Annual weeds are much more abundant than perennials, but they are much easier to control. One way to decrease the annual weed population in an area to be seeded is to allow the weeds to germinate prior to seeding, kill them, and then sow the grass. Using this weed-control procedure, called the *stale seedbed technique*, it is necessary to prepare the seedbed in the normal fashion until it is ready for seeding. At this point, irrigation is begun and continued for 7 to 10 days. If the weather is favorable, the weeds will sprout and can then be easily destroyed with a *contact herbicide* (one which kills the plant when sprayed on the leaves and stem). Be sure the herbicide selected does not leave a residue in the soil that will harm the grass. Care should be taken not to disturb the soil afterward, since this will cause new seeds to sprout. If the soil has become hard, a very light ¼-inch (6-millimeter)-deep raking may be used to break the surface crust. The grass can then be seeded, covered with a mulch, and irrigated. Resulting turf will generally contain 50 percent or less of the usual annual weed population. With some turf species, a *preemergence herbicide* (one that controls weeds as they germinate) can also be used at planting.

Some weeds will always be present after the grass has emerged, despite all precautions. These can be controlled by pulling or, after the lawn has been mowed once, by the application of a *postemergence herbicide* (one that controls a weed that has already germinated). Several postemergence herbicides will kill broadleaf (nongrass) weeds in turf; however, controlling emerged grassy weeds must still be done by hand.

Recommendations for specific herbicides are given in Table 12-2 or can be obtained from a local nurseryman, county Cooperative Extension Agent, or Farm Advisor. Like all pesticides, herbicides are dangerous to humans, pets, the environment, and desirable plants if used improperly. Read the label thoroughly and follow all directions precisely. Particular care should be taken to prevent herbicides from blowing to areas adjacent to the lawn (such as shrub beds), where damage to plants may occur.

Grading. Grading is generally done by the construction crew so that the ground slopes evenly away from the building at not less than a 1 percent grade. However, in some cases this has not been done or has been done improperly, or grade changes are called for in the landscape plan. In these cases care should be taken to remove the existing topsoil, grade the subsoil layer, and then replace the topsoil. When the topsoil is missing, it should be replaced with purchased topsoil of good quality applied to a depth of 4 inches (10 centimeters).

Soil Amending. Depending on the condition of the topsoil, soil amending can be needed to alter pH, improve fertility, or increase or lessen water-holding capacity with organic matter.

Ideally, a soil test should be made, and the proper pH- and fertility-correcting chemicals added as described in Chapter 5. However, it is usually sufficient to assume the pH is in an acceptable range and give a general fertilizer application of 2 pounds (1 kilogram) per 1000 square feet (100 square meters) each of nitrogen, phosphorus, and potassium. The fertilizer should be worked throughout the topsoil layer.

Amending the soil with organic matter need not follow any formula. Any of the amendments listed in Chapter 5 can be used, remembering the basic rule that a large amount of amendment is usually needed to cause any appreciable change in the soil.

Surface Preparation. The surface on which seed is to be sown should be free of clods and as smooth as possible. Clods can usually be broken up by rototilling or, in a small area, by hand. A light raking will remove remaining clods and provide the final surface.

Finally the area should be rolled with a lawn roller to compact the soil surface. Lawn rollers can usually be rented and are filled with water to provide the necessary weight.

Table 12-2 *Lawn Herbicides*

Name	Time of application	Weeds controlled	Remarks
For newly seeded lawns			
Glyphosate (Kleenup®, Roundup®)	Anytime prior to planting	All	Useful for killing all existing weeds
Siduron	Immediately following seeding	Emerging annual grasses	Water immediately after treatment
Bromoxynil	After grass has emerged and weeds are in the 2- to 4-leaf stage	Many broadleaf weeds	Larger weeds require higher rates
For established lawns			
Benefin	Spring, before weeds germinate	Crabgrass and other annual grasses	Does not leach readily; do not use on bentgrass
Dacthal	Spring, before weeds germinate	Crabgrass and other annual grasses	Less effective on heavy soils
Bensulide	Spring, before weeds germinate	Crabgrass and other annual grasses, a few broadleaves	
2,4-D amine	When weeds are actively growing	Most annual and perennial broadleaves	Do not apply when windy; do not use on St. Augustine-grass
Glyphosate (Kleenup®, Roundup®)	When weeds are actively growing	All	Must be applied *only* to the weed (spot treatment) with a paintbrush
DSMA	Postemergence	Young crabgrass, chickweed, oxalis, Dallis grass, sandbur	Do not use on St. Augustine-Bahiagrass, or centipede-grass; apply 1–3 times 5–7 days apart
Aminotriazole	When weeds are growing rapidly	Annual and perennial weeds	*Spot treatment only;* will kill turf if contacted; brush on with paintbrush
Triclopyr	When weeds are actively growing	Annual and perennial broadleaf weeds	Use on cool season grasses only

Seeding and Mulching

Seed Selection. Selecting the proper variety or mixture of grass for the area is important to growing an attractive lawn and should be done carefully. Selected grass species are listed in Table 12-1; because of the wide variety of grasses grown throughout the country, a discussion of the merits and drawbacks of all is not possible. The state Cooperative Extension Service or Farm Advisor's Office will be able to assist in the choice.

Buying seed is not simply a matter of finding the greatest amount of seed for the money. Information on both *germination rate* (percentage of the seeds which will germinate under ideal conditions) and purity (percentage of the seeds of the specific variety or mixture being purchased) should be noted. The label will give the germination test date, which should be within the current year. Otherwise the seed may be more than a year old and have a decreased germination rate.

The aim in buying seed is to obtain the highest germination rate and purity at the best price. In no circumstances should purity or germination rate be sacrificed in favor of low price. Problems of weeds and poor germination will far outstrip any small savings in seed cost.

Other factors to be considered include the amount of sunlight the area receives and the soil condition. Some seed mixtures are specially formulated for shade, poor soils, wet soils, and so on. Choosing a suitable seed mix can avoid problems later.

Seeding Techniques. When seeding any area greater than 30 by 30 feet (9 × 9 meters), a mechanical seeder (Figure 12-2) should be used to assure even coverage. However, for small areas hand seeding is satisfactory. No matter which method is chosen, the seed label should be carefully followed as to the quantity of seed planted in an area. Since grass seed is lightweight and easily blown by the wind, seed should be sprinkled on a windless day or during early morning or late afternoon, when the wind is likely to be at a minimum.

When seeding (or fertilizing with granular fertilizer), a pattern such as the one shown in Figure 12-3 will assure even coverage. To avoid missing strips between passes, the wheel of the spreader should be run just inside the previous wheel track. A steady walking pace is necessary to prevent double dosing and insufficient coverage problems.

Figure 12-2 Two types of mechanical seeders for distributing seed or fertilizer: hand spreader (top), and push type (bottom). (*Photos by [top] Rick Smith and [bottom] Vocational Education Productions, California Polytechnic State University. San Luis Obispo, CA*)

Mulching. Mulches are applied to newly seeded areas to hold in moisture and to control erosion and washing away of the seed during rainfall or irrigation. Unlike mulches used for weed control, mulch over seed must be applied in a thin layer so the germinated seedlings can grow up through it. Coarse sawdust, shredded bark, peat moss, wood shavings, and straw are several mulches for this purpose; they should be applied so that the ground remains visible through the mulch (Figure 12-4).

Postseeding Care

Irrigation. Frequent irrigations to keep the soil surface moist are the key to obtaining good germination.

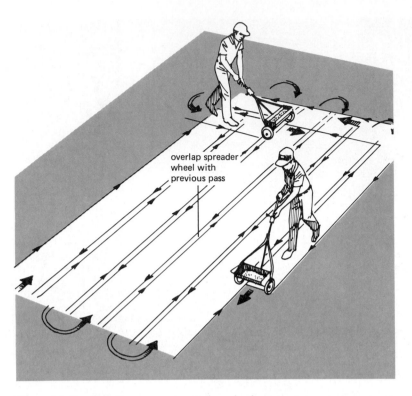

Figure 12-3 A push-type mechanical seeder being run in a pattern to assure even coverage.

Figure 12-4 Straw mulch applied to a newly seeded lawn. (*Photo by George Taloumis*)

Figure 12-5 Laying sod. (*Photo courtesy of Vocational Educational Productions, California Polytechnic State University, San Luis Obispo, CA*)

Drying out even once during the germination and initial establishment stages can kill the majority of the seedlings and necessitate reseeding. Watering may be needed as often as three or four times a day to keep the soil surface moist. As the roots penetrate down into the ground, less frequent but longer waterings should be made; watering every two to three days will be sufficient.

Fertilizing. A light topdressing of about ½ pound of nitrogen per 1000 square feet (0.25 kg/100 square meters) of lawn is recommended as the first fertilizing of turfgrass seedlings. The fertilizer should be applied only when the majority of the seedlings are 1½ to 2 inches (4–5 centimeters) high and should be immediately watered in.

Mowing. The first mowing should take place as soon as the seedlings exceed the normal cutting height. Afterward, mowing should be done regularly.

Sodding

Sod consists of long strips of turf cut from the ground with the roots intact (Figure 12-5). It is heavy due to the weight of attached soil and is delivered on pallets to the site where it will be used. A new method of sod production is now being used by some growers in which the sod is grown in an organic medium without soil. This produces a lightweight, good-quality sod which is easier to lay at a cost of only a few cents more per square foot. Sodding can be done any time during the growing season, and it creates a high-quality lawn immediately.

Buying Sod

Sod is expensive and should be selected carefully. An ideal sod is thick and uniform, free of weeds and diseased areas, and has only a small layer of thatch. You should always ask to see a sample of the sod you are purchasing. Specific arrangements for its delivery can then be made to allow adequate time for preparing the site in advance.

Site Preparation

The same methods of site preparation that precede seeding should also precede sodding. Negligence in site preparation may initially be covered up by a healthy sod, but the sod will not remain healthy if the soil into which it sends new roots has not been thoroughly prepared.

Laying Sod

Sod must be laid as soon as possible after it is harvested. If it is stacked or left in rolls for more than a day or two, it will develop heat and disease problems and deteriorate quickly. If the sod must be left stacked, the outside edges should be lightly watered; the sod should not be heavily watered.

When laying sod, the strips should be staggered (Figure 12-5) so the ends do not form a line which will be noticeable later. Avoid stretching the sod, since it will shrink later and leave bare strips between the pieces. Cut the sod with a sharp knife to fit tightly around sprinklers and other obstructions and tightly against lawn edges. Gaps between the sod will fill with weeds and become a maintenance problem.

After the sod is laid, the area should be rolled to ensure contact between the sod and the prepared soil beneath. Roll with a water-filled lawn roller in the opposite direction the sod was laid.

Care after Transplanting

Immediately after being laid, sod should be watered thoroughly enough to wet both the sod itself and the first several inches (7–10 centimeters) of soil beneath. During the first two weeks it should be lightly irrigated once each day during the warmest period to minimize transpiration during reestablishment of the roots. Avoid walking, playing, or other heavy traffic for a couple of weeks.

After the sod has rooted it will benefit from occasional light application of fertilizer. Mowing should begin when the grass has grown sufficiently to require it. The mowing height should be set relatively high until the sod is well established.

Lawn Establishment by Stolonizing and Plugging

Stolonizing and plugging are used to establish new lawns of vigorous grasses such as Bermuda and creeping bentgrass. Stolonizing consists of spreading small stem sections or sprigs of the turf variety over a prepared area and then rolling or lightly covering with soil. The stolons root quickly and form a turf in several months.

In plugging, small 2- to 4-inch (5- to 10-centimeter)-diameter pieces of sod are transplanted into a prepared area 6 to 14 inches (15–35 centimeters) apart (Figure 12-6). The closer the plugs are spaced, the sooner the area will fill in. Warm-season grasses such

Figure 12-6　A plug of turfgrass ready to be planted in a lawn. (*USDA photo*)

as Bermudagrass, zoysiagrass, St. Augustine-grass, and buffalograss are the species most commonly established by plugging.

Lawn Renovation

Renovation is used when a lawn is in poor condition but has not deteriorated to the point where it should be completely reestablished. The poor condition may be due to insufficient soil fertility, weed invasion, thatch accumulation, soil compaction, or any combination of these and other causes. Although renovation can be started at any time during the growing season, late summer is best for cool-season grasses and late spring for warm-season grasses.

Determining the cause of the poor growth is the first step in renovation. If there is excessive weed growth, footpaths, or pale color, neglect may be suspected. However, if only isolated areas are dying, other causes should be suspected. Excess shade, poor drainage, diseases, and insects may be problems. Sending a soil sample for analysis will indicate nutritional or pH problems.

The next step should be weed control, by hand or with the chemicals recommended in Table 12-2. Excess thatch should be removed, if necessary, using a machine rented for this purpose (see Dethatching).

Aeration of the soil can be done with a rented machine or a hand aerator. Finally, bare spots should be reseeded, followed by fertilizing as indicated by the soil test and a deep, thorough irrigation.

Given minimum care afterward, the turf should remain in good condition. If additional problems develop, or if renovation does not rejuvenate the lawn, consult a specialist.

Lawn Maintenance

Mowing

Mowing is the most frequent maintenance chore in caring for a lawn, and it must be done correctly for the turf to look its best.

Mowing Frequency

Mowing is essentially a pruning operation done on grass. It removes a portion of the photosynthesizing part of the plant, which in turn lessens the supply of carbohydrates going to the roots for growth and respiration. Thus roots are as affected by mowing as the blades of the grass.

Mowing should be as frequent as necessary to remove only 30 percent of the blade length at any one time. Cutting that removes more than 30 percent of the blade is detrimental to the roots because of the abrupt drop in carbohydrate flow.

Mowing Height

The height at which grass should be maintained varies with the species and may range from ½ to 4 inches (1.3–10 centimeters). Table 12-3 lists suggested mowing heights for grasses grown throughout the United States.

Mowing at a height less than suitable for a particular grass is called *scalping* and leaves the turf an unattractive brown stubble. Occasional scalping will not permanently harm grass. But if repeated on a susceptible area (such as a berm top), it will eventually kill the turf.

Returning Clippings

Many people use grass catchers to remove clippings during mowing. Although this does give a neater surface, it is not essential to turf maintenance. Returned clippings supply additional nutrients when they decay and reduce fertilization requirements. Clippings need only be removed if they are overly heavy and long and would inhibit passage of light to the grass underneath.

Mower Operation

Mowing only when grass is dry is suggested for two reasons: first, because the mower is less likely to clog with soggy clippings and the clippings will distribute evenly rather than in clumps; second, because grass blades are more susceptible to disease invasion through the cut surface when wet.

The importance of sharpening mower blades so they cut grass cleanly is often overlooked. A clean cut

Table 12-3 *Preferred Cutting Heights of 19 Turfgrasses as Determined by the Resulting Turfgrass Quality and Vigor* [a]

Relative cutting height	Cutting height (in.)	Turfgrass species
Very close	0.2–0.5	Creeping bentgrass
		Velvet bentgrass
Close	0.5–1	Colonial bentgrass
		Annual bentgrass
		Bermudagrass
		Zoysia grass
Medium	1–2	Buffalograss
		Red fescue
		Centipedegrass
		Carpetgrass
		Kentucky bluegrass
		Perennial ryegrass
		Meadow fescue
High	1.5–3	Bahiagrass
		Tall fescue
		St. Augustinegrass
		Fairway wheatgrass
Very high	3–4	Canada bluegrass
		Smooth brome

*Extracted from James B. Beard, *Turfgrass: Science and Culture*, Prentice-Hall, Englewood Cliffs, NJ, 1973.

minimizes the damaged area and reduces tip browning of the blades. It also decreases moisture loss and eliminates bruised blades, which are a site of disease entry.

Irrigation

As for irrigation of any plant, watering of grass should be done according to need rather than on a fixed schedule which does not take weather conditions into account. The aim is to apply water just before wilting, wet the top 6 to 8 inches (15–20 centimeters) of soil where the roots grow, and wait until wilting approaches before watering again.

Determining exactly when wilting is about to occur is difficult. One technique called *footprinting* is advocated by Dr. James Beard of Michigan State University. Walk across the grass, then observe how quickly the turf stands upright again. Turf with adequate water will recover rapidly, while turf approaching wilting will recover very slowly.

Where rainfall is insufficient, a permanent sprinkler system can be installed before surface preparation and seeding. Pop-up sprinkler heads, which are below cutting level normally but are elevated above the turf by water pressure during operation, are the most popular. They can be used in conjunction with plastic pipe to create a durable inexpensive system easily installed by a homeowner. To minimize water loss by evaporation, sprinklers should be used late in the afternoon or early in the morning. The former is preferable when disease problems are common because the foliage will not remain wet and susceptible to infection through the night.

Subsurface irrigation is a relatively new technique in use primarily on golf courses. Underground pipes spaced closely throughout the turf area ooze water at a slow rate. The water is moved by capillary action to the grass roots. The main advantages of a subirrigation system are the elimination of evaporation and runoff. Disadvantages include the possibility of clogging and buildup of salts. In addition, not all soils have sufficient capillary action to utilize the system.

Fertilization

Choosing the Fertilizer

Since the goal in fertilizing a lawn is the production of vegetative growth, the fertilizer chosen should be high in nitrogen. Fertilizers with two to four times as much nitrogen as either phosphorus or potassium are suggested. In addition, it is best to buy fertilizers with part of the nitrogen in a form that is immediately available and another part in a slow-release form.

Combinations of fertilizer plus herbicide, fungicide, or insecticide are widely sold for use on lawns. Their use eliminates separate applications of each chemical, but extra care must be taken to apply them at the proper rate.

Timing and Rate of Fertilizer Applications

The frequency at which fertilizers should be applied to a lawn depends on whether the goal is to establish a young lawn or to maintain an older one, on the soil type, and on the weather conditions. The following recommendations are based on a loam soil with irrigation or precipitation every one to two weeks. In lighter soils or with more frequent irrigation, fertilizer should be applied more often.

Young Lawns. Establishing lawns require frequent fertilizer for rapid growth. As a general guide, a ½ to 1 pound per 1000 square feet (0.25–0.5 kg/100 square meters) application of immediately available nitrogen should be applied when grass reaches 2 inches (5 centimeters). This should be followed by applications at the same rate every three to four weeks during the first growing season. In cold-winter climates applications should be stopped about one month before the onset of winter dormancy.

Established Lawns. Established lawns require less frequent fertilization to maintain an attractive appearance. Two fertilizer applications per year, one in early spring and one in fall, are normally sufficient. Two pounds (1 kilogram/100 square meters) of nitrogen per 1000 square feet is recommended at each application.

In areas with mild winter temperatures grass frequently does not have a dormant period and will grow throughout the year. Accordingly, an additional fertilizer application will be needed in midwinter to supply nutrients for healthy growth.

Fertilizer Application Methods

Fertilizer is normally applied to lawns in granular form using a broadcast and water-in method. A fertilizer spreader is best to assure even coverage over large lawns. Hand broadcasting or applying liquid fertilizer with a hose-end applicator is practical for small areas. The latter method is used by some commercial lawn care companies and is valuable for very sandy soils with a poor nutrient-holding capacity. A diluted solution can be mixed and applied in place of plain water at recommended intervals.

Dethatching

Thatch is a layer of stems and decaying organic matter which accumulates between the turf surface and the soil (Figure 12-7). The accumulation occurs when the rate at which the turf sloughs off dead parts is greater than the rate at which they decompose.

Although a thin layer of thatch is not harmful, any layer greater than ½ inch (1–2 centimeters) can cause problems. The heavily thatched area may dry out fast, diseases and insects may invade, and there may be

Figure 12-7 A thatch layer in turfgrass. (*Photo courtesy of The Lawn Institute*)

Figure 12-8 Dethatching with a vertical mowing machine. (*Photo courtesy of the Cooperative Extension Service of the University of California*)

decreased tolerance to heat, cold, and drought. Vigorous grass varieties are the most susceptible to thatch accumulation because of their rapid growth rate.

Excess thatch is detected by examining a cross section of turf. It can then be removed mechanically by a "vertical mower" (Figure 12-8). This machine cuts and pulls the turf up in narrow strips. The material is then raked off and the lawn watered and fertilized. Timing of the mowing should allow at least 30 days of good growing conditions afterward for recovery of the turf. Early spring is suitable on most grasses.

For small lawn areas, renting a vertical mowing machine may be impractical. Hard raking with a stiff metal rake will loosen and remove the thatch.

Preventing future buildup of thatch involves altering the care of the turf to balance the accumulation and decomposition rates. If irrigation or fertilization has been excessive, it should be reduced to slow the growth of the turf. Liming may be needed since the pH of thatch is generally too low for the optimum decay rate. Finally, topdressing the turf with soil will hold in the moisture necessary for decomposition and provide a source of microorganisms for the breakdown process.

Aeration

Aeration is needed when a lawn has been compacted by foot or vehicular traffic. Compaction is particularly prevalent in clay soils. Under compacted conditions, the movement of oxygen and other gases into and out of the soil is restricted. Water will soak in very slowly or run off. Turf roots will eventually begin to die.

Aeration removes small cores or slices of grass and soil from the turf area, penetrating to a depth of up to 4 inches (10 centimeters) from the soil surface. The remaining soil then can expand into the open area and relieve the compaction problem. Engine-driven aerators can be rented if a large area is to be treated. For spot treating or small lawns, a home aeration tool, which is used like a shovel, is more practical.

Weed Control

Weeds in a lawn are undesirable because they detract from its even appearance. Their varying textures and colors make them stand out, and their tendency to grow faster than the surrounding turf increases the mowing frequency. Seasonal weed species (crabgrass, for example) crowd out desirable turf and then die, leaving bare spots in the lawn. When weeds flower and set seed, they are even more conspicuous in a lawn and cause additional problems. Seed heads of common Bermuda grass, for example, generally will not be cut by reel-type mowers but must be removed by hand. Others have seeds that are spiny (bur clover and puncture vine) and will injure bare feet.

The objective of lawn weed control should be to keep weed populations within a tolerable range, not necessarily to have a flawless lawn. A few weeds seldom detract from the lawn enough to require control measures. In fact, many homeowners prefer a lawn containing a combination of plants including clover and ground ivy.

A healthy lawn will generally not have severe weed problems. However, due to disease, insects, or poor cultural practices, weeds sometimes become established in large numbers. Thus the first line of defense against weeds is to keep the grass healthy. Correct mowing height and frequency, fertilization, and insect and disease control will prevent weed seedlings from becoming established and even enable the grass to outcompete established weeds. For the few remaining, hand pulling is a sound control method. Pulling after a rain or irrigation will help remove plants with taproots.

Occasionally, weeds such as crabgrass become a severe problem in even the healthiest of lawns. Preemergence or postemergence herbicides can then be utilized. Crabgrass is most often controlled with a granular herbicide (often combined with a fertilizer) applied preventatively in spring before the crabgrass germinates. Weeds escaping preemergence treatment can be hand pulled later or sprayed with a selective postemergence herbicide that kills only the weeds without harming the grass. Several applications may be necessary for complete control. Table 12-2 lists herbicides used for weed control in lawns.

Lawn Substitutes

Dichondra

A number of plant species will tolerate some foot traffic and can be planted as grass substitutes. Dichondra is the best known of these, although its use is limited to the warmest climates of the country.

Dichondra (Figure 12-9) is an extremely attractive grass substitute with a flat, matlike habit of growth. It is restricted to areas where the winter temperature does not drop below 25°F (−4°C) and is tolerant of the very warm summer temperatures such as those found in the Southwest.

A dichondra lawn must be established in spring or summer for soil temperatures to be sufficient for seed germination and rapid growth. Soil preparation, sowing, and postseeding care are essentially the same as for grass. Mulches of coarse sawdust, shredded peat moss, or other relatively fine material are preferred over straw because the creeping growth habit of the dichondra will not readily obscure a straw mulch.

Plug planting can be used to establish a dichondra lawn and is useful when only a small area is to be planted. In large areas the expense is prohibitive. As with plug planting of grasses, the area should be thoroughly prepared beforehand to facilitate rapid spreading. Flats of dichondra can be cut into plugs 2 inches (5 centimeters) square or larger and transplanted 6 to 12 inches (15–30 centimeters) apart.

Figure 12-9 Dichondra, a popular lawn plant in California and Florida. (*Photo by Rick Smith*)

The dichondra lawn requires regular watering and frequent fertilization to remain attractive. From this standpoint, it is not a low-maintenance plant. Fertilizing every two to three weeks is best, using a high-nitrogen fertilizer at 1 pound per 1000 square feet (0.5 kg/100 square meters) of area.

Mowing frequency depends on growing conditions. Shade and abundant moisture increase dichondra lushness and height, and mowing may be required once per month. In full sun, growth will stay low, and mowing will never be required.

Primary weeds that invade dichondra are oxalis, chickweed, and miscellaneous lawn grasses such as Bermuda grass and crabgrass. Herbicides to control them are recommended in Table 12-4.

Table 12-4 *Dichondra Herbicides*

Name	Time of application	Weeds controlled	Remarks
Newly seeded lawns			
Bensulide	Preemergence at seeding	Most annual weeds	Follow label directions
Napropamide	Preemergence at seeding	Most annual weeds	Follow label directions
Established lawns			
All of the above	Above	Above	Follow label directions
Monuron	Preemergence or when weeds are in the two-leaf stage	Most annual grass and broadleaf weeds, oxalis	Use exact rate; irrigate immediately after application
Neburon	Preemergence	Most annual grass and broadleaf weeds	Safer than monuron, lasts longer in the soil; available only in weed and feed mixtures
DSMA	When weeds are young	Crabgrass	Follow directions carefully; 1–3 applications 5–7 days apart are required
Fluazifop-butyl	When weeds are actively growing	Annual and perennial grasses except annual bluegrass	Check label for current registration
2,4-D	When weeds are growing actively	Broadleaf weeds	*Spot treatment only;* brush onto leaves on broadleaf weeds with a paintbrush; will kill dichondra if contacted

Other Lawn Substitutes

Dichondra is the only widely used lawn substitute for large areas. However, a number of uncommon plants will tolerate limited walking such as a front yard might receive. Most will be hard to find, and all are established by plugging. It is advisable to try a small area before using the species to cover an extensive lawn area.

Table 12-5 lists a number of lawn-substitute plants along with their temperature hardiness and ideal

Table 12-5 *Lawn-Substitute Plants*

Common name	Botanical name	Cold hardiness	Remarks
Corsican sandwort	*Arenaria balerica*	−5 to 5°F (−21 to −15°C)	Needlelike leaves; grows up to 3″ (7 cm) high; best in shade with moist soil; produces white flowers in May
Mountain sandwort	*Arenaria montana*	−20 to −10°F (−29 to −23°C)	Grows 1″ (3 cm) high with grasslike leaves; tolerant of poor soil and flowers profusely in spring with white blooms
Moss sandwort	*Arenaria verna*	−35 to −20°F (−37 to −29°C)	Bright green needlelike foliage; mat-forming, and develops lumps unless thinned occasionally
Chamomile	*Chamaemelum nobile*	−20 to −10°F (−29 to −23°C)	Often sold under the name *Anthemis nobilis* and used commonly as a lawn plant in Europe; withstands drought and emits a pleasant odor when stepped on; fernlike; bright green leaves and good traffic tolerance; daisylike white flowers
Mock strawberry	*Duchesnea indica*	−10 to −5°F (−23 to −21°C)	2″ (5 cm) high with yellow flowers in spring; strawberry-like leaves
Beach strawberry	*Fragaria chiloensis*	−20 to −10°F (−29 to −23°C)	Native plant good for sandy soils and spreading by runners; white flowers and small edible fruits in spring
Colchis ivy	*Hedera colchica*	−10 to −5°F (−23 to −21°C)	Relatively rare ivy with heart-shaped leaves which tightly overlap over the ground; shade-tolerant
Mazus	*Mazus reptans*	−35 to −20°F (−37 to −29°C)	Deciduous plant, although leaves hang on long in fall; plant grows 1″ (3 cm) tall with 1″ (3 cm) leaves; purple flowers in May; prefers moist soil and will grow in partial shade
Dwarf lily turf	*Ophiopogon japonicus*	5–10°F (−15 to −12°C)	Dark evergreen foliage is grasslike, to 6″ (15 cm) tall; covers an area rapidly, grows in shade or sun, and produces lilac to white flowers in summer; good in sandy soil
Garden lippia	*Phyla nodiflora*	0°F (−18°C)	Sometimes sold under the name *Lippia repens* or *Lippia canescens*; has a matlike growth habit with gray-green leaves and is good in heat and sun; mow occasionally to maintain an even surface; very drought- and traffic-tolerant
Rusty cinquefoil	*Potentilla cinerea*	−35 to −20°F (−37 to −29°C)	Strawberrylike leaves and yellow flowers on a plant 2 to 4″ (5-10 cm) high
Self-heal	*Prunella vulgaris*	−35 to −20°F (−37 to −29°C)	Native to temperate Northern Hemisphere; leaves are semievergreen, lying flat on the ground; purple flower spikes in summer; best in damp soil and spreads rapidly
Pearlwort	*Sagina subulata*	−20 to −10°F (−29 to −23°C)	Mat-forming with needlelike leaves, similar to *Arenaria*; grows up to 4″ (10 cm) tall and is good in shady areas
Creeping thyme	*Thymus serphyllum*	−35 to −20°F (−37 to −29°C)	Very flat growing (less than 1″; 3 cm); evergreen; will withstand poor soil, shade, drought; small purple flowers in summer; many species of thyme are sold under this name, including *Thymus lanuginosus* and *Thymus lanicaulis*, and since these grow taller, they are less tolerant of foot traffic
Creeping speedwell	*Veronica repens*	−10 to −5°F (−23 to −21°C)	Grows to 4″ (10 cm) tall, with a mosslike appearance and shiny foliage; bluish flowers in spring
Pennyroyal	*Mentha pulegium*	−5°F (−21°C)	Very flat-growing, and has a strong mint scent when stepped on

growing conditions. Many of the plants mentioned here are suited to growing in shade, where grass is particularly difficult to maintain.

Selected References for Additional Information

Ali, A. D., and C. L. Elmore. *Turfgrass Pests.* University of California Cooperative Extension Publication 4053. 1989.

MacCaskey, Michael. *All About Lawns.* San Francisco: Ortho Books, 1985.

Rice, Robert P., Jr. *Nursery and Landscape Weed Control Manual.* 2nd ed. Fresno, CA: Thomson Publications, 1992.

Sunset Magazine and Book Editors. *Lawns and Groundcovers.* Menlo Park, CA: Sunset, 1989.

Time-Life Books Staff. *Lawn and Garden.* Menlo Park, CA: Sunset Publishing, 1994.

Turgeon, A. J. *Turfgrass Management.* 3rd ed. Englewood Cliffs, NJ: Prentice Hall, 1990.

University of California Cooperative Extension. *Guide to Turfgrass Pest Control.* Leaflet 2209. 1988.

Vengris, Jonas. *Lawns.* 3rd ed. Fresno, CA: Thomson Publications, 1982.

*Vocational Education Productions. *Professional Turf Management.* San Luis Obispo, CA: Vocational Education Productions of California Polytechnic State University.

*Indicates a video.

13

Diagnosing and Treating Outdoor Plant Disorders

The diagnosing of some plant disorders is easy; for example, wilting usually indicates lack of water, and caterpillars found chewing through a leaf are proof of an insect infestation. But as often as not, a problem is hard to diagnose, and a systematic approach to solving the problem should be used.

Identifying the Affected Plant

The first step in diagnosing a plant problem is identifying the plant that is affected. A common name is sufficient for fruits, nuts, or vegetables, since those names are standard throughout the country. However, with ornamentals, the botanical name is often essential.

Identifying the plant will narrow the list of possible causes considerably, since species vary in their susceptibility to different maladies. A species often will be prey to the same problem repeatedly, and a knowledgeable nurseryman can often diagnose a problem given only the plant name and the symptoms.

Determining the Cause of the Problem

Plant problems are almost always attributable to either a parasitic organism or an environmental condition. Parasitic organisms include insects, rodents, and microorganisms such as bacteria and fungi. Environmental conditions include heat, drought, cold, and the like. Each of these causes of plant injury will be discussed along with its typical symptoms in the following sections.

Insects Damaging to Plants

Insects pass through several stages in their maturation process, and in one or more than one of these stages, they may damage plants. Butterflies are the best-known example of this phenomenon. They begin life as eggs laid on a host plant. The eggs hatch into a larval or caterpillar form, and during this stage the insect feeds on and causes damage to plants. After reaching full size, the caterpillar makes a cocoon (pupates), later emerging as the adult butterfly, a nectar-feeding insect form. However, although the butterfly itself is harmless, it lays the eggs for the next generation of caterpillars.

This egg-larva-pupa-adult sequence (Figure 13-1) is common to many insects, with the larva stage often referred to as a worm or grub. The adult will vary, with adult beetles, flies, or moths emerging as commonly as butterflies.

Damage to plants by insects and related pests is almost always from feeding of two types: chewing or sucking. Chewing insects include many larvae (such as caterpillars and grubs) as well as grasshoppers, boring insects, snails, and beetles. These insects eat through plant parts rapidly, becoming larger and more damaging as they mature. Sucking insects, on the other hand, do not chew holes. They feed on the sap either by inserting their mouthparts into the phloem or by rasping a small area and feeding on sap oozing from the wound. Sucking pests include insects such as aphids and scales, and mites. Their feeding damage is harder to diagnose.

Damage by chewing insects is relatively easy to identify. Irregular holes (Figure 13-2) appear on leaves, or a stem may be severed completely. The insects may

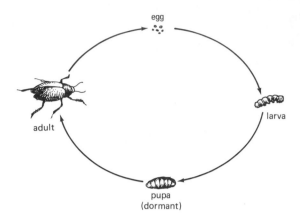

Figure 13-1 The typical maturation sequence of a metamorphosing insect.

still be present and eating or may have left droppings on the plant.

If insects are suspected but not present, an effort should be made to determine what insect is causing the damage, for this affects the control to be applied. The plant should first be inspected under the leaves and near buds and feeding sites. If no insects are present, the pests may be in the soil at the base of the plant. Many insects hide there during the day, coming out to feed at night. For the same reason, checking the plants after dark will often reveal the insect.

Sucking insects present a different detection problem. They are usually small (some microscopic), and their damage varies. Typical symptoms include puckering and bleaching of the leaves (Figure 13-3). A hand lens is helpful in detecting these pests.

Recognizing damage by insects feeding on roots is not as easy as recognizing aboveground infestations. Grubs and other larvae are most often the culprits, chewing the roots of a plant until the root system is insufficient to sustain the foliage. The resulting symptoms are the same as for any condition harming the roots: yellowing, wilting, and eventual death of the top portions. When digging up the plant, signs of insect damage (and often the insects themselves) will be apparent.

The following sections include most of the chewing and sucking pests found in home gardens.

Caterpillars

These chewing insects consume amounts of leaves equaling several times their body weight each day, and sometimes even change color depending on what they have eaten!

Grubs and Borers

Grubs and borers are usually the larvae of beetles. They live underground eating plant roots (grubs) or burrow into stems and fruits (borers). Some, such as the Japanese beetles, are destructive in both larval and adult stages.

Figure 13-2 Damage due to a chewing insect, in this case a caterpillar. Similar damage is caused by slugs and snails. (*USDA photo*)

Figure 13-3 The leaf on the bottom is puckered from a sucking insect feeding on it. The top leave is normal. (*USDA photo*)

Leaf Miners

Leaf miners are miniscule worm-like insects which tunnel between the upper and lower leaf surfaces. Their feeding develops scribblelike white patterns on affected leaves.

Grasshoppers

Grasshoppers are chewing, defoliating insects throughout their life cycle. They have no distinctly different larval, pupal, and adult stages; young grasshoppers resemble the adults.

Beetles

Beetles damage large numbers of plants by chewing and are particularly troublesome in late summer.

Aphids

Aphids are small, sucking insects generally found clustered on the growing tips of plants. Their feeding causes disfigurement of the leaves and buds (Figure 13-4).

Snails and Slugs

These pests feed at night or during wet weather, chewing holes in leaves, flowers, and fruits. Because they are not insects, they cannot be controlled with most insecticides.

Scale and Mealybugs

These sucking insects spend their lives in colonies attached to leaves or stems. Scales (Figure 13-5) look

much like raised spots, and mealybugs (see Figure 18-2) like dots of cotton. Their unique appearance stems from the waxy secretion that oozes from the back of the insect after it has settled and begun feeding. Once protected in this manner, the pests are unaffected by most insecticide sprays.

Figure 13-4 Typical symptoms of aphid damage to a delphinium (note distorted new growth). (*USDA photo*)

Mites

Spider mites cluster on the underside of plant leaves, sucking juices and often give a pinprick pattern to the top surfaces. In heavy infestations, some species spin fine webs. Since the insects are nearly microscopic, a magnifying glass and strong light are helpful in diagnosing a mite problem. Mites on fuchsias can cause deforming as in Figure 13-4.

Thrips

Thrips are sucking pests only slightly larger than mites. They attack leaves or flowers and cause damage in the form of streaks, small bleached spots, or distorted leaves (Figure 13-7).

Gall-forming insects

Both mites and aphids as well as other insects and diseases, can cause the abnormal swellings known as galls. Although unsightly, most are not seriously deleterious to plant health.

Figure 13-6   Spider mites and their damage. (*USDA photo*)

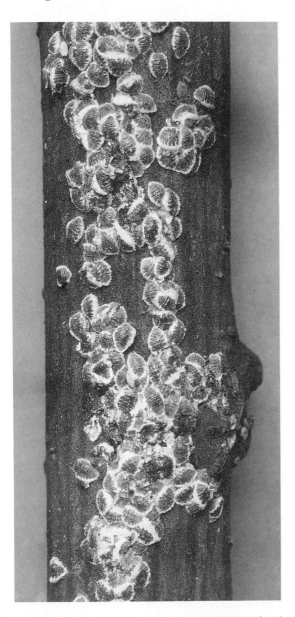

Figure 13-5   European elm bark scale. (*USDA photo*)

Figure 13-7   Cuban laurel thrips on an ornamental fig (*Ficus* sp). (*Photo courtesy of the Cooperative Extension Service of the University of California*)

Nematodes (Eelworms)

Nematodes are microscopic or nearly microscopic worms. Most are soil dwelling, burrowing into roots and decreasing their ability to translocate. This results in yellowing and general lack of vigor. Root nematodes can sometimes be seen with a magnifying glass, but one of the most common types, the root-knot nematode, is identified by the characteristic swellings formed on roots by its feeding (Figure 13-8).

Damage from foliar nematodes is less common. But some species emerge from the ground and crawl up to the leaves on a film of water such as from dew or rain. They then enter the leaves through the stomata and begin feeding. Typical symptoms include death of irregular sections of leaves and spotting.

Microorganisms Damaging to Plants

Several types of microorganisms damage plants, causing diseases that kill the foliage or rot the roots. These microorganisms can be divided into four types—fungi, bacteria, viruses, and mycoplasmas. Each is discussed in the following sections.

Figure 13-8 Root-knot nematode damage to a tomato. (*USDA photo*)

Fungi

Fungi are single- or multicelled plants that lack chlorophyll. Therefore they rely on other sources for carbohydrates and parasitize green plants. A typical fungus has an extended threadlike body which increases in size by cell division. At maturity, a reproductive system is formed that produces dustlike spores which spread the fungus.

Fungi are responsible for more plant diseases than any other type of microorganism, but they are also the most easily treated. The diseases vary widely, but most can be classified into one of the types listed in Table 13-1.

Bacteria

Bacteria are single-celled, lack chlorophyll, and frequently grow in colonies. They are the cause of relatively few plant diseases, but diseases caused by them are difficult to treat. Leaf spots, rots, cankers, blights, and wilts (Table 13-1) are all caused by bacteria. In addition, bacteria can cause galls.

Viruses

Viruses (Figure 13-3) are among the smallest microorganisms, and only a few cause plant diseases of importance to the home gardener (Table 13-2).

They are spread between plants by insects. The insect feeds on a virus-infected plant, ingests the virus, moves to an uninfected plant, and spreads the infection. Aphids, mites, leafhoppers, and nematodes can spread plant viruses. The virus may eventually be fatal to the plant, although death may be quite slow. To date, virus diseases are untreatable.

Mycoplasmas

Mycoplasmas are very similar to bacteria and were long thought to be a small bacteria type. Of the 30 types that are known, only a few cause plant diseases. Symptoms are similar to those of virus disease and are difficult to diagnose.

Large Animal Pests Damaging to Plants

A number of rodents and similar animals cause damage in the garden. The more common ones and their symptoms are discussed in the following sections.

Rabbits

Rabbits are troublesome in many areas of the country. They nibble vegetable gardens and the tender

Table 13-1 *Fungus Diseases of Cultivated Plants*

Disease	*Symptoms and parasitizing pattern*
Mildew (Figure 13-9)	The fungus grows on the surfaces of plant leaves and stems. The threadlike body can be easily seen with the naked eye as a powdery white covering. Infected leaves eventually wither and drop, and the whole plant dies.
Wilt	Fungi invade and multiply in the phloem and xylem of the plants, clogging them and restricting translocation. The entire plant or only a section will wilt and not recover.
Rot	Roots, underground storage organs, and the lower portions of stems are attacked by these soil-inhabiting fungi. Cells in infected areas die and decay, disrupting absorption and translocation and causing the death of the plant.
Leaf spot (Figure 13-10)	Blotches or spots appear on the foliage and sometimes stems or flowers, spreading rapidly to cover the plant. The spots are usually very distinctive in appearance, having, for example, a dark center surrounded by rings of different-colored tissue. The spots gradually reduce the photosynthetic area of the leaves and cause the death of the plant.
Rust (Figure 13-11)	The fungi which cause rusts are distinct by virtue of requiring two different species of host plants to complete the life cycle. Often one host will be a cultivated plant and the other a native plant or weed. Common symptoms of rusts are reddish-brown spots on the foliage and stems. On some species, the fruiting body which bears the spores is a 1-inch mass of orange gelatinous horns with a bizzare appearance.
Canker (Figure 13-12)	Relatively uncommon disease affecting woody plants. Fungi cause a sunken bark area, which girdles the branch or trunk and eventually kills that portion of the plant above the infected area.
Smut	Uncommon disease most often encountered on corn. Invaded portions swell grotesquely, eventually bursting open and dispersing a multitude of dusty black spores.
Witches'-broom	Mainly a disease of woody plants and not frequently encountered. Fungi infect the entire plant, but symptoms appear on small branches as unrestrained twiggy growth resembling a bird nest. The disease is sometimes fatal.
Blight	Usually affects woody plants and causes quick wilting and death of young and growing tissues such as flowers and twigs. Can spread downward to kill more mature portions of the plant.

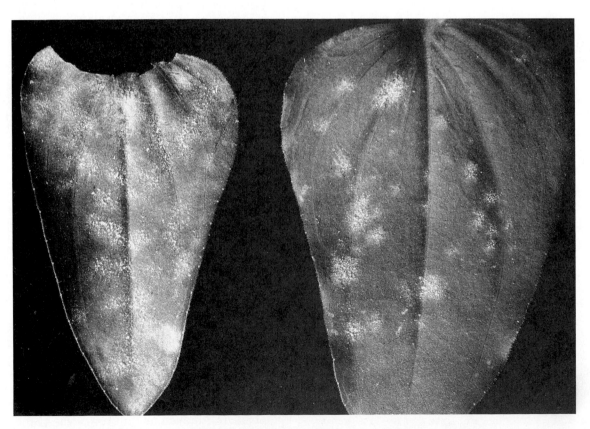

Figure 13-9   Mildew on zinnia leaves. (*Photo courtesy of Dr. R. E. Partyka*)

Figure 13-10 Leaf spot disease on strawberry leaves. (*USDA photo*)

spring growth of many ornamentals including peony and dahlia.

Mice

Mice are a problem chiefly in the winter when food is scarce, burrowing under mulch or snow layers to feed

Figure 13-11 Rust fungus disease on a rose leaf. (*Photo courtesy of Vocational Education Productions, California Polytechnic State University, San Luis Obispo, CA*)

on the bark of fruit trees. The feeding can girdle the tree and cause its death.

Pocket Gophers

These burrowing rodents create a nuisance by digging holes and tunnels throughout the garden (Figure 13-14). They also feed on roots and can kill a large plant in a day or two. Pocket gopher damage can be easily diagnosed; a plant will wilt suddenly and, when lifted, will come out of the ground easily and have no roots.

Moles

Moles also burrow, but feed mainly on insects they find in the soil. The damage results from their burrows, which leave a raised pattern throughout an area with mounded exit holes. Moles are particularly damaging to lawns because they burrow through grass roots, causing the death of the turf above.

Deer

Deer are an increasing problem in rural areas. They eat twigs from shrubs or trees and will also attack vegetable and flower plantings. They usually feed late at night.

Figure 13-12　A bark canker along the upper part of the stem. Normal, healthy bark is at the bottom. (*Photo courtesy of Dr. R. E. Partyka, Columbus, OH*)

Figure 13-13　Watermelon mosaic virus on summer squash. (*Photo courtesy of Dr. O. W. Barnett, Clemson University*)

Table 13-2　*Typical Virus- and Mycoplasma-Caused Plant Diseases*

Disease	Causal organism	Symptoms
Stunt	Virus or mycoplasma	Slowed or stopped growth under favorable growing conditions and with no other apparent cause
Curly top	Virus	Causes twisting and distortion of foliage; symptoms are similar to weed-killer injury
Mosaic	Virus	Mottled light and dark areas on foliage or fruit
Yellows	Mycoplasma	Yellowing of foliage, stunting, and witches'-broom shoot growth under good growing conditions and due to no other apparent cause

Birds

Birds are a problem where fruit plantings are grown. They will eat cherries, raspberries, and other species.

Dogs

Dog urine is toxic to plant foliage, causing browning or yellowing. Male dogs frequently damage the lower leaves of shrubs, and females leave circular spots on turf.

Environmental and Other Nonparasitic Causes of Plant Injury

Frost Damage

Plants differ in their susceptibility to frost according to age and species: young tissues are more tender than mature ones; woody plants are more frost-resistant than herbaceous ones. Accordingly, a light frost may strike sporadically, damaging one plant and not the next. A key symptom is sudden overnight wilting, particularly on the uppermost leaves and buds.

Winterkill

Winterkill is the partial or total death of normally hardy perennial plants due to cold temperatures. In minor cases a few branches will die back, and the plant will resprout from buds on the lower stems or crown. With complete winterkill, both tops and roots die. This occurs during very cold winters when temperatures fall below the critical temperature for root survival.

Figure 13-14 Gopher damage. (*Photo courtesy of Vocational Education Productions, California Polytechnic State University, San Luis Obispo, CA*)

Winterkill manifested as "burning" of evergreens is a common occurrence in cold climates. Leaf tissues become brown, and since evergreens do not refoliate yearly, the appearance of the plants can be permanently marred. Burning is usually due to lack of soil moisture during the winter combined with continuing water need and drying of the needles by winter winds. Watering and mulching are therefore recommended during the winter when the soil is not frozen.

Drought

Wilting is the main symptom of drought, and watering revives a plant in a drought condition. Drought-struck plants will usually revive completely unless their leaves are already leathery or crispy. Even beyond that stage, many species will regenerate from the roots after complete death of the aboveground parts.

Poor Drainage and Flooding

Poor drainage, in which roots remain in water-saturated soil, will kill most plants. Those native to bog and other lowland areas are exceptions. Death is due to lack of oxygen. The length of time a plant can remain in standing water without injury varies, but most will tolerate several days to a week of occasional flooding.

Lack of Soil Nutrients

Lack of adequate soil nutrients will stunt and bleach plants but will not usually kill them very quickly. Lack of a required element will cause different symptoms on different species, and this makes nutrient deficiencies difficult to diagnose. However, probably 80 percent of all deficiencies are due to a lack of nitrogen, phosphorus, potassium, magnesium, or iron. Descriptions of deficiency symptoms for these elements are given in Table 13-3.

Unavailability of Soil Nutrients Due to pH

Even if an essential element is present, a plant may be unable to absorb it due to an excessively high or low soil pH. Iron and manganese are most commonly

Table 13-3 *Deficiency Symptoms for Commonly Lacking Nutrients*

Element	Symptoms
Nitrogen	General leaf yellowing, showing up first on older leaves; deciduous trees defoliate earlier in fall
Phosphorus	General yellowing of foliage, with young expanding leaves and shoots frequently showing purple discoloration
Potassium	Older leaves die at the margins, and there is reduced shoot growth and leaf size; the marginal burn may be preceded by yellowing and is most severe late in the growing season
Iron	Yellowing of areas between the veins on young leaves
Magnesium	Retarded growth followed by yellowing of areas between veins on old leaves; yellowing spreads later to young leaves

affected by pH. Symptoms will be the same as for normal deficiency (Table 13-3).

Excess Soil Salts

A nutrient excess is frequently called a *salt buildup*. Nitrogen can be the primary damaging element if the toxic buildup stems from overfertilization, but compounds present in normal irrigation water and alkali soils can be as damaging.

Symptoms of excess salts are wilting and browning of leaf margins. This is due to death of the roots from dehydration by salts, and the effect is often indistinguishable from that of drought. The cure is by leaching, as described in Chapter 11.

Excess Heat (Heat Scorch)

Excess heat damage (see photo, Figure 1-4) frequently occurs on plants growing in paved areas or on south-facing walls; it is due to the reflection of sun-generated heat from these surfaces. The leaf margins and areas between the veins brown, and succulent new growth may die. Tolerance to excess or reflected heat varies among species.

Soil Compaction

Constant foot or equipment traffic causes soil compaction around plants. The air spaces normally present in the soil are compressed, and water absorption and air penetration are inhibited. Plants in compacted soil grow slowly, and portions may die back. In severe cases, the entire plant may die.

Improperly Applied Chemicals

Pesticide sprays can injure plants if applied improperly, in excess concentration, or to plants for which they are not intended. The lawn weed killer 2,4-D is among the most frequently damaging. Spray drift can carry the chemical to neighboring areas, and curling and distorting of plant foliage (Figure 13-15) will result. Similar damage can appear when one sprayer is used for both weed killer and insect or disease sprays. Only a small amount of weed killer residue is needed to cause injury.

Chlorosis of plant foliage is a symptom of weed killer injury by chemicals such as Simazine. This weed killer injury can be distinguished from nutrient deficiency by the suddenness of the symptoms: weed killer injury will appear in two to three weeks, whereas nutrient deficiency will progress over several months. There are no practical remedies for weed killer injury, and the chances of its killing the plant depend on how

Figure 13-15 2,4-D herbicide damage on squash leaves. (*Photo courtesy of Marge Coon*)

much chemical was absorbed. Often the foliage will be marred for one or two seasons, but the plant will grow out of the damage and regain its former appearance.

Insect and disease sprays applied in excess concentrations or on plants for which they are not approved show different symptoms. Sudden browning of foliage to which the chemical was applied is the most common and may be followed by defoliation. A single defoliation from chemicals, disease, or insects is seldom fatal.

Grade Changes

Raising the soil level over the roots of plants more than a few inches (5 centimeters) can be fatal because water and oxygen flow to the roots will be restricted. For this reason, tree wells are necessary when grade changes are planned. They should be made as large as practical, extending to the branch tips if possible. Supplemental drain tiles may need to be installed (Figure 13-16).

Air Pollution

Air pollution damage is a problem only in a few areas of the country. It affects only susceptible plants, such as conebearing evergreens. The damage takes many forms, but usually involves some browning and death of the foliage. Pollution damage is difficult to diagnose and should only be suspected if the area frequently accumulates smog.

Allelopathy

Allelopathy is the injury of one plant by another via the excretion of toxic substances by the roots. Both black

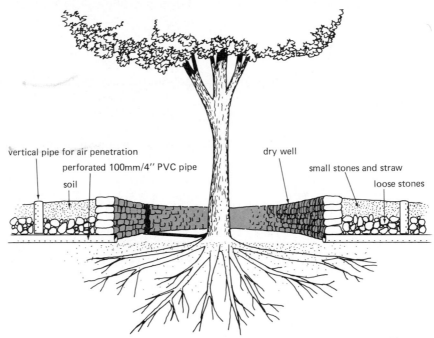

vertical pipe for air penetration

perforated 100mm/4″ PVC pipe

soil

dry well

small stones and straw

loose stones

Figure 13-16   A tree well with drainage tiles.

walnut and eucalyptus trees are allelopathic to some other species.

Treating Plant Disorders

The steps in restoring a sick plant to health relate directly to the cause of the disorder. For environmental problems, there are two choices: correct the condition harming the plant, or find a plant that will tolerate the situation. Excess salts, drought, soil compaction, and pH problems, for example, can all be corrected. Air pollution, winterkill, and reflected heat problems, on the other hand, cannot be eliminated, but plant selection can be changed.

Animal damage to plants can be dealt with by either preventing the damage or removing or killing the animal. Fences will keep out rabbits, dogs, and deer, although an 8-foot (2.4-meter) fence will be needed to prevent intrusions by deer. Blood meal may also work as a deer and rabbit repellent. Netting or noise-making devices are most successful with bird problems. Girdling of trees by mice and rabbits can be avoided by pulling away mulch layers and enclosing the trunk with a tall collar of tin or hardware cloth (Figure 13-17).

Damage by moles, gophers, and ground squirrels is harder to prevent. Poison baits are the best recourse against pocket gophers and ground squirrels, but moles are less likely to accept baits. Trapping with spe-

cially designed mole traps is frequently successful, but to be effective the trap must be placed in a main burrow used repeatedly by the animal. When moles infest a lawn, it is often because the lawn contains grubs, a

Figure 13-17   A hardware cloth collar to protect the base of a tree from rodent damage. (*Photo by George Taloumis*)

favored food. Applying an insecticide drench will destroy the food source, and the moles will retreat.

Treating Microorganism-Caused Diseases

Plant disease control is both cultural and chemical, and an understanding of both is essential to restoring and maintaining plant health.

Cultural Disease Control

Cultural control involves using plant care techniques that minimize disease infection and spread.

Watering. Water is a key factor in plant disease attack and spread. Many fungi spores lie dormant until in the presence of water, germinating only then to invade the host plant. Accordingly, keeping water off plants susceptible to foliage diseases lessens the chances of infection. Excessive soil water similarly favors the growth of disease organisms causing root rot, and overwatering will trigger that disease.

Splashing water is often a factor in foliage disease spread, carrying the microorganisms from infected to uninfected plants. Careful watering, with the water applied only to the soil, can decrease disease problems on susceptible species. The timing of watering can also reduce or enhance disease spread. The foliage of plants watered late in the day may stay wet overnight and be susceptible to disease. Plants watered from morning to midday dry quickly and are less vulnerable.

Sanitation. Sanitation involves removing all disease-harboring materials to reduce the population of microorganisms. Many disease organisms have a seasonal cycle with dormant and active periods. The overwintering site for many foliage disease organisms is the fallen leaves that were infected during the summer. Accordingly, leaf and twig litter under disease-infested plants should be cleared away promptly in fall to prevent it from being a source of infection next season. These infested plant parts should be burned or hauled away, not composted.

Weed removal is another sanitation measure. One disease will frequently attack several related plants. If one of those plants is a weed, it can be a source of infection for the other plants. As another example, rust fungi require two plants to complete their life cycle, one of which is often a weed. Without a weed host nearby, the rust will not be able to attack the cultivated species.

Pruning. Pruning can be a valuable control technique for localized diseases such as gall, wilt, and canker. The infected portion of the plant should be pruned out and burned as soon as it is detected, cutting back several inches (5-10 centimeters) into healthy tissue. Between cuts and after pruning, the pruners should be dipped in alcohol to avoid spreading the microorganisms by the blades.

Roguing. Roguing involves pulling out diseased plants to stop the spread to others. It is valuable for annual plants or when only a few plants are infected. In addition, it is the only control for virus diseases and the most practical for mycoplasmas and most bacteria.

Crop Rotation. Crop rotation is usually associated with vegetable plants, but it can apply to flowers as well. It involves planting a different species in an area each year instead of the same one repeatedly. Rotation is a valuable control for soilborne diseases, most of which are hard to control by chemical means. By using crop rotation, the parasitic microorganisms growing in conjunction with a species will be depleted yearly because they are not able to prey on the new crop.

Resistant Cultivars and Species. Planting species and cultivars resistant to disease is one of the soundest disease-control techniques. Many vegetables and flowers have resistance to diseases imparted by breeding, and the small extra amount these cost is well spent. Landscape plants are similarly bred for disease resistance, although some have natural disease and insect resistance. Planting these ornamentals in lieu of disease-susceptible ones can greatly lessen maintenance.

Chemical Disease Control

Chemicals are very helpful with some disease problems and nearly useless with others. Leaf spots, rust, mildew, smut, and blight diseases can usually be chemically controlled, but wilts, rots, cankers, witches' brooms, and all virus diseases are only marginally controllable or uncontrollable by chemicals. Fortunately, most commonly encountered diseases fall in the controllable group.

Fungicides. These are the most common disease-controlling chemicals, and most can be used on a variety of plants for numerous diseases. The container label and Cooperative Extension Service publications will give information regarding which diseases the product controls.

Although most fungicides are used as foliage sprays, a few combat soil-borne diseases and are applied to the root zones of plants. These are called *fungicidal soil drenches*. Still fewer are applied by one of the methods just described but are absorbed into the plants instead of simply forming a protective covering. These are called *systemics* and are effective on wilts when few other chemicals are.

Most fungicides are only *protectants* which stop the germination and growth of fungal spores but do not actually kill the fungus. A protectant is most useful in the early stages of a fungus infection before the pathogen becomes established enough to do significant damage to the plant. For this reason, monitoring the health of plants is important to catch disease infections before they become epidemics. In addition, protectants must be reapplied frequently in order to continue to give protection to a plant. New growth which occurs after spraying, for example, will not be protected and an additional spraying will be needed.

A few fungicides are classified as *curatives* (chemotheraputants). Curatives kill existing fungus completely as well as preventing germination and growth of spores newly landed on the plant.

Bactericides. These are generally antibiotics such as streptomycin and tetracycline. They are less common than fungicides and are used only on a few bacteria- and mycoplasma-caused diseases including blights. Copper sprays may also control bacteria.

Seed and Bulb Treatments. Often a home gardener will buy a packet of seeds that are oddly colored (usually pink or blue) because of treatment with a fungicide before packaging. The treatment will have been applied by dusting or quickly dipping the seeds (followed by immediate drying) in liquid fungicide. The fungicide gives protection against damping off.

If damping off has been a frequent problem, seeds can be home-treated before planting by shaking them in a bag with a teaspoon of powder fungicide recommended against damping off before planting. If bulbs are bought and appear to have mold growing on them (normally penicillium), they can be treated by dipping them in fungicide mixed to proper application strength before planting. The dip will slow or stop the development of the organism in the soil.

Treating Nematode and Insect Problems

Cultural Controls

The cultural controls for combating insect problems are less successful than those for combating diseases but can sometimes be effective. One strategy is to re-move the insect hiding places. Accordingly, weed control and removing litter at the base of a plant are two controls. Pruning off of an insect-infested portion of a plant is also used.

Supplying plants with adequate nutrition and other requirements for healthy growth is another cultural control technique. Healthy plants are less susceptible to both insect and disease problems than those under stress.

Physically removing large insects such as worms by hand picking during the day or the night, depending on when they are active, can be effective. A strong stream of water can be used to wash off aphids.

Chemical Controls

Chemicals used to control insect pests are most often applied to plants as sprays or soil drenches and protect the plant by killing the insects. However, baits, which attract insects and kill them, and repellents are two other chemical controls.

Sprays. Surface sprays either kill on contact or are absorbed by insects as they walk across a sprayed leaf. These sprays are effective on most pests but are not useful for those with protective coverings such as mature scales and mealybugs.

Stomach poisons applied as foliage sprays are useful on chewing insects, which ingest the poison as they feed on the plant. They are less effective on sucking insects, which feed on the untreated sap.

Dormant oils are the third type of spray and are used on dormant deciduous plants. Applied in early spring, they are the primary means of combating scales and mealybugs. The oil coats trunks, branches, and twigs and excludes air, suffocating insects and their eggs. Later in the season, light oils safe for foliage can be used.

Soil Drenches. Soil drenches combat underground pests. They are contact or systemic and are usually the same chemicals as surface sprays but in weaker concentrations.

Systemics. Systemic insecticides are absorbed into a plant and kill only those pests that feed on the plant. They can be applied as foliage sprays or soil drenches (penetrating by root uptake) and, occasionally, by injection or implantation into the trunks of trees (see Chapter 18). Some insecticides have both contact and systemic action.

Baits. Baits are the most commonly used control material for snails, slugs, sowbugs, centipedes, and sometimes rodents. They act as stomach poisons, and careful attention should always be paid to placing them where other animals will not eat them.

Repellents. Few repellents are available commercially except for rodent repellents. However, homemade solutions of varying effectiveness can be made by following the instructions in organic gardening books.

Choosing Pesticides

With the numerous chemicals sold for controlling insect and disease pests, it is difficult for a gardener to know which to apply for a particular pest problem. The criteria for selecting a chemical should be that it controls the pest satisfactorily and is as nontoxic as possible both to people and to beneficial organisms.

The best sources of the toxicity and uses of pesticides are government publications (Cooperative Extension, Farm Advisor, United States Department of Agriculture) and the label of the material. The label will state "Danger," "Warning," or "Caution" in bold letters. "Danger" indicates that the chemical is very toxic to humans, "Warning" that it is moderately toxic, and "Caution" that it is not very toxic.

Table 13-4 lists the LD_{50} of many common insecticides, fungicides, and bactericides. LD_{50} is an abbreviation for the phrase "lethal dose killing 50 percent of a population." It indicates how much of a pesticide is required to kill 50 percent of a population of laboratory animals and is expressed in milligrams per kilogram of total body weight of the animals. Therefore the larger the number, the more of the material was needed to kill the animals and the *less* toxic the material. For comparison purposes, the LD_{50} values of salt, caffeine, and aspirin are included in the table, but this comparison in no way implies that pesticides less toxic than these materials should be used injudiciously.

Organic Gardening

Organic gardening is a popular way of gardening that does not permit the use of any synthetically manufactured granular fertilizers, insecticides, disease sprays, and the like. Organic gardeners contend that such materials work against rather than with nature, upsetting the natural balance of organisms in the environment and creating more problems than they solve. In lieu of conventional manufactured chemicals, they advocate mainly cultural solutions to plant disorders and, when necessary, pesticides, repellents, and baits that occur naturally in the environment. They contend that by virtue of being naturally occurring, chemicals do less

harm to the environment and at the same time pose little threat to humans.

The cultural controls discussed for diseases and insects are essentially the same as those used by organic gardeners. In addition, the following pest control methods are advocated by organic gardeners, although the effectiveness of several has not been proven experimentally.

Companion and Repellent Planting

In companion planting, relatively insect-resistant plants are grown next to susceptible ones to repel potential pests from the susceptible plant. Marigold roots, for example, exude a substance that repels nematodes, and onions and garlic are thought to ward off a number of pests. Garlic has been used successfully on some bacterial and fungal diseases.

Trap Crops

Trap planting is believed to work in the opposite manner as repellent planting. Insects often have definite plant preferences. In theory, planting a more attractive plant nearby will lure potential pests away from the principal crop. Dill supposedly can attract hornworms away from tomatoes, and Japanese beetles are reportedly lured by zinnias.

The effectiveness of trap crops is questionable. It is possible that the presence of two attractive crops compounds rather than alleviates a pest problem.

Traps

Traps can attract insects by providing food or shelter. Molasses, beer, and rotting fruit are reputed feeding lures which attract insects and then drown or stick them in the trap as they feed. Since many insects spend the day under plant refuse, a substitute hiding area can be made by laying a board on the ground. Slugs, sowbugs, and so on will congregate there and can then be destroyed. More sophisticated light and electrocuting traps can also be purchased. Traps can also be used to detect the presence of small numbers of insects so that controls can be used at the most effective time.

Barriers

Barriers prevent insects from reaching a plant. Flypaper, for example, can be placed around a tree trunk to stop insects crawling up from the ground. Wood ashes scattered around the base of a plant will repel snails and slugs. Vegetable garden transplants can be protected

Table 13-4 *Oral LD$_{50}$ Values of Selected Garden Chemicals*

Chemical and associated trade names	Use	LD$_{50}$[a]
Acephate (Orthene®)	insecticide	700
Anilizine (DYRENE®)	protective fungicide	2710
Aspirin	—	750
Azadirachtin (BioNEEM®)	insecticide (growth regulator)	5000
Bacillus popillae milky spore disease	insecticide for Japanese beetle larvae	5000
Bacillus thuringiensis BIOTROL, THURICIDE®, DIPEL	insecticide	no toxicity at tested doses
Bayleton	curative fungicide	363
Bendiocarb	insecticide	40
Benomyl (BENLATE)	curative fungicide	9590
Caffeine	—	200
Captan (ORTHOCIDE®)	fungicide	9000–15,000
Carbaryl (SEVIN®)	insecticide	500–850
Chlorothalonil (DACONIL®, BRAVO®)	protective fungicide	10,000
Chlorpyrifos (LORSBAN®, DURSBAN®)	insecticide	97–276
Copper	protective fungicide, bactericide	low toxicity
DCPA (DACTHAL®)	herbicide	3000
Diazinon	insecticide	300–400
Dienochlor (PENTAC)	miticide	3160
Dimethoate (CYGON®)	insecticide	215
Disulfoton (DI-SYSTON®)	insecticide	2–12
Dursban	insecticide	96
Endosulfan (Thiodan®)	insecticide	23
EPTC (EPTAM®)	herbicide	1600
Gasoline	—	150
Hexakis (Vendex®)	miticide	2630
Kelthane	miticide	100–1000
light and heavy oils	insecticide	15,000
Malathion	insecticide	1000–1375
Mancozeb (Dithane®)	protective fungicide	4500
Metaldehyde	insecticide, snail, slug killer	600–1000
Metasystox R	insecticide	65
Methoxychlor	insecticide	6000
Oftanol	insecticide for soil insects	20
oryzalin (SURFLAN®)	herbicide	10,000
oxadiazon (RONSTAR®)	herbicide	8000
PCNB (TERRACHLOR®)	soil fungicide	1650–2000
Pyrethrum	insecticide	820–1870
Resmethrin	insecticide	4240
Rabon	insecticide	4000–5000
Rotenone	insecticide	50–75
Ryania	insecticide	750–1200
Salt	—	3320
Simizine (PRINCEP®)	herbicide	5000
Soap (Safer Soap®)	insecticide	very low toxicity
Steinernema carpocapsae beneficial nematodes (Exhibit®)	insecticide	none
Streptomycin	bactericide	9000
Sulfur	protective fungicide	17,000
Tetramethrin	insecticide	4640
Thiophanate-methyl	curative fungicide	15,000
Triforine (Funginex®)	curative fungicide	16,000

[a]Class 1: highly toxic (1–50); class 2: moderately toxic (50–500); class 3: low order of toxicity (500–5000); class 4: very low toxicity (5000+).

from cutworms with a paper collar held together with a paper clip.

Biological Control

Biological control uses beneficial insects and microorganisms to control harmful insects. A number of insects use other insects as food or for part of their reproductive cycle. These beneficial insects can reduce the population of pests to a level at which their damage does not justify additional control. Among these are ladybugs, praying mantises, fireflies, lacewings, spiders, and many wasps.

Beneficial insects can be encouraged in the garden by minimizing the use of broad-spectrum chemicals that kill both pests and beneficials (insects that either do no harm or are actually beneficial in a garden). Buying and releasing predators is only slightly effective, since the beneficials disperse over a wide area instead of remaining congregated on one property.

Bacteria, fungi, nematodes, and viruses infect and kill insects, but research in this area has been slow, and only a few are available for home use. Ultimately, these may prove to be one of the most valuable forms of insect control.

One disease of insects is *Bacillus thuringiensis,* a bacterium that kills the larvae of moths and butterflies. Cabbageworms, cutworms, corn borers, and tent caterpillars are among the insects controlled. The bacteria are applied as a spray as soon as worms are detected. After ingesting the bacteria, the worms stop feeding and are dead two to four days later. The dead insects then serve as sources of infection for future worms. *Bacillus thuringiensis* is completely harmless to plants and to animals other than the larvae.

An insect growth-regulating chemical, Enstar®, is being used to combat infestations of whiteflies, mealybugs, and aphids. In addition to killing eggs and preventing maturation of developing insects, it also sterilizes adults. At stronger concentrations, it will kill adults outright.

A third biological control is the beneficial nematode species *Steinernema carpocapsae.* This nematode is active in the soil against the larvae of borers, weevils, chafers, and against cutworms and fungus gnats. However it is completely harmless to non-insect species. The nematode enters the body of the host insect and multiplies in the digestive track, killing the larvae.

Homemade Insect Sprays

Recipes using common household substances for insect control are also available. Buttermilk, soap and water, turpentine on rags, and various oils are suggested by organic gardeners as insecticides and repellents.

Botanical repellents can be made at home from a number of common plants. Their effectiveness has never been thoroughly investigated. Onions, garlic, hot peppers, tomato leaves, nettles, and horseradish are all reputed to have excellent qualities when mixed in water and applied as a spray.

Integrated Pest Management

Integrated pest management is an intermediate course between spraying indiscriminately with insecticides and rejecting them completely, as organic gardeners do. It combines many controls—cultural, biological, and chemical—to prevent and treat plant problems. Emphasis is placed on learning as much as possible about the pest or disease problem—the life cycle, host plants, most favorable environment, and so on—and using the information to formulate a control plan. Integrated pest management also involves establishing an amount of damage to plants that can be tolerated and accepted, as opposed to assuming that plants be completely pest and disease free. In an integrated program insecticides are used only when cultural and biological controls are ineffective, and they are used when they will be maximally effective.

When an insecticide is necessary, it is chosen carefully with regard to effectiveness, toxicity to beneficial insects, persistence in the environment, and danger to humans. It is unnecessary, for example, to spray a long-lasting, toxic insecticide for pest control when a less toxic and less persistent one will work as well. Similarly, one should not use a multipurpose spray that will kill both harmful and beneficial insects when a systemic drench would kill only insects damaging the plant.

While it would be difficult for a gardener to be familiar with the life cycles of all the insects and diseases of plants, he or she can use integrated pest management to a limited extent by following these steps.

1. *Determine the cause of the problem.* Decide whether it is due to an insect, disease, rodent, nutritional deficiency, or other source by analyzing symptoms, inspecting for pests, and using a reference book for additional information where necessary.

2. *Determine whether the harm being done is sufficient to justify control.* Set a level of tolerable damage. Mildew attacking annuals in late fall should, for example, be more tolerable than mildew on plants in spring, since the plants will soon be frosted anyway. Treatment might not be justified at all. In the same way, if lettuce has already been severely chewed by insects, spraying may be superfluous because the lettuce will not recover.

3. *Use cultural and biological controls, if practical and available, to treat the problem.* Hand picking a few cabbageworms or trapping slugs in beer is more ecologically sound than spraying and can be nearly as effective. Similar cases can be made for using *Bacillus thuringiensis* on caterpillars and for hosing off aphids and mites with water.

4. *Apply a pesticide as a last resort or if it is the only available control.* Choose the chemical carefully so that it is as nontoxic, nonpersistent, and specific to the pest as possible. For example, miticide should be used for mites, not an all-purpose combination. The chemical should be applied to the damaged plant. Spraying unaffected neighboring plants is not preventative, it is overkill and unnecessarily pollutes the environment.

Applying Pesticides to Outdoor Plants

Applying chemical pesticides safely and correctly will maximize their effectiveness.

Application Equipment

Dusters

Dusting (Figure 13-18) is less common than spraying with liquid, but some gardeners prefer it simply because a duster does not need to be emptied and cleaned after every use. The disadvantages are that it leaves an

Figure 13-18 A duster for applying pesticide. (*Photo by Rick Smith*)

Figure 13-19 An atomizer sprayer. (*Photo by Rick Smith*)

unattractive residue and that not all pesticides are formulated to apply as dusts. Unless a pesticide states that it can be used as a dust, it should not be.

Atomizer Spray

Many pesticides are now sold in plastic pump spray bottles in premixed form. The bottle holds only about a pint (0.5 liter) of spray.

Compressed-Air Tank Sprayer

The tank of the sprayer is partially filled with pesticide, and the remainder contains compressed air. Pulling the trigger on the spray wand releases the pressure and dispenses pesticide in a fine spray. A tank sprayer is probably the best all-around sprayer for the home gardener. The wand makes it possible to reach both high into trees and beneath the foliage of low-growing plants (Figure 13-20).

Hose-End Sprayer

This sprayer uses concentrated pesticide in the container and dilutes it with water under pressure from the hose. However, because it proportions pesticide into water relatively imprecisely, it is not recommended for general pesticide applications.

Pesticide Safety

The importance of safety precautions in storing, mixing, applying, and disposing of pesticides cannot be overemphasized. As potentially deadly chemicals, they merit respect and careful use.

Figure 13-20 A gardener using a compressed-air tank sprayer. (*USDA photo*)

Storing

1. Store pesticides out of reach of children and animals and at a moderate temperature. A locked storage cabinet is suggested.

2. Store them only in the original containers. Although it may seem practical to share a bottle of pesticide with a neighbor, don't do it. Storing pesticides in bottles other than their original containers is one of the main causes of poisoning.

Mixing

1. *Read the label* before opening the container. It is your best source of information on the pesticide, its toxicity, its uses, and so forth.

2. Do not use kitchen measuring utensils for pesticides. Keep a separate set expressly for pesticide use.

3. Do not mix pesticides unless listed on the label. Although many are compatible, some are not and can seriously damage plants.

4. Never use a pesticide at stronger than the recommended rate. Doing so pollutes the environment unnecessarily, will not make the chemical any more effective, and could damage the plant.

Application

1. Think about whether the pesticide will be a danger to pets. Cover pet food and water bowls and fish ponds when applying pesticide nearby.

2. Do not spray on windy days. The pesticide will blow, which will make it less effective and may also cause it to come into contact with you.

3. Wear protective clothing: a long-sleeved shirt, trousers, socks, and shoes.

4. Avoid breathing the pesticide. The lungs take up pesticide quickly. Stand back from the plant being sprayed or use a respirator, and do not smoke.

5. If pesticide is spilled on the skin, wash it away immediately with soap and water. For spills on clothes, change, and launder the garments separately from your regular wash.

6. For ground spills of small quantities, absorb the spilled material with cat litter or paper towels and then flood the area with water to dilute the pesticide to a minimally toxic level.

7. Apply the pesticide only to plants on which it is *registered* on the label. Registration means that the pesticide has been tested and found to be safe on that species. Especially do not use pesticides on vegetables and fruits unless they are listed on the label, and wait the *harvest interval* (the number of days after application of the pesticide before the vegetable can be safely eaten) if an interval is given.

Disposal

Try to mix exactly the amount needed, but if some pesticide remains after spraying, several alternatives can be followed for disposing of this material.

1. Spray on additional plants.

2. Seal the excess in an unbreakable container, and put it in the trash.

Do not pour excess pesticide down drains, and do not burn the containers.

Selected References for Additional Information

Agrios, G. N. *Plant Pathology.* 3rd ed. New York: Academic Press, 1988.

*Autrey, Russell. *Herbicide Use and Safety in the Landscape.* San Luis Obispo, CA: San Luis Video Productions.

*Indicates a video.

*Autrey, Russell. *Pesticide Safety in the Landscape.* San Luis Obispo, CA: San Luis Video Productions.

*Clokey, Joe, and Jim Harrigan. *Integrated Pest Management (Horticulture).* San Luis Obispo, CA: San Luis Video Productions.

Pierce, Pamela K. *Controlling Vegetable Pests.* San Ramon, CA: Ortho Books, 1991.

Pirone, Pascal P. *Diseases and Pests of Ornamental Plants.* 5th ed. New York: Wiley, 1978.

University of Arizona College of Agriculture. *Homeowner's Guide to Outdoor Pesticide Safety.* Tucson: University of Arizona College of Agriculture.

Ware, George W. *Complete Guide to Pest Control: With and Without Chemicals.* 2nd ed. Fresno, CA: Thompson Publications, 1988.

Ware, George W. *The Pesticide Book.* Fresno, CA: Thompson Publications, 1993.

Yepsen, Roger B., Jr. *The Encyclopedia of Natural Insect and Disease Control.* Emmaus, PA: Rodale Press, 1984.

Part III

GROWING PLANTS INDOORS

Introduction

No other branch of horticulture has undergone the rise in popularity that indoor plant growing has during the past three decades. Almost every home now displays at least a few indoor plants, whether they are hanging baskets, large freestanding specimens, or windowsill pots. Indoor plant enthusiasm remains constant because many people enjoy being surrounded by plants in their homes, at work, where they shop, or where they dine. Lush green foliage creates a cool, tranquil feeling and mixes with any decor from modern to traditional. Indoor plants link the outdoor world of nature and the indoor domain of people in a mutual living space.

Many commercial buildings, more now than ever before, are designed with atriums, skylights, and artificial lighting to accommodate the light requirements of plants. Commercial *interiorscaping* or interior landscaping is a full-fledged multibillion dollar industry with a technological base.

Significant scientific advancements have been made during these decades to make houseplant growing more successful for the average person. Specifically, new cultivars have been selected or generated from tissue culture mutations. These new plants are much more durable and better able to survive—even thrive—in less light and less humidity than the original cultivated species. They have also been selected for pest resistance, although this has proven to be a more elusive quality than environmental adaptability. Also, new species never commercially produced or even cultivated until relatively recently have been introduced.

In response to consumer preferences, much research has focused on developing new and more dependable flowering plants, ones with longer lasting flowers in a wider range of colors. A florist pot plant (see Chapter 14) that would have been expected to last two weeks indoors several decades ago now will last months in good condition. People can throw out their Christmas poinsettias on the 4th of July!

Lastly plants have been selected for compactness, which has also been induced by growth retardants. Compactness improves *keeping quality,* the ability of a plant to maintain an attractive appearance over time in unfavorable growing conditions.

Beyond the aesthetic considerations, recent research has shown that plants can improve indoor air quality and presumably human health as well. In the early 1970s the energy crunch hit full force and with it came new building designs that restricted ventilation with outdoor air in order to avoid constant heating or cooling of fresh air. While this did conserve energy, it also led to the creation of so-called *sick buildings* in which people experienced a much greater than normal rate of upper respiratory infections, rashes, headaches, and allergy symptoms than people in older unremodeled buildings with more air exchange. Pollutants in these sick buildings include mold and bacteria; cleaning solvents; tobacco smoke; vapors from inks, paints, and plastic; and fibers from upholstery and carpet. Because of inadequate air exchange, these toxins accumulate to levels that make the air inside these buildings even more polluted than the outside air of a large industrial city. Better ventilation is, of course, the obvious solution, but this ignores the energy problem, so methods of *scrubbing* the air through better filtration and recently through the use of indoor plants are being investigated.

Dr. B. C. Wolverton, principal investigator for NASA, has explored the pollutant-removing capabilities of plants in conjunction with NASA's interest in using plants for air scrubbing in sealed, orbiting space stations. Dr. Wolverton tested about 20 species of potted houseplants for their abilities to remove benzene and trichloroethylene gases from the atmosphere of a sealed chamber. He found that at initial concentrations of 0.119 and 0.435 parts per million, *Spathiphyllum* 'Mauna Loa' removed 80 percent and 23 percent respectively. Although *Spathiphyllum* was the most effective species, all species in the experiment removed some of the gases.

Continuing research focused on the role of potting soil, and particularly the microorganisms it contains, in pollutant absorption. Results showed that the soil/root complex alone was capable of purifying air, removing 20 percent of benzene and 9.2 percent of trichloroethylene. Two years of continuing experiments confirmed that plants and soils continuously exposed to air containing toxic chemicals actually *improve* over time in their ability to clean the air. *Aglaonema* 'Silver Queen' initially removed 47.6 percent of benzene in a given time but after six weeks of intermittent exposure was able to remove 85.8 percent in the same amount of time. This phenomenon was attributed to the well-known genetic adaptability of soil microorganisms to improve their detoxifying potential with exposure to toxic agents.

The plant roots were also acting as detoxifiers. Tests with *Epipremnum aureum* (pothos) in a system using forced-air circulation through the root zone in conjunction with carbon filtration showed that initial benzene and trichloroethylene levels of 0.235 and 0.140 ppm respectively dropped to effectively zero in two hours. Overall the research concluded that the root/soil complex was the most efficient part of indoor plants for removing these typical toxic compounds.

The Indoor Environment

A plant must adapt to many environmental conditions if it is to live indoors. These are light, temperature, humidity, water quality, containerization, and air circulation.

In the final analysis, plants will always be at a disadvantage indoors because human comfort will always be the primary consideration. Temperature will be controlled to the liking of people, as well as light intensity, air circulation, and other factors. The conditions that control the health of plants sharing that living space are considered secondarily.

Light

Probably the most important factor is reduced light. Light intensities comfortable for people to live and work in are insufficient for many species of plants. If plants do not receive sufficient light to manufacture the carbohydrate required for respiration, they will eventually die.

Temperature

Outdoors there is usually a 5 to 15°F (3–8°C) difference between day and night temperatures, with the temperature dropping slowly after sunset. These cooler night temperatures reduce the rate of respiration and prevent the rapid depletion of stored carbohydrate reserves generated during the daylight hours and necessary for growth and maintenance. However, plants grown indoors are often not exposed to cool night temperatures. As a result, their carbohydrate reserves are used more quickly, and the plants have little or none to use for new growth.

Insufficient Humidity

Insufficient humidity is a third factor contributing to the less than ideal conditions indoor plants experience. Home heating creates dry air conditions indoors in most parts of the country. This causes tropical plants to lose water more rapidly by transpiration than they would in their native environment. The plants absorb water to replenish that which has been lost, but often the water is transpired more rapidly than it can be replaced. The leaf tips then turn brown, and the plant develops other symptoms indicative of insufficient humidity.

Water Quality

The quality of the water used on indoor plants is also unlike that which they would receive outdoors. Under natural conditions rainwater supplies moisture. It has relatively few impurities and, generally, a neutral pH (neither acid nor alkaline reaction). But tap water is used indoors. Depending on its source and prior treatment, it may contain numerous chemicals such as chlorine, sodium, or fluoride. In addition, it may be acidic or basic, which will affect the availability of nutrients from the growing medium over time.

Containerization

Plant roots restricted to a pot live under unnatural growing conditions. The nutrients and water available to the plant are limited by the small amount of medium in the pot and must be more carefully regulated than for plants outdoors.

Some physical characteristics of the potting medium also change by being restricted. Soil that is loose and drains quickly outdoors may pack down and drain slowly in a pot. Consequently, garden soil is not often suitable for houseplant use unless it is mixed with other materials (Chapter 15).

Air Circulation

In the outdoor environment air movement is fairly constant due to changing temperatures and natural winds. This circulation moves oxygen and carbon dioxide over the leaves of the plants, aiding in photosynthesis and respiration and drying foliage after rainfall. But indoor air can be relatively stagnant during the winter months when windows remain closed. Any fresh air is likely to come through the opening of outside doors and may be 40°F (22°C) colder than room temperature. Accordingly, plants in the direct path of doors often suffer undiagnosed illnesses from rapid air temperature changes.

Conclusion

With so many modified environmental factors to cope with, it would seem that growing plants indoors would be difficult. But the size and superb condition of many tropical plants growing in homes disprove this. Their owners obviously understand the requirements of plants and meet those requirements. Their skill can be acquired by anyone through reading, observing, and, most importantly, gaining experience in growing many plants.

Selected References for Additional Information

Brown, Ross. "Are Plants the 'Magic Bullet' to Cure Sick Buildings?" In *Interior Landscape Industry,* March 1989, pages 50-55.

Marchant, Brent. "Interim NASA Report Reveals Impact of Roots in Cleaning Air." In *Interior Landscape Industry,* October 1989, pages 12-14.

Wolverton, B. C., Anne Johnson, and Keith Bounds. "Final NASA/ALCA Research Results." In *Interior Landscape Industry,* January 1990, pages 12-16.

Wolverton, B. C., and Douglas L. Willard. "NASA/ALCA Test Update." In *Interior Landscape Industry,* June 1989, pages 8-12.

14

Indoor Plant Maintenance

Purchasing Indoor Plants

With plants, as with any other purchase, it pays to shop around. There are bargains and lemons among indoor plants, and the way a plant looks in the store is not necessarily the way it will look after two or three months under home growing conditions.

One of the biggest disappointments can occur with purchases of flowering plants. Such plants are usually grown in greenhouses under bright light intensities which trigger flower induction. Under indoor growing conditions the light intensity is often not strong enough to cause new buds to set, so after its present buds are exhausted, the plant is not likely to bloom again. Table 14-1 lists some reliable houseplants that bloom indoors without the use of supplemental lights.

However, even though most flowering plants will not rebloom in the house, it is still possible to have color in indoor plants by either purchasing ones with *variegated* foliage (striped or marked with yellow, white, pink, or red) or houseplants currently in bloom that will continue to develop the current crop of buds even though the light level is insufficient to set new ones. These can bloom for several months to the better part of a year and are well worth the investment (see section on florist plants).

When you are considering buying a houseplant, take the time to examine the plant carefully. A well-grown houseplant will be compact and have richly colored foliage, strong new growth, and a full crop of leaves. The stem should be strong enough to easily support the weight of the top and have no weak or discolored areas, especially at the base. An insect-infested plant is never a good buy, so check the growing tips, leaf undersides, and potting medium for hiding pests or signs of their damage.

The price charged for the plant will vary according to where it is purchased, species, cultivar, and,

sometimes, proximity to the large indoor plant-growing areas of Florida and California. Mass merchandising outlets such as supermarkets and discount stores normally have the lowest prices, florists the highest, and local plant shops and greenhouses are in between. However, mass merchandising outlets usually have the least selection. Also, if the plants have been displayed in the store for long, they may have been improperly cared for and be in poor condition.

The price charged for a plant is often a reflection of the ease with which it is propagated and the time required to grow it to selling size. Wandering Jew and coleus, for example, propagate very easily from cuttings and grow rapidly. They are among the least expensive indoor plants. Palms, on the other hand, are grown from seed, and kentia palm seed takes an average of 8 to 12 months just to germinate.

Before transporting a newly purchased plant, take the time to wrap it in paper (folding the leaves upward) or enclose it in a paper bag before taking it outdoors. It will be less likely to break, and in below-freezing weather the paper will act as insulation against the sudden temperature drop.

Then find out the plant's growing requirements for light, watering, humidity, potting medium, and the like. Check a reliable reference, and then try to match the ideal growing conditions as best you can in your home.

Acclimatization and Conditioning

Plant "shock" is a common occurrence in newly acquired houseplants. It usually results from moving plants directly from the production area to home growing conditions without a conditioning period. The sudden change to lower light and humidity disturbs the

Table 14-1 *Selected Houseplants That Flower Without Artificial Lighting*

Common name	Botanical Name
Bromeliad	*Aechmea, Billbergia, Nidularium,* and so on
Begonia	*Begonia* spp.
Spider plant	*Chlorophytum* spp.
Coleus	*Coleus* × *hybridus*
Crocus*	*Crocus* spp.
Orchid	*Cypripedium, Cattleya,* and so on
Crown of thorns	*Euphorbia milii* cultivars
Rose-plaid cactus	*Gymnocalycium* spp.
Amaryllis	*Hippeastrum* hybrids
Hyacinth*	*Hyacinthus orientalis* cultivars
Pincushion cactus	*Mammillaria* spp.
Grape hyacinth*	*Muscari* spp.
Daffodil*	*Narcissus* spp.
Geranium	*Pelargonium* spp.
Swedish ivy	*Plectranthus* spp.
Primroses	*Primula obconica* and *P. malacoides*
Crown cactus	*Rebutia* spp.
Easter cactus	*Rhipsalidopsis* spp.
African violet	*Saintpaulia ionantha* cultivars
Strawberry begonia	*Saxifraga stolonifera*
Christmas cactus	*Schlumbergera bridgesii* cultivars
Thanksgiving cactus	*Schlumbergera truncata* cultivars
Christmas cherry	*Solanum pseudocapsicum*
Spathe flower or Peace lily	*Spathiphyllum* spp.
Cape primrose	*Streptocarpus saxorum* cultivars
Tulip*	*Tulipa* spp.

*These are spring-flowering bulbs that can be *forced* in winter on a windowsill, then planted outside in spring.

metabolism of the plant and temporarily halts growth. Leaf drop may occur, and some plants may even die.

Properly grown houseplants are "conditioned" or "acclimated" prior to sale. The light and humidity are gradually lowered to the levels commonly found in indoor conditions, and watering and fertilization are less frequent. Plants that have not been acclimated often betray the fact by their appearance: they are bright green and extremely lush appearing, with much new growth and dense succulent or soft foliage. Basically, they are extremely healthy and vigorous plants and would continue to grow well in a greenhouse. But they are not able to adjust well to indoor conditions.

It is possible to acclimate a tropical plant that was not conditioned before sale. First, give the plant the maximum amount of light and humidity available and keep the soil moist. This will go as far as possible toward approximating the former growing environment and minimize shock. Then gradually move it further from the light and decrease the humidity and water to the recommended range over several months. This long

period of adjustment will allow the plant to acclimate to an indoor growing environment, but even then some foliage drop should be expected. After acclimatization the plant should grow slowly but satisfactorily provided the correct light level and other requirements are met.

Maintenance Activities

Routine maintenance is an essential part of houseplant care if the original health and beauty of plants are to be preserved. Two major aspects of maintenance, watering and fertilizing, are significant enough to require discussion in individual chapters, but the less commonly practiced maintenance activities discussed here are no less important.

Inspection

Regular inspection for pests and diseases is a part of maintenance that is commonly overlooked, to the decline and ruin of many beautiful plants. Its purpose is much the same as annual checkups for people: to detect and cure minor health problems before they become bigger and more difficult to control. Inspection of houseplants should take place every one to two weeks and should consist of a general search of the plant for any signs of poor health. New growth, in particular, should be carefully scrutinized under bright light for signs of pests such as spider mites (see Chapter 18); a small magnifying glass is helpful for this purpose. Likewise, more mature leaves should be examined both above and underneath, where insects frequently hide. The leaf axils are another favored spot of concealment for pests.

At the same time, the general condition of the plant should be evaluated. If young growth is spindly or many lower leaves are beginning to yellow, the growing conditions should be altered to remedy the problem before the plant deteriorates any further.

The soil surface and roots should be checked for signs of soil-dwelling insects. Possibly toxic fertilizer accumulations (with their characteristic white coating) may be noticeable.

Plant inspection may appear to be time-consuming, but it actually will take only a few minutes. The time will be well spent if an insect infestation can be detected before the pests spread to other plants.

Cleaning

Cleaning is the second aspect of maintenance. Outdoors, rainfall regularly washes the leaves of plants. It removes the dust and dirt that block out light and decrease photosynthesis. But indoors most water is applied directly to the soil, and so the leaves are never cleansed. Periodic washing or dusting is a substitute for natural cleansing by rainfall and should be performed whenever the leaves show an appreciable amount of dust. In addition, it will help remove insects and their eggs, lessening pest problems.

Several methods may be used to clean the leaves, and the one chosen should be determined by personal preference and the size of the plant. Small plants are perhaps most easily cleaned by the spray nozzle found on most kitchen sinks. Excess water will conveniently drain in the sink, and leaves should be sprayed thoroughly on both the tops and the undersides.

A shower is very useful for large-specimen plants. Place the plant in the tub, and adjust the water to approximately room temperature. Then leave the plant under a low-pressure spray for about 5 minutes, turning the plant occasionally to make sure all the leaf surfaces are adequately washed. If there is a particularly stubborn residue on the plant, the leaves can be sponge cleaned with a drop or two of detergent in a bowl of water, but the plant should be thoroughly rinsed afterward.

Many indoor gardeners find that carrying large plants to the shower every month is inconvenient, if not completely impossible, so they use a damp cloth or sponge for cleaning instead. This method is satisfactory provided the leaves are smooth and not easily damaged. It is especially recommended for areas in which the water has an appreciable mineral content. This "hard" water will leave unsightly residue on sprayed or showered plants, so wiping should be substituted.

Dusting is another way of cleaning large-leaved plants. Only a very soft duster, such as a feather or lamb's wool duster, should be used.

Leaf Shines

Many commercial "plant shines" or "leaf cleansers" are now available in aerosol sprays, fiber-tipped applicators, and foil-encased towelettes. All claim to improve the appearance of foliage and enhance the looks of plants. Most garden books, though, contend that they are bad for the health of the plants because they cause "pore" clogging. Neither is completely correct.

Most plant shines and leaf cleaners will not harm houseplants. However, a few will, especially on young foliage, and this is most often attributable to an aerosol propellant. The pore-clogging claim is not valid: first, because the "pores" (technically stomata; see Chapter 2) are not like human pores and do not respond in the same manner to clogging. In fact, commercial nurseries often use chemical sprays called *antidesiccants* or *antitranspirants* during transplanting. These antidesiccants are solely for the purpose of sealing the stomata

and preventing moisture loss. Second, most stomata are found on the undersurfaces of leaves, where leaf shine is not applied. Consequently, the majority will not be affected.

Nor are leaf shines or cleaners necessarily the best way to cleanse plant leaves. Their chemical makeup varies greatly among the different brands. Some leave an artificial-looking gloss and an oily residue that attracts dust, while others leave no residue and only a slight shine. People who prefer to use a leaf shine or cleaner should compare several brands before deciding.

Homemade leaf shines such as milk or salad oil were commonly used in the past. Salad oil cannot be recommended because of the resulting greasy film. Milk does not have this objectionable characteristic and is safe and inexpensive.

The hair-covered leaves of tropicals such as purple velvet plant (*Gynura aurantiaca*) and panda plant (*Kalanchoe tomentosa*) can be either cleaned with a spray of water or dusted with a soft paintbrush. Leaf shines should not be used, and neither should hand wiping, which can flatten or break off the hairs. When rinsing the hairy-leaved members of the African violet family, always use water at room temperature. Gesneriads are extremely sensitive to temperature changes. Water more than 10°F (5°C) cooler than room air temperature will cause yellow splotching on the leaves wherever it touches. The discoloration that results is caused from the destruction of leaf chlorophyll and is permanent (see Figure 17-9).

Whereas some indoor tropicals have hair-covered leaves, others possess a natural filmlike coating that should not be rubbed or covered over with leaf shine. Bromeliads are one example. The foliage of many bromeliad species is covered with flaky silverish scales (Figure 14-1). The scales are decorative, but their function is highly practical. They are cells for absorbing water directly through the leaves.

Cacti and succulents frequently have a waxy cuticle that rubs off easily with the fingers. The coating serves as an insulation layer against moisture loss and is an adaptation factor that helps them survive in arid environments. As a general rule, leaf polishes or cleansers should not be used on cacti and succulents.

Turning

Normally plants indoors receive their light from a nearby window. The leaves facing the light receive more solar energy causing the plants to develop more in the direction of the light. To prevent plants from becoming lopsided, it is advisable to turn them a quarter turn every few weeks so that a different section of foliage faces the light.

Figure 14-1 White bands of water-absorbing scales on *Aechmea* bromeliad. (*Photo by Rick Smith*)

Pruning

Pruning is commonly regarded as an outdoor plant maintenance chore. Consequently indoor plant pruning is often neglected, resulting in many poor-looking houseplants.

Pruning has several functions. First, it causes branching and new growth, producing a fuller, more attractive plant. Second, it removes dead, insect-infested, or diseased plant parts to halt the spread of the organisms.

Pinching

The most commonly practiced form of pruning for indoor plants is a technique called *pinching* (Figure 14-2). Pinching consists of breaking off up to about 3 inches (8 centimeters) of the growing tips.

The meristems of plants are known to produce a number of chemicals, including a plant hormone called *auxin* (see Chapter 3). This hormone constantly moves down the stem of the plant from its source of manufacture. It inhibits growth of buds lower on the stem and prevents branching. This hormonal effect on growth is termed *apical dominance,* indicating that the top bud (the apical meristem) "dominates" the growth of the buds below it. Buds directly below the growing point are kept fully dormant, since the auxin is in strongest concentration there. As auxin continues moving down the stem, the concentration decreases. The lowest buds on the stem receive auxin in least concentration and sometimes overcome dormancy and grow into shoots.

Auxin manufacturing rates and movement vary between species, accounting for why some plants branch naturally when others do not. Pinching removes the source of auxin manufacture on plants that will not branch of their own accord.

When a growing tip is removed, it is nearly always the buds nearest the tip that begin growing (Figure 14-2). Although sometimes only one will grow, usually two to three will surge into growth at the same time, producing several new leafy shoots. Since they are also growing tips, they soon begin auxin manufacture, keeping the buds below dormant, and the cycle continues.

Pinching must be done regularly on some species to promote constant branching and to develop a full plant. As new shoots lengthen, the tips are removed continually to make the plant full and dense.

Leafy, vining houseplants most often require regular pinching to prevent a "stringy" appearance. Ideally they should be pinched frequently after potting to develop a large number of cascading stems. The most common vining and upright-growing houseplants requiring frequent pinching are listed in Table 14-2.

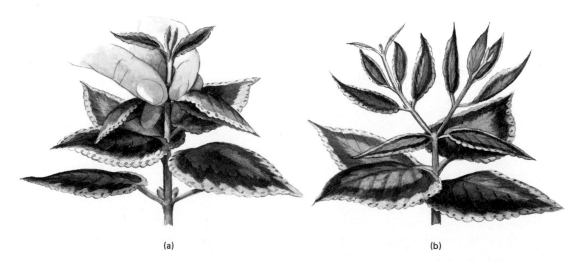

(a)   (b)

Figure 14-2 (a) Pinching a coleus. (b) One month later after branching has occurred.

Table 14-2 *Most Common Houseplants Requiring Frequent Pruning*

Common name	Botanical name
Coleus	*Coleus* × *hybridus*
Goldfish vine	*Columnea* spp.
Tahitian bridal veil	*Gibasis geniculata*
Purple passion vine	*Gynura aurantiaca* 'Purple Passion'
Pocketbook plant	*Hypocyrta* spp.
Polka-dot plant	*Hypoestes phyllostachya*
Bloodleaf	*Iresine* spp.
Aluminum plant	*Pilea cardierei*
Swedish ivy	*Plectranthus* spp.
Christmas cherry	*Solanum pseudocapsicum*
Wandering Jew	*Tradescantia* or *Zebrina* spp.

Renewal Pruning

Renewal pruning is used to rejuvenate a plant that has become unattractive and bare-stemmed through drought, insufficient light, or some other cause. It involves removing all or nearly all of the foliage, leaving only 1- to 2-inch (3- to 5-centimeter) stems (Figure 14-3). In several weeks new growth will begin from the leafless stubs. With regular pinching *and* improved growing conditions, the plant will become attractive again in a few months.

Renewal pruning is a drastic measure and should be employed only if the plant is basically healthy, with a strong root system and stored carbohydrate reserves. For plants in a severely weakened condition, a different treatment should be used. First, the source of the problem should be corrected. This will encourage healthy growth and will often be sufficient to break the dor-

mancy of lower buds. After the plant is partially refoliated, the remaining unattractive portions can be cut back. By allowing some foliage to remain, further carbohydrate depletion will be prevented.

Pruning by Thinning

Not all indoor plants are pruned by pinching. Many houseplants such as spider plants grow in a form called a rosette (see Chapter 2) and produce new foliage from a short section of stem at soil level called the crown. Others, such as most ferns, send up new leaves from an underground *rhizome* system (underground stems) and botanically speaking have no "stems" aboveground at all, only leaf petioles. Each new shoot that develops is a single leaf, and the dormant buds are found at the crown at or just below soil level. Pinching a leaf back will never cause branching. Instead, it should be cut off at the base to promote new growth from the crown.

In pruning these types of houseplants, usually only a few leaves are removed at a time as needed to improve the appearance of the plant. However, in cases in which the entire plant is unsightly or heavily infested by insects, renewal pruning may be advisable. As with renewal pruning of houseplants having stems, all the foliage is cut off just above ground level, and the plant is allowed to refoliate under improved growing conditions and care.

Plants That Seldom Require Pruning

A moderate percentage of houseplants seldom require any pruning except removal of dead leaves. Pinching or renewal pruning of these plants will often destroy their

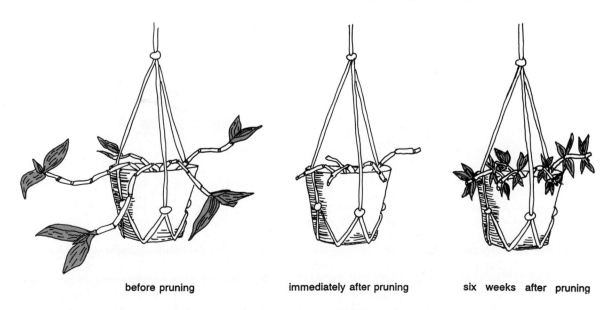

before pruning   immediately after pruning   six weeks after pruning

Figure 14-3 Stages in renewal pruning.

Table 14-3 *Selected Plants Requiring Minimal or
No Pruning*

Common name	Botanical name
African violet	*Saintpaulia*
Bromeliad	*Aechmea, Neoregelia, Vriesia* etc.
Cactus	*Cereus, Gymnocalycium,* *Mammillaria* etc.
Chinese evergreen	*Aglaonema commutatum* cultivars
Dracaena	*Dracaena*
Dumbcane	*Dieffenbachia* spp.
Norfolk Island pine	*Araucaria heterophylla*
Orchid	*Cattleya, Paphiopedilum*
Palm	*Chamadorea, Howiea, Phoenix* etc.
Rosette-type succulents	*Agave, Aloe, Haworthia* etc.
Snake plant	*Sansevieria* spp.
Spathe flower	*Spathiphyllum*
Umbrella plant	*Brassaia*

form and create an unnatural appearance. Selected houseplants that require minimal pruning maintenance are listed in Table 14-3.

Pruning Flowering Houseplants

Flowering houseplants follow the basic rules for pruning which apply to other houseplants. For many, however, the flowering period can be lengthened by removing the flower heads and seed pods as soon as the blooms fade. If the seed pods are not removed, energy will be channeled into seed production rather than into the formation of additional flower buds.

Exceptions to this practice are plants in which the fruit is an ornamental feature, such as Christmas cherry. Blooms should also be left if one is trying to raise seeds to grow additional plants. Table 14-4 lists plants from which blooms should not be removed.

Pruning for Disease and Insect Control

Pruning for the control of diseases and insects is not a common practice because in most cases the entire plant is infected before the problem is discovered. However,

these problems often do start on one or two leaves or a shoot tip. Quick removal of the unhealthy parts can be preferable to a series of chemical sprays. Pruning is particularly valuable in dealing with houseplant foliage diseases because chemical cures are rarely available. When pruning for this purpose, all diseased portions should be cut back to healthy areas.

The clipping of browned leaf tips is commonly used to improve the appearance of houseplants such as dracaenas and palms. Although temporarily effective, the leaf tips usually brown again within weeks.

Transplanting and Repotting

Transplanting (or repotting, as it is sometimes called) is a routine part of indoor plant maintenance. It is performed repeatedly from the seedling or cutting stage through maturity.

Primarily it is to provide crowded roots with additional potting medium from which to absorb water and nutrients. Less often, transplanting is used as a general tonic or cure-all for an undiagnosed plant ailment suspected of being caused by the roots or the potting medium.

Most people repot more often than necessary, switching to bigger pots when frequent fertilization would have been sufficient. Although unnecessary repotting is seldom harmful, it is a considerable amount of extra work.

Timing of Repotting

Several factors must be evaluated to determine if a plant should be moved to a larger container. First, does the plant look top-heavy in its pot and tip over easily? A plant with a large foliage mass will grow best if given an adequate volume of potting medium to support that top growth. Keeping it in an undersized pot will stunt its growth.

Next, does the plant require frequent watering to keep the soil moist, and does it wilt within a few days of watering? Roots that are overcrowded extract avail-

Table 14-4 *Plants from Which Faded Flowers and Fruits Should Not Be Removed*

Common name	Botanical name	Remarks
Bromeliad	*Aechmea, Nidularium,* and other genera	Colorful fruit often follows blossoming
Asparagus fern	*Asparagus* spp.	Ornamental red berries follow bloom
Spider plant	*Chlorophytum* spp.	New plants form on flower stalks
Cactus	*Mammillaria, Rebutia,* and other genera	Red or orange fruits frequently follow blossoming
Christmas cherry	*Solanum pseudocapsicum*	Red fruits form from white blooms
Chinese evergreen	*Aglaonema commutatum* cultivars	Flowers followed by red berries; seldom flowers indoors
Wax plant	*Hoya carnosa* cultivars	Next season's blooms from old flower stubs

able water and nutrients very quickly, necessitating more frequent watering and fertilization.

Third, is the health of the plant declining with no change in care or growing location? Some houseplant media pack and crust on the surface after a period of months or years. Repotting in fresh media may be advisable.

In extreme cases of underpotting, roots will be found growing on the surface of the medium. Roots may also grow vigorously at the base of the pot, pushing the plant out of the container. This phenomenon is occasionally encountered in dracaenas and spider plants.

Some indoor gardeners use the presence of roots growing from the drainage hole or circling the root ball as an indicator for repotting. This method is not always reliable. Roots grow most vigorously in areas that hold both moisture and air. The soil near drainage holes and around the inside surface of unglazed clay pots provides these conditions most ideally. Accordingly, roots will grow there in abundance, while the bulk of the potting medium is undisturbed. However, if large numbers of roots have grown in circles, repotting is justified because the bulk of the root system is not in contact with the potting medium.

The season chosen for repotting indoor plants is not important. Most plants go through seasonal dormancy and growth spurts like outdoor plants, and transplanting during either of these periods is acceptable. Most transplanting is performed in spring or summer because the increased vigor due to increased light and humidity makes plants outgrow their pots quickly during these periods.

Transplanting of blooming plants should be restricted to a period when they are not flowering, since it is common for a plant to drop buds and flowers prematurely when the roots are disturbed. Repotting blooming plants immediately after flowering is a reliable practice. It provides room for new root growth and, consequently, better foliage growth before the next flowering period.

Unpotting

The amount of difficulty in removing a plant from its pot depends on the container and plant involved. Plants in 8-inch (20-centimeter) diameter or smaller-sized containers can usually be unpotted by a technique known as *knocking out*. To do this the medium should be moist (but not freshly watered) to keep it from falling away from the roots. Supporting the top of the soil with one hand and the plant stem between the fingers, the pot is turned upside down, and the edge lightly tapped on a counter or table (Figure 14-4). The soil and roots should slide out intact. If they do not,

Figure 14-4   Knocking a spider plant out of its pot. (*Photo by Rick Smith*)

harder tapping, shaking, or running a knife around the rootball may be necessary. Only in very difficult cases should a plant be loosened from its pot by pulling on the stems. Root or stem damage is likely to result.

Unpotting a plant that has been growing in a narrow-mouthed pot is always difficult. A portion of the roots will nearly always be lost when the root ball is pulled through the narrow opening. As with knocking out, the potting medium should be moist and the plant dug and eased through the narrow opening. Laying the pot on its side is often an aid in easing the roots out. The medium will fall to the bottom side, and a tool can be inserted in the space above to pull the roots free.

Flooding is another way to unpot a plant from a narrow-mouthed container. The pot should be submerged in water and a stream of water directed at the mouth to loosen the medium and flood it from the roots. Eventually enough will flood away to allow the roots through the opening with minimal damage. Because unpotting by flooding causes less root damage, it is valuable for delicate or highly prized specimens.

Unpotting a large plant is often a two-person job. Since it is too large to be held upside down and knocked out, the usual procedure is to lay the plant on its side and ease the roots out horizontally. Foliage breakage can be minimized by folding the leaves upward and wrapping the top of the plant snugly with newspaper tied in place with string.

A good time to inspect the roots is after removing a plant from its container. Healthy young roots are firm and usually white. (A few are other colors, such as the orange of *Dracaena* roots.) But they generally should not be brown or mushy. Such roots are dead and decaying and should be trimmed off before the plant is repotted. Soil insects may also be detected in a repotting operation.

Roots that have circled the root ball should have special treatment at repotting. If repotted in that condition, they may continue to wrap around the soil ball, never branching into the fresh soil and its available nutrients. Wrapped roots may be loosened by pulling them away with the fingers (Figure 14-5) or by cutting several places around the root ball with a knife. This, like pinching, causes branching and the development of new *feeder roots*.

Selection of the New Pot

When selecting a new pot for transplanting, both size and drainage should be considered. The new pot should preferably be only 1 to 2 inches (3-5 centimeters) deeper and wider than the previous pot. This additional area will hold an adequate volume of fresh potting medium for new root growth. If a larger pot is chosen, it is likely that a large portion of the new potting medium will stay waterlogged because the roots have not grown through it.

Drainage is a characteristic of the potting medium, but where drainage water is deposited is determined by the pot. A container with a drainage hole or holes allows excess water to pass out of the pot. In a pot without holes, water filters through the potting medium and excess eventually collects in the bottom. This excess water keeps the medium at the base of the pot saturated, excluding oxygen necessary for root respiration. For this reason, pots with holes are preferable unless watering is carefully regulated.

A water reservoir can be created in holeless pots to keep drainage water away from the roots. Stones or gravel at least a half-inch (1 centimeter) in diameter should be poured in a 1-inch (3-centimeter) or deeper layer in the bottom of the pot. The layer is then covered with nylon net, a sheet of sphagnum moss, or screen to keep out potting medium and leave air spaces open for water. In repotting large specimens, consideration should be given to the weight that gravel adds. In such cases styrofoam "peanuts" packing material can be substituted.

It is a common misunderstanding that gravel in the bottom of any pot, having holes or not, will im-

Figure 14-5 Loosening the roots of a pot-bound plant. (*Photo by Kirk Zirion*)

prove the drainage. The gravel layer does *not* improve the drainage. Actually it worsens it by decreasing the depth of soil in the pot because drainage is a function of both the depth of the pot and of the components that make up the medium. (Medium components' effects on drainage are discussed in Chapter 15 and the physics of water drainage based on pot height are discussed in Chapter 17.) A gravel layer only provides a reservoir in pots from which the water has no way of escaping. Gravel is unnecessary for pots with holes. The only improvement these pots need is a single piece of broken clay pot (round side up) or screen over the drain hole to keep in the potting medium.

Repotting

Repotting is a simple process. The container is filled with potting medium to a level that will permit the plant to be growing as it was previously.

A plant should never be set deeper than it was originally growing with the idea of improving its appearance. This exposes the stems to soil fungi and bacteria which can invade the tissues and kill the plant.

After firming the medium lightly, the plant should be watered to settle out air spaces and lessen shock. Watering may cause the medium to settle somewhat, and it can then be refilled around the top. Enough room should be left to make watering convenient. A space of ½ to 1 inch (2–3 centimeters) is sufficient.

Double Potting

Double potting (Figure 14-6) involves inserting a clay or plastic pot with drainage holes inside a holeless decorative pot. The spaces between the two pots are filled with sphagnum moss, gravel, or other material.

Potting by this method provides the advantages of a free-draining pot with the beauty of a decorative one.

Posttransplanting Care

Sometimes plants *shock* after transplanting. They wilt and may remain wilted for days or weeks. To lessen shock, keep newly transplanted plants out of direct sunlight for a few days. Delicate plants can also be enclosed in a plastic bag for a week or two following transplanting. This increases the relative humidity while the roots are reestablishing and regaining their normal absorption potential.

In the rare cases where bagging is ineffective and the plant remains wilted, pruning may be necessary. One-quarter of the foliage should be cut off and the plant should be observed for several days. If it remains wilted, more foliage will need to be removed.

— filler material

Figure 14-6 Double potting.

Plants with Special Requirements

Bonsai

Bonsai is the art of growing and training trees to be miniature versions of the form they achieve in nature (Figure 14-7). It originated before the eleventh century in China, and some bonsai are hundreds of years old. However, a bonsai need not be old to have the appearance of age. Instead, aging is accomplished by pruning, training, and restricting the soil volume and fertilizer available to the roots.

Bonsai are not difficult to create and care for, but they do require a larger investment of time than many other houseplants. The first step in creating a bonsai is to select the plant. Visit a local nursery and choose an easy-to-grow plant such as pyracantha, pine, or juniper in a 1-gallon can. A large, bushy plant is not necessarily the best choice. Instead, purchase a plant with an irregular form, less foliage, and even some dead wood, since this will help create an aged appearance.

The next step is to select the front or primary viewing angle of the bonsai, the angle from which it is most attractively viewed. If the trunk has a natural curve, the viewing angle should emphasize it. Next the two lowest major branches are located and all branches growing below them removed flush with the trunk. The ultimate height of the bonsai is then decided, the top removed to that point, and the two lowest branches shortened to be one-half the length of the trunk.

Next the upper branches are chosen. They should alternate up the trunk and not necessarily be the strongest branches. Smaller, thinner branches may appear more natural. The goal is to space the branches evenly both up and around the trunk and to trim them gradually shorter as they progress upward to mimic branch growth in nature.

Wiring can be done at the same time as pruning. Copper wire of a 10 to 20 gauge is wrapped in spiral fashion around the trunk and branches, which can then be bent to the desired shape. The wiring can be removed in several months, and the tree will retain the new form. Bottommost branches are preferably wired to point either horizontally or downward and branches progressing up the trunk point sequentially at slightly more vertical angles (Figure 14-7).

The next step is to remove the bonsai from its pot and examine the roots. If there are predominately large (anchorage) roots and few fibrous (absorptive) roots, the root ball will need to be pruned to fit a training pot half the depth of the present container and grow in this pot for several months to develop fibrous roots. If not, and it is immediately planted in a very shallow bonsai

(a) the plant in its
original state
     (b) after pruning and
training the top
     (c) after potting

Figure 14-7 Steps in bonsai making. (*Photo from* Bonsai: The Art and Technique *by D. Young, 1985. Reprinted by permission of Prentice Hall, Englewood Cliffs, NJ*)

pot, the shock of root removal may be too great and might kill the plant.

The final container will need to have its drainage holes covered with screen wired into place, and if it is suspected that the bonsai will need to be wired through the roots to be stable in its pot, an additional long strand of wire can be secured through the drainage hole onto a short peg with the ends of the wire left hanging over the edge of the pot. The root ball should then be flattened, and long, thick roots pruned off to make a root ball that will fit the new pot with one inch (2–3 cm) of clearance on the sides and the bottom. Normally a bonsai is potted slightly off center, and this must be taken into account when shaping the root ball. After filling the pot to 1 inch depth (2–3 cm) with potting medium (a sterilized mix of equal parts soil, sphagnum peat moss, and coarse sand), the bonsai can be placed in the pot and checked for height. At this point the trunk base should be uncovered to show the point of origin of the first major root. This should be positioned to be slightly above the top of the container with the bulk of the root ball sloping downward to the lip. At this point wiring the tree to the pot for stability can be done if necessary by crossing the wires over the root ball and twisting them taut to form a knot near the trunk. The wire and knot should be covered with moss, which is laid over the entire root area as a final touch.

Watering of bonsai will depend on the species. Because of the limited soil volume, bonsai dry out relatively quickly. They must be watered frequently to prevent death from drought but should not be kept constantly moist since this would encourage undesired, vigorous growth. Watering every day during warm weather is not unusual.

Fertilizer for bonsai should be in liquid or water-dissolving form (see Chapter 15) applied from spring through fall. A dilute solution applied once per month is usually adequate, since the objective is to maintain the plant at its present size, not to encourage growth.

Bromeliads

Members of the family Bromeliaceae (Figure 14-8) are termed *bromeliads*. In nature they are tree-dwelling (epiphytic) plants but they are not parasites as they use the trees only for anchorage. As houseplants they are dependable and easy to care for. All are valued for their foliage, and many have long-lasting flowers and colorful fruits.

Bromeliads survive under a wide range of light levels from bright to dim, but the foliage will show the best coloration under the proper light level for the species. As a general rule, plants in the genera

Figure 14-8 *Neoregelis carolinea* 'Tricolor,' a popular bromeliad for indoor growing. The leaves are banded green and white. At flowering, the center cup becomes red. (*Photo courtesy of Marge Coon, Vista, CA*)

Ananas, Dyckia, Orthophytum, and *Tillandsia* are bright-light-requiring. *Aechmea, Neoregelia,* and *Billbergia* are moderate- to bright-light-requiring, and *Cryptanthus, Guzmania, Nidularium,* and *Vriesea* are moderate- to-dim-light-requiring.

The water requirements of bromeliads differ from those of other houseplants. Potted types use soil moisture but also frequently have a center cup formed by the leaves which should be kept half (not completely) full of water. Both potted and plaque-mounted species absorb water efficiently through their foliage, and misting is beneficial. Since misting is the only water source for plaque-mounted bromeliads, daily spraying is advisable when there is little humidity in the air.

Bromeliads should be fertilized with a water-soluble fertilizer mixed at the strength recommended for foliar application. The fertilizer should be applied to the medium of potted specimens and also to the cup (Figure 14-8). Misting of both potted and mounted species with the solution is recommended.

Potted bromeliads generally have small root systems and require repotting infrequently. Use the Cornell epiphytic mix discussed in Chapter 15 or another similar light-weight medium. Table 14-5 is a list of bromeliads particularly suitable for growing as houseplants.

Cacti and Succulents

Confusion persists about the difference between cacti and succulents. Basically a cactus is a member of the *Cactaceae* family and a succulent is any plant that is fleshy and contains a large quantity of water in its leaves or stems. Almost all cacti are succulents; however, obviously, not all succulents are cacti. Many other genera contain succulent plants.

The presence of thorns or spines does not necessarily classify a plant as a cactus. Many succulents have spine-like organs, and only careful examination and a fairly detailed knowledge of plant anatomy will enable one to classify a plant as a member of the *Cactaceae* with certainty.

Cacti (Figure 14-9) are classified as either *xerophytic* (adapted to dry conditions) or *epiphytic* (tree dwelling in tropical forests). The former includes all desert cacti and all commonly grown succulents. The epiphytic cacti include Christmas cactus, Easter cactus, orchid cactus, and related species. The care for the two types differs greatly. Xerophytic cacti require bright light, a coarse, fast-draining potting medium, and infrequent watering for most of the year. The epiphytes grow best in moderate light and a richly organic and constantly moist potting medium.

Table 14-5 *Selected Bromeliads for Indoor Growing*

Common name	Botanical name
Burgundy vase plant	*Aechmea* 'Burgundy'
Vase plant	*Aechmea maculata*
Urn plant	*Aechmea fasciata*
No common name	*Aechmea gamosepala*
No common name	*Aechmea maculata*
Patriotic plant	*Aechmea mertensii*
Pineapple	*Ananas comosus*
Queen's tears	*Billbergia nutans*
Muriel Waterman	*Billbergia* 'Muriel Waterman'
Earth star	*Cryptanthus* cultivars
Cryptbergia	*Cryptbergia rubra*
Orange star	*Guzmania lingulata*
Flasklike neoregelia	*Neoregelia ampullacea*
Blushing bromeliad	*Neoregelia carolinae*
Fireball	*Neoregelia* 'Fireball'
Black Amazon bird nest	*Nidularium innocentii* 'Innocentii'
No common name	*Nidularium regelioides*
Pink quill	*Tillandsia cyanea*
Blushing bride	*Tillandsia ionantha*
No common name	*Tillandsia xerographica*
Flaming sword	*Vriesea splendens*
Sword plant	*Vriesea* hybrids

Improper watering is a major cause of death in cacti and succulents. Xerophytic species should be watered infrequently from fall through spring, their dormant period. The medium should be allowed to dry out thoroughly between waterings, and if possible, they should be kept at a cool 45 to 65°F (7–18°C) temperature. As soon as new growth begins, the watering frequency should be increased so that the medium is rewet as soon as it approaches dryness.

Because of their sensitivity to overwatering, xerophytic cacti should be transplanted only when actively growing in spring or summer. They should be repotted in a dampened media in lieu of postpotting watering.

Transplanting spiny cacti and succulents can be difficult; however, they require repotting only every two to four years. The process can be made easier by handling the plants with a band of folded newspaper, as shown in Figure 14-10.

Table 14-6 lists cacti and succulents suitable for growing as houseplants. When purchasing xerophytic cacti, beware of large cacti at bargain prices. They are often dug from the desert and their roots severed. Such plants are difficult to reroot and usually die over several months. They may also be endangered protected species.

Ferns

Most ferns (Figure 14-11) are moderate- to low-light-requiring houseplants. During winter, if light intensities are poor, they can be placed directly in an east, north, or west window. But in summer they should be either moved back from east and west windows or given protection by a sheer curtain.

With the exception of a few genera such as *Pellaea* and *Cyrtomium,* ferns will not survive complete drying of the potting medium. The medium should contain some moisture at all times. A quick-draining but moisture-retentive potting medium is best for root growth. Most prepackaged houseplant media are suitable. Humidity level is more critical for ferns than other houseplants, and a 30–60 percent relative humidity range is necessary for the healthy growth of most species. Table 14-7 lists selected ferns recommended for indoor growing.

Figure 14-9 A group of succulents. (*Photo by Rick Smith*)

Figure 14-10 A folded newspaper aids in transplanting a cactus. (*Photo by Rick Smith*)

Table 14-6 *Selected Cacti and Succulents for Indoor Growing*

Common name	Botanical name
Cacti	
Sea urchin cactus	*Astrophytum asterias*
Bishop's cap	*Astrophytum ornatum*
Silver torch	*Chamaecereus silvestri*
Peanut cactus	*Cleistocactus strausii*
Golden barrel cactus	*Enchinocactus grusonii*
Rainbow cactus	*Echinocereus pectinatus* var. *neomexicanus*
Orchid cactus	*Epiphyllum* hybrids
Fire barrel	*Ferocactus acanthodes*
Rose-plaid cactus	*Gymnocalycium mihanovichii* cv. *friedrichii*
Agave cactus	*Leuchtenbergia principis*
Pincushion cactus	*Mammillaria* spp.
Ball cactus	*Notocactus* spp.
Bunny ears	*Opuntia microdasys*
Crown cactus	*Rebutia* spp.
Easter cactus and others	*Rhipsalidopsis* hybrids
Christmas cactus and others	*Schlumbergera* hybrids
Succulents other than cacti	
Agave	*Agave angustifolia*
Aloe	*Aloe* spp.
Ponytail	*Beaucarnea recurvata*
Jade plant	*Crassula argentea* cultivars
Watch chain	*Crassula lycopodioides*
String of buttons	*Crassula perforata*
Hen and chickens	*Echeveria* hybrids, *Sempervivum* spp.
Crown of thorns, pencil plant, and others	*Euphorbia* spp.
Ox tongue	*Gasteria* spp.
Wart plant	*Haworthia* spp.
Wax plant	*Hoya carnosa* cultivars
Panda plant, air plant, and others	*Kalanchoe* spp.
Mother-in-law's tongue/ Snake plant	*Sansevieria trifasciata* cultivars
Burro tail and others	*Sedum* spp
Wax ivy, string of beads, and others	*Senecio* spp.

Florist Plants

The species traditionally called florists plants (Figure 14-12) such as chrysanthemum, poinsettia, and forced bulbs are handsome, long-lasting additions to the indoor environment. Recent improvements in keeping quality have greatly extended the anticipated indoor life expectancies of many species. In addition, the range of species used for indoor color and their flower colors have also increased dramatically in recent years. In pre-

Figure 14-11 A cultivar of Boston fern. (*Photo by Andrew Schweitzer*)

vious years selection was limited to perhaps a dozen major species and a plant was expected to last only a few weeks; now fifty or more popular species are available and most will last from one to several months indoors with minimal care.

The life of the flowers can be extended considerably with correct care in the home. The plant should be placed in bright light but preferably out of direct sunlight and drafts. The potting medium should be kept moist.

Table 14-7 *Selected Ferns for Indoor Growing*

Common name	Botanical name
Maidenhair fern	*Adiantum raddianum*
Maidenhair fern	*Adiantum tenerum*
Bird's nest fern	*Asplenium nidus*
Mother fern	*Asplenium bulbiferum*
Holly fern	*Cyrtomium falcatum* cultivars
Boston fern	*Nephrolepis exaltata* cultivars
Boston fern	*Nephrolepis duffii*
Button fern	*Pellaea rotundifolia*
Green cliff brake	*Pellaea viridis*
Hart's tongue	*Phyllitis scolopendrium*
Common staghorn fern	*Platycerium bifurcatum*
Rabbit's foot fern	*Polypodium aureum* cv. *areolatum*
Rabbit's foot fern	*Polypodium aureum* 'Undulatum'
Climbing bird's nest fern	*Polypodium punctatum*
Holly fern	*Polystichum tsus-simense*
Cretan bracken	*Pteris cretica* cultivars
Australian bracken	*Pteris tremula*
Ladder bracken	*Pteris vittata*
Tongue fern	*Pyrrosia lingua*
Leatherleaf fern	*Rumohra adiantiformis*

Figure 14-12   A selection of florist flowering plants, clockwise from top: Kalanchoe, gerbera daisy, exacum, cyclamen, and chrysanthemum. (*Photo by Andrew Schweitzer*)

Gesneriads

The family Gesneriaceae includes many outstanding houseplants such has episcias, African violets, and gloxinias. Many popular flowering houseplants are found in this family (Table 14-8).

Table 14-8   *Selected Gesneriads for Indoor Growing*

Common name	Botanical name
Columnea	*Columnea* 'Chanticleer'
	Columnea 'Early Bird'
	Columnea 'Mary Ann'
Flame violet	*Episcia* 'Moss Agate'
	Episcia 'Acajou'
	Episcia 'Cygnet'
	Episcia 'Jinny Elbert'
Nautilocalyx	*Nautilocalyx forgettii*
Candy corn plant	*Nematanthus wettsteinii*
African violet	*Saintpaulia ionantha* cultivars
Slipper plant	*Sinningia pusilla* cultivars
Cape primrose	*Streptocarpus saxorum*
	Streptocarpus 'Wiesmoor hybrids'
	Streptocarpus 'Constant Nymphs'
	Streptocarpus 'Diana'
	Streptocarpus 'Fiona'
	Streptocarpus 'Karen'
	Streptocarpus 'Marie'
	Streptocarpus 'Louise'
	Streptocarpus 'Paula'
	Streptocarpus 'Tina'
	Streptocarpus 'Margaret'

Most gesneriads are of tropical origin and should be grown at temperatures from 55 to 80°F (13–27°C). Nighttime temperatures in the lower half of this range and daytime temperatures in the upper half will provide the best conditions for growth. Light should be moderate to bright to achieve flowering, but long periods of strong sunlight in summer should be avoided. Less intense winter sunlight is acceptable. Gesneriads are also known to grow well with artificial lighting as the only light source.

The potting medium of gesneriads should remain slightly moist at all times. All water applied should be at room temperature since cold water can cause permanent leaf spotting.

The humidity range most acceptable for gesneriad growing is 50-80 percent. They can be successfully grown in less humidity provided other growing conditions are met satisfactorily. Because of the humidity desirable for raising gesneriads, they are particularly suited to terrarium growing.

Hanging Plants

The care of hanging plants differs from the care of those set on tables, windowsills, or the floor for several reasons. First, the air temperature is usually warmer around hanging plants. Heat will rise to the top of a room, and consequently the air surrounding a plant hung at eye level may be 5°F (3°C) warmer than the air at floor level. There is also greater air movement around a suspended plant since there is no supporting surface to block air flow.

Because of these environmental differences, hanging plants lose moisture from both the medium and the leaves at an accelerated rate. More frequent watering will thus be required. Since many hanging plants are drought-sensitive ferns, it is important to check these plants frequently for moisture. Table 14-9 lists a number of excellent houseplants for hanging baskets.

Orchids

Although many orchids are epiphytes, a number are terrestrials. Selected species of both types grow well as houseplants (Table 14-10). The anatomy and growth pattern of many orchids are unusual. Monopodial orchids (Figure 14-13) have a rosette form, or leaves only on opposite sides of the stem, giving them a two-dimensional appearance. *Sympodial orchids* (Figure 14-13) have storage organs called *pseudobulbs*, with each pseudobulb bearing one leaf. These orchids grow by extension of rhizomes and generally gain one new pseudobulb and one new leaf per year.

Orchids (Figure 14-14) fall into the following three growing-temperature categories:

Cool	(45–58°F)	(7–14°C)
Intermediate	(55–70°F)	(13–21°C)
Warm	(62–80°F)	(17–27°C)

However, as an all-round temperature range, 56 to 62°F (13–17°C) night temperature and 62 to 80°F (17–27°C) day temperature are acceptable.

Epiphytic orchids are potted in osmunda fiber, bark, or similar lightweight potting medium. Watering should remoisten the potting medium just before it completely dries. Since water drains through these loose media freely, care must be taken to ensure that the entire medium is wet. Watering thoroughly with a sink sprayer will accomplish this, as will submersion watering (Chapter 17).

Terrestrial orchids may be grown in an all-purpose potting medium suitable for foliage plants.

Table 14-9 *Selected Plants Suitable as Hanging Baskets*

Common name	Botanical name
Lipstick vine	*Aeschynanthus* spp.
Rattail cactus	*Aporocactus flagelliformis*
Asparagus fern	*Asparagus* spp.
Begonia	*Begonia* spp.
Rosary vine	*Ceropegia woodii*
Spider plant	*Chlorophytum* spp.
Coleus	*Coleus × hybridus*
Cissus vine	*Cissus* spp.
Goldfish vine	*Columnea* spp.
Episcia	*Episcia* spp.
Creeping fig	*Ficus pumila* cultivars
Bridal veil	*Gibasis geniculata*
Purple passion vine	*Gynura aurantiaca* 'Purple Passion'
English ivy	*Hedera helix* cultivars
Wax plant	*Hoya carnosa* cultivars
Pocketbook plant	*Nematanthus* spp.
Fern	*Nephrolepis, Adiantum, Davallia,* and other genera
Arrowhead	*Syngonium podophyllum cultivars*
Trailing peperomia	*Peperomia* spp.
Heart-leaf philodendron	*Philodendron scandens oxycardium*
Swedish ivy	*Plectranthus australis* and *oertendahlii*
Easter cacti	*Rhipsalidopsis gaertneri* and *rosea*
Strawberry begonia	*Saxifraga stolonifera*
Devil's ivy	*Epipremnum aureum* cultivars
Trailing sedum	*Sedum morganianum ×* *rubrotinctum* and other species
German and wax ivies	*Senecio macroglossus* cultivars
Wandering Jew	*Tradescantia* and *Zebrina* spp.
Christmas cactus	*Zygocactus truncatus*

Table 14-10 *Selected Orchids for Indoor Growing*

Common name	Botanical name
Pine Pink	*Bletia purpurea*
No common name	*Brassavola cucullata*
Lady-of-the-night	*Brassavola nodosa*
No common name	*Brassia maculata*
No common name	*Calanthe vestita* cultivars
Tulip cattleya	*Cattleya citrina*
Summer cattleya	*Cattleya gaskelliana*
Autumn cattleya	*Cattleya labiata* cultivars
No common name	*Coelogyne cristata* cultivars
No common name	*Coelogyne massangeana*
No common name	*Cycnoches ventricosum* cv. chlorachilon
No common name	*Dendrobium aggregatum*
Spice orchid	*Epidendrum atropurpureum*
No common name	*Gongora armeniaca*
No common name	*Hexisea bidentata*
No common name	*Laelia anaceps* cultivars
No common name	*Laelia autumnalis* cultivars
No common name	*Laelia rubescens*
No common name	*Lycaste aromatica*
Tiger orchid	*Odontoglossum grande*
Columbia buttercup	*Odontoglossum cheirophorum*
Dancing-lady orchid	*Odontoglossum ornithorhynchum*
Butterfly orchid	*Oncidium papilio*
Dancing-lady orchid	*Oncidium sarcodes*
Moth orchid	*Phalenopsis equestris*
No common name	*Pholidota chinensis*
Coral orchid	*Rodriguezia secunda*
No common name	*Stanhopea oculata*

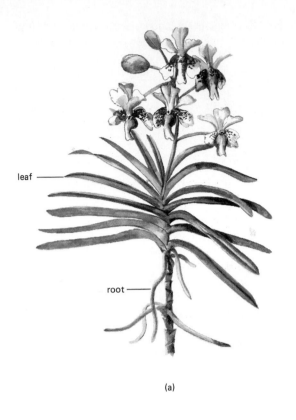

leaf

root

(a)

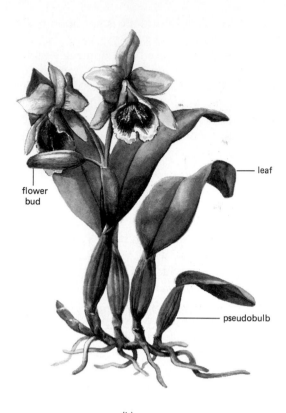

leaf

flower bud

pseudobulb

(b)

Figure 14-13 The monopodial (a) and sympodial (b) growth forms of orchid.

Figure 14-14 Cattleya orchids. (*Photo courtesy of Vocational Education Productions, California Polytechnic State University, San Luis Obispo, CA*)

They should be watered whenever the medium becomes nearly dry.

When repotting epiphytic orchids, a moist medium is more convenient. The bark or osmunda should be soaked in water overnight prior to use in repotting. The orchid should be gently eased from its previous pot and the old medium cleaned from its roots. Next, any dead roots should be cut away and the plant repotted in the fresh medium. Finally, staking should be done, if necessary. The orchid should not be watered until the medium dries.

The light requirements of orchids can be met by placing the plants in a south-facing window during the fall and winter months, and in an east or west window* during the spring and summer when light intensities become brighter. Orchids receiving insufficient light will be surprisingly attractive, with sleek, bright green leaves, but they are unlikely to flower.

Orchids grow best in a humid atmosphere, which is the most difficult growing condition to meet indoors. Misting will improve the humidity for a few minutes; it is more important as a source of moisture that is absorbed by the foliage. Other methods for increasing household humidity are discussed in Chapter 17.

Fertilizer should be applied to orchids every other week while they are in active growth. The fertil-

_____

*Avoid west-facing windows in the sunny West unless a net curtain is used.

izer should be of a liquid or dissolvable powder type and applied to the medium, leaves, and pseudobulbs at dilute strength.

Selected References for Additional Information

Bond, Rick. *All About Growing Orchids.* San Ramon, CA: Ortho Books, 1988.

*Brooklyn Botanic Garden. *Bonsai.* Brooklyn, NY: Brooklyn Botanic Garden.

Catteral, Eric. *Begonias, The Complete Guide.* London: Crowood, 1992.

Davenport, E. *Ferns for Modern Living.* Kalamazoo, MI: Merchants Publishing, 1977.

Graf, A. B. *Exotica IV.* 12th ed. New York: MacMillan, 1986.

Hodgson, Larry and Susan M. Lammers. *All About Houseplants.* San Ramon, CA: Ortho Books, 1994.

*Interior Landscape Industry. *Introduction to Plant Maintenance.* Chicago: Interior Plant Industry.

Koreshoff, D. *Bonsai: Its Art, Science, History and Philosophy.* Beaverton, OR: Timber Press, 1984.

*Krome Communications. *Indoor Plant Maintenance* (12 tapes). Pittsburgh, PA: Krome Communications.

Lamb, E., and B. Lamb. *Pocket Encyclopedia of Cacti in Color.* New York: Sterling, 1980.

Rentoul, J. N. *Growing Orchids,* Vols. 1, 2, 3, and 4. Beaverton, OR: ISBS/Timber Press, 1982–1986.

*Vocational Education Productions. *Foliage Plant Production.* San Luis Obispo, CA: Vocational Education Productions.

Wall, Bill. *Bromeliads.* New York: Sterling, 1990.

Wilson, L. *Bromeliads for Modern Living.* Kalamazoo, MI: Merchants Publishing, 1977.

Young, Dorothy. *Bonsai: The Art and Technique.* Englewood Cliffs, NJ: Prentice-Hall, 1990.

*Indicates a video.

15

Potting Media and Fertilizers

Indoor Plant Potting Media

The word *medium* (plural, *media*) is not familiar to most people when used in relation to indoor plant growing. A medium is any material used for rooting or potting plants. It is soillike in that it performs the same functions as soil does for plants outdoors: supporting the roots and acting as a reservoir for moisture, nutrients, and air. But a medium can be a mixture of many ingredients, some manufactured and some naturally occurring. It need not be a natural material like soil. In fact most commercially available media used for houseplants contain no soil at all.

Potting media have several advantages over most pure soils for growing houseplants. First, they are made up of materials that create large pore spaces, and they are therefore less likely to pack around the roots of plants. Therefore, danger of root death due to air exclusion is largely eliminated. Second, they have a loose structure that drains well. Water passes through quickly (the definition of "good" drainage), yet some moisture is retained. Third, they are usually free of disease organisms, weed seeds, and insects which could damage indoor plants. Finally, they are lighter in weight, which makes moving large plants easier and decreases shipping costs.

Overall, potting media are superior to natural soils for growing plants in pots and are the best choice for rooting and growing indoor plants.

Potting-Media Components

Many brands of premixed potting media are available. They are sold under the name "houseplant soil" or sometimes "cacti soil" or "African violet soil," although most contain no soil at all. They are completely satis-

factory for indoor plant growing, although they can be relatively expensive when used in quantity. A few, particularly the types sold for cacti, could be improved by the addition of more coarse sand. When purchasing a premixed potting medium, read the label for assurance that the product has been sterilized. If it has not, it will need to be home sterilized.

Premixed potting media do not have one set formula or list of ingredients. They vary widely among manufacturers, depending on the available materials.

Any potting medium will generally consist of two to three types of components. The first is *organic, plant-derived material*. Its water-absorbing and -holding capacity will be great, and it will frequently have a spongy structure resistant to packing. In addition, its nutrient-holding ability or "cation-exchange capacity" will be substantial, enabling it to function as a reservoir for nutrients.

The second component is called the *coarse aggregate*. The aggregates improve the drainage of the medium because the angular shape of each particle wedges them apart, creating large or *macro pore spaces* (Figure 15-1). These pore spaces facilitate the free flow of air and water through the medium and increase the drainage rate. In general, coarse aggregates have a hard structure and poor water-holding and cation-exchange capacities.

A third optional component is soil. It is used to hold and supply nutrients and acts as a reservoir of water in addition to the organic material. Soil particles are usually smaller than the coarse aggregate or organic material particles. Consequently, they fit close to each other and other materials, forming *micro pore spaces* (Figure 15-1) which restrict the flow of air and water. However, these micropores are efficient at retaining water due to a phenomenon known as *capillary action*. The concepts of pore space, capillary action, and cation exchange are discussed in more detail in Chapter 5.

275

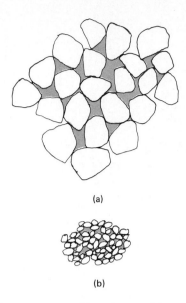

(a)

(b)

Figure 15-1 Macro (a) and micro (b) pore spaces.

The following are the most common ingredients used in making prepackaged mixes (Table 15-1). They are divided into three categories.

Organic, Plant-Derived Materials

Peat Moss. Sold under the names Canadian peat moss, sphagnum peat moss, and just peat moss, this bog plant is harvested, dried, and used in the United States and Canada in enormous quantities each year. In its raw state, it can be bought in mats or bags, is light brown, and has a stringy texture (Figure 15-2). It is also commonly available shredded or "milled," with fibers broken down so they mix easily with other ingredients.

Peat moss is the main ingredient in many prepackaged houseplant mixes, often making up 50 percent of the volume of the medium. It is even used alone for some purposes such as air layering and rooting cuttings.

When dry, peat moss is a very fluffy, lightweight material. When wet, it is capable of holding a large volume of water without packing down, therefore providing plant roots with the ideal combination of moisture and air.

Peat moss is usually very acidic, with a pH range of 4.0 to 5.0. This range is too low for most houseplants, so pH-increasing materials are normally added when peat moss is part of the potting medium. One of

Table 15-1 *Common Ingredients of Houseplant Media*

Ingredient	Type	Remarks
Compost	Organic, plant-derived	A combination of decayed leaves, twigs, grass, and so on made at home; must be sterilized before use
Activated charcoal	Organic, plant-derived from partially burned wood, but functions as a coarse aggregate	Absorbs odors, chemicals from the soil for several months before becoming saturated; can be reactivated by heating in an oven at 350°F (180°C) for about 30 minutes; black
Coarse sawdust	Organic, plant-derived material	Available only in certain areas of the country; should be coarse (1/8"; 3 mm) if possible
Ground bark	Organic, plant-derived material	By-product of lumber mills like coarse sawdust; holds water, nutrients
Bark chunks	Organic, plant-derived material	By-product of lumber mills; holds moisture, nutrients; used for orchids
Sphagnum peat moss	Organic, plant-derived material	Holds water and nutrients well, acid pH, breaks down very slowly; water repellent when dry.
Leaf mold	Organic, plant-derived material from partially decomposed leaves	Flaky texture; decomposes in the medium, releasing nutrients; must be sterilized before use
Humus or muck soil	Organic, plant-derived material	Dark brown and highly decomposed with a powdery feel; can be difficult to rewet if allowed to dry out; also called Michigan peat
Sand	Coarse aggregate	Inert (not chemically reactive); use only coarse grades and not ocean sand
Aquarium gravel	Coarse aggregate	Heavy, inert, and very coarse; excellent for use with cacti and succulents; various colors
Perlite	Coarse aggregate	Lightweight and coarse; contains fluorides which are toxic to some plants; white in color
Vermiculite	Coarse aggregate; also a water- and nutrient-holding component	Lightweight, holds water and nutrients; useful as a coarse aggregate, but it packs down over time; beige and shiny
Scoria or lava rock	Coarse aggregate	Found in natural deposits near volcanos; relatively lightweight with a pleasing dark color

Figure 15-2　Milled (left) and unmilled (right) sphagnum moss. (*Photo by Rick Smith*)

the few disadvantages of peat moss is that if it is ever allowed to dry out thoroughly, it will be very difficult to rewet. Water will run over the fibers rather than being absorbed. The moss or medium must be gradually rewet by submersion or bottom watering or by the use of a chemical "wetting agent" or surfactant. For home use a drop of detergent in a gallon of water will serve this purpose for occasional use.

Leaf Mold. Leaf mold (Figure 15-3) is a second possible component of prepackaged houseplant media. Like peat moss, it is derived from plants and is made up of partially decayed leaves and twigs. It is dark brown with a flaky texture. Leaf mold is capable of holding water and nutrients, although not as much as peat moss. Leaf mold may be used in premixed potting media in addition to or as a complete substitute for peat moss.

Figure 15-3　Leaf mold. (*USDA photo*)

Leaf mold is a natural product in the decomposition of leaves. It is seldom sold as a pure product but can be made at home by composting fallen leaves (Chapter 5). It can also be gathered from the ground in deciduous woodlands.

Humus Peat or Muck. This ingredient is actually a soil type. It is found in low-lying areas and is composed largely of organic material in the form of highly decayed vegetable matter such as leaves, stems, and roots. It is black and fluffy when dry and has a fine, powdery feel. It is commonly available under the name of garden peat or Michigan peat, although it is not peat at all. Muck continues decomposing rapidly when used in a potting medium. Its fineness may create aeration problems.

Do not confuse humus peat with sphagnum peat. The latter is a lighter-brown fibrous material and is the more expensive of the two.

Bark. Shredded or ground bark is occasionally used as an organic potting medium component. Fir, redwood, pine, hemlock, and oak are all satisfactory sources of bark. Bark of incense cedar and walnut should be avoided since it may contain toxic substances.

Compared to peat moss, bark has less water-holding capacity and a higher pH. However, pH-altering chemicals are still needed to raise the pH to an acceptable level. Nitrogen fertilizer is generally included in bark-based potting media to compensate for nutrients temporarily lost as the bark decomposes.

One type of bark derived from pines has been shown to be mildly antibiotic, suppressing the growth of fungi in the potting medium. However, this property does not render it significantly superior to bark from other species.

Hydrating Gel Granules. These synthetic acrylic copolymer granules or spikes are an occasional component of potting media and are sold for use in sandy soils as well (Figure 15-4). They absorb more than 100 times their weight in water, swelling to become ¼ inch (0.5 cm) diameter pieces of nondecomposing gel. Gel granules are most useful as "insurance" in potting media for ferns and similar moisture-loving species which suffer serious damage or die if they dry out. Their value is in their extreme moisture-holding capacity.

The other use for gel granules is as a component of a medium to lessen watering frequency. To this end they are employed in potting mixes for use in commercial interiorscapes to decrease maintenance costs. They might also be useful for home use for people who have large numbers of plants or who travel frequently for extended periods.

(a)

(b)

Figure 15-4 (a) Hydrating gel granules. This brand is sold premixed with a microencapsulated fertilizer (see fertilizer section). (*Photo courtesy of Gromax, Salinas, CA*) (b) The root zone of a plant with hydrated gel granules, showing the roots penetrating the granules. (*Photo courtesy of JRM Chemical Co., Inc., Cleveland, OH*)

Coarse Aggregates

Sand. Sand is a naturally occurring material used occasionally in premixed potting media, although heavy shipping weight discourages frequent use. It is primarily quartz which is not chemically reactive in the medium. Not all grades of sand are satisfactory for use in potting media. In most cases only coarser grades known as builder's or mason sand are used in potting media (see exception under University of California Mix 'C').

Perlite. This aggregate (Figure 15-5) is commonly used in commercial media because of its extremely light weight. It is made by heating volcanic rock until it expands and then grinding the resulting material into irregular chunks ¹⁄₁₆ to ⅛ inch (1.5–3.0 millimeters) in diameter. Like sand, perlite does not hold appreciable amounts of water or nutrients and does not react in the potting medium. In appearance and size it is very similar to white aquarium gravel.

One of the disadvantages of perlite is its tendency to float to the top of a potting medium when a plant is watered. Another is that, due to its light weight, it may not provide enough base weight for larger plants, and they may tend to fall over when the medium is dry. Cacti potted in a largely perlite medium may tend to fall out of their pots since large-bodied cacti frequently have sparse root systems which provide little anchorage.

Perlite has also been proved to contain levels of fluoride that are toxic to some houseplants. As a precaution, it should be rinsed in water three times before it is used. Perlite-containing potting media should be avoided for members of the Liliaceae family since they are particularly susceptible to fluoride injury.

Vermiculite. Vermiculite (Figure 15-6) is made by a heat expansion process similarly to perlite, but a micalike material is used in lieu of volcanic rock. The resulting

Figure 15-5 Perlite. (*Photo by Kirk Zirion*)

Figure 15-6 Vermiculite, an occasional component of potting media. (*Photo courtesy of Vocational Education Productions, California Polytechnic State University, San Luis Obispo, CA*)

Figure 15-7 Horticultural charcoal. (*Photo by Kirk Zirion*)

Figure 15-8 Scoria, a coarse aggregate used in potting media. (*Photo by Kirk Zirion*)

material is shiny, silvery-brown in color, lightweight, and with a spongy feel. The pH is neutral to slightly alkaline, in the range of 7.0 to 7.5. Vermiculite naturally contains appreciable amounts of potassium and magnesium, two nutrients essential to plant growth.

Vermiculite is unusual among the coarse aggregates in that it is an active part of a medium. It has a good capacity for holding water and nutrients, although the size of the individual particles still facilitates good drainage.

Several size grades of vermiculite are manufactured, although not all of them may be available in garden centers. The fine grade is useful for seed germination, and the medium for rooting cuttings or adding to media. The coarse grade is occasionally used for packing plants for shipping. The main disadvantage of vermiculite is its tendency to compress and lose its drainage-enhancing quality with time or when it is watered frequently. Nonetheless, vermiculite is still a popular material for potting media.

Charcoal. Charcoal (Figure 15-7) in small chunk form (such as used for aquariums) is sometimes used as a potting media additive in place of other aggregates. It can be beneficial in some instances in that it absorbs and holds odors, toxic pesticides, and other undesirable chemicals in the short run, but its use is not practical because of its relatively expensive cost and short-term effectiveness.

Scoria or Lava Rock. Scoria (Figure 15-8) is a natural volcanic rock ground to a size of ¹⁄₁₆ to ⅛ inch (1.5–3.0 millimeters). It is gray or black and intermediate in weight between perlite and sand. Where readily available, it is an inexpensive and satisfactory coarse aggregate.

Soils

Loamy Soil. Loamy soil is a well-balanced combination of the basic soil particles sand, silt, and clay in conjunction with a moderate amount of humus or other organic matter. It is the rich topsoil found in many successful gardens. Loamy soil holds water well, does not pack or crust when dry, and can be easily crumbled in the hands when it is moderately damp. It is an excellent potting medium additive even in relatively large quantities but should not be used alone.

Clay Soil. Clay soils are composed largely of tiny plate-like particles which pack together closely, forming many micro pore spaces. Clay soils are very water- and nutrient-retentive but poorly aerated. They are seldom added to premixed potting media but can be used in restricted amounts when balanced carefully with organic and coarse aggregate components.

Sandy Soil. If the sand it contains is coarse enough, a sandy soil may substitute in part for the coarse aggregate portion of a potting medium. However, fine sandy soils are less widely recommended since the particles create relatively small pore spaces.

Silty Soil. This soil has a floury feel when dry and has poor aeration like clay. It has neither the nutrient-holding capacity of clay nor the aggregate function of sandy soil. It should not be used in houseplant media.

Homemade Potting Media

Many people make their own potting media for various reasons: first, buying premixed potting media in bags becomes expensive if there are more than a few plants in the household. Second, premixed media are often formulated with more thought to eye appeal and shipping costs than plant growth. They can be heavily composed of dark, rich-looking humus but lacking essential aggregates like sand.

There is no one formula or list of ingredients that must be adhered to when making a potting media at home. The mix composition is largely determined by the availability of different ingredients in the area and their relative cost. The organic part of the mix need not necessarily be made up of sphagnum moss, for example. Leaf mold gathered from the floor of a woods, compost, or bagged humus can be substituted in whole or in part. Likewise, leftover aquarium gravel, crushed charcoal, cinders, or sand from a gravel pit or lake can be used in place of more expensive perlite and vermiculite. Ocean sand contains salt even after thorough washing and should not be used.

In addition to the bulk ingredients used in homemade media, other ingredients will sometimes be needed to alter pH or improve fertility. Although in quantity they constitute a very minor component, they are essential parts of a good potting medium.

pH Alteration

Many potting media formulas call for additions of small amounts of powdered dolomitic limestone to a mix with a peat moss, leaf mold, or bark base. The limestone counteracts the acidity of the organic component to give the media the slightly acid pH at which most tropicals grow best. It also contains quantities of calcium and magnesium needed for growth.

Fertility Improvement

Many fertilizer materials, both natural and synthetically made, are recommended for incorporating into homemade potting media. The natural materials such as bone meal and dried manure decompose slowly and re-

lease their nutrients slowly into the potting medium. Plants are supplied constantly with fertilizer over several months. Other dry chemical fertilizers such as ammonium nitrate, super phosphate, or a typical complete fertilizer of nitrogen, phosphorus, and potassium are less expensive but effective for a shorter period of time.

Incorporated fertilizers are largely optional except when bark is used as a potting medium component (see previous section on bark). Occasional fertilization should supply all the nutrients required by plants.

Potting Media Formulas

The following media formulas are useful for the majority of houseplants and show the variation that is acceptable when mixing a growing medium. They can be taken either as bases from which to experiment or as standard formulas to be followed exactly. For the experienced gardener, the feel of the medium may indicate its suitability, with the desired product being loose, with a rich supply of organic matter to produce a spongy feel.

Soilless or Artificial Media

Mix 1. An adaptation of the foliage plant mix developed by Cornell University:

> 4 qt (4 liters) shredded sphagnum moss
> 2 qt (2 liters) vermiculite
> 2 qt (2 liters) perlite
> 1 tbsp (15 ml) ground dolomitic lime
> 1 tsp (5 ml) 10-10-10 granular fertilizer.

Mix 2. University of California Mix "C," using fine sand:

> 1 part (by volume) fine sand
> 1 part (by volume) peat moss.

Mix 3. Bark-based mix:

> 3 parts milled pine bark
> 1 part sphagnum peat moss
> 1 part sand.

For mixes 2 and 3, add for each cubic foot (35 liters) of mix:

> 6 oz (200 g) dolomitic lime
> 2 oz (65 g) hydrated lime
> ¼ tsp (1–2 ml) fritted trace elements.

Soil-Containing Media

Mix 1. For use with heavy soils that are clay-based:

1 part (by volume) soil

2 parts (by volume) organic matter

2 parts (by volume) coarse aggregate.

Plus for each bushel (35 liters), add 6 to 8 oz (200-250 g) ground or dolomitic limestone.

Mix 2. For use with medium-weight soils such as loams:

1 part (by volume) soil

1 part (by volume) organic matter

1 part (by volume) coarse aggregate.

Plus for each bushel (35 liters), add 6 to 8 oz (200–250 g) ground or dolomitic limestone.

Mix 3. For use with light, sandy soils:

1 part (by volume) soil

1 part (by volume) organic matter.

Plus for each bushel (35 liters), add 6 to 8 oz (200–250 g) ground or dolomitic limestone.

Mix 4. For stretching commercially prepared mixes:

2 parts (by volume) premixed potting medium

1 part (by volume) loamy garden soil of good quality

1 part (by volume) coarse aggregate.

Specialized Potting Media. The formulas just listed are sufficient for most houseplants. However, a few plants require special media, whereas others will do better with slightly different media.

Very organic medium—for ferns and African violets:

3 parts shredded sphagnum moss

1 part sand.

Xerophytic cacti and succulent medium—fast draining and coarse:

1. 1 part coarse sand or aquarium gravel
   1 part topsoil
   1 part peat moss
2. 1 part premixed commercial medium
   1 part coarse sand.

Table 15-2 *Acid-Loving Houseplants That Commonly Show pH-Caused Chlorosis*

Common name	Botanical name
Azalea	*Rhododendron* spp
Camellia	*Camellia japonica*
Citrus	*Citrus* spp.
Climbing fern	*Lygodium palmatum*
Flame of the woods	*Ixora coccinea*
Gardenia	*Gardenia jasminoides*
Heath, heather	*Erica* spp.
Hydrangea (blue)	*Hydrangea macrophylla*
Sasanqua	*Camellia sasanqua*
Sweet potato	*Ipomaea batatas*
Venus fly trap	*Dionaea muscipula*

Acidic medium—for plants requiring a low pH (5.0–5.5) (see Table 15-2):

1. 3 parts sphagnum peat moss
   1 part vermiculite
   1 part sand.
2. Pure sphagnum peat moss.

Rooting media—fast draining with a moderate moisture-holding capacity (not for xerophytic cacti):

1. Pure vermiculite (medium grade)
2. 1 part perlite
   1 part shredded sphagnum moss.

Seedling media—fine texture and a moderate moisture-holding capacity:

1. 1 part vermiculite (fine grade)
   1 part finely milled sphagnum moss.
2. 1 part sand
   1 part premixed commercial medium
   1 part finely milled sphagnum moss.

Epiphytic orchid media—extremely fast draining and well aerated:

1. Pure bark chunks (½ inch; 1–2 centimeters) in diameter.
2. Pure osmunda fiber (Figure 15-9).

Epiphytic bromeliad medium—fast draining:

1. 1 part sphagnum peat moss
   1 part bark.
2. University of California "C" Mix.

Figure 15-9 Osmunda fiber derived from the osmunda fern. The pole at the top is used to support climbing vines. Loose fiber at the bottom is used for potting orchids and other epiphytes. (*Photo by Rick Smith*)

Mixing Potting Media

For the most thorough mixing, the ingredients for a houseplant medium should be damp but not wet. Mixing ingredients that are completely dry will cause dust to fly as well as the finer particles to settle to the bottom of the medium.

Dampening sphagnum peat moss can be difficult when it is fully dry. It can be allowed to soak in a bucket overnight and then squeezed out to the correct dampness the next day. As an alternative, hot water will wet it quickly.

Sterilizing Potting Media

Sterilizing houseplant media is done to prevent the introduction of unwanted disease organisms, insects, and weed seeds. It is particularly important for seedling and rooting media, where soil-borne diseases are common. Not all media require sterilization to be considered safe to use. Vermiculite and perlite are disease-free because they are heat-treated in the manufacturing process. However, all other ingredients including peat moss, leaf mold, soil, and sand require sterilization before use.

Heat sterilization is the most common soil sterilization method. The damp mix is placed in a large baking pan, covered with foil or a lid, and baked at about 180°F (85°C) for 30 minutes or more until heated to the center. It is then allowed to cool and sealed in a plastic bag to prevent recontamination.

Soil-containing and soilless media give off an odor during baking which some people find objectionable. The unpleasantness can be avoided if the medium is baked in a sealed roasting bag instead of a pan.

Any fertilizer that is to be included in the potting medium should be added after it is sterilized. Some fertilizers are affected by hot temperatures and will break down and release all their nutrients in the baking process.

Indoor Plant Fertilizers

The growing medium in which a plant lives is a reservoir for the nutrients it needs for growth. As these nutrients are absorbed or leached from the medium, they must be replaced with fertilizers.

Major Nutrients for Houseplant Growth

Many nutrients in varying amounts are needed by plants, but the ones used in the greatest amounts are nitrogen, phosphorus and potassium, and to a lesser extent calcium, magnesium, and sulfur. They are termed *plant macronutrients.*

Nitrogen. Nitrogen (abbreviated N) is the nutrient used in greatest quantity by houseplants. Plants absorb nitrogen constantly and use it for manufacturing many chemical compounds within the plant, including chlorophyll and amino acids. Without a constant supply of nitrogen, leaves are unable to synthesize chlorophyll and become pale yellow.

Phosphorus. Phosphorus (abbreviated P) is the second essential nutrient for plant growth. It plays a part in many plant activities including energy transfer. Phosphorus is particularly essential for proper flower and fruit formation.

Potassium. Potassium (abbreviated K) is the third element nutrient for plant growth. It is involved in carbohydrate translocation and protein synthesis. Potassium also enhances coloration, strengthens stems, and aids in strong root development.

Calcium. Calcium is needed for cell division and expansion. An insufficient supply of this element also slows root growth and can cause death of the meristem.

Magnesium. Magnesium is a constituent of chlorophyll and, like potassium, is necessary for protein synthesis.

Sulfur. Sulfur is a component of many amino acids and two of the B vitamins found in plants.

Of the six elements considered macronutrients, nitrogen, phosphorus, and potassium are most generally included in houseplant fertilizers. Each element is included as part of a chemical salt that can be easily shipped and stored. Nitrogen, for example, can be included as calcium nitrate ($Ca(NO_3)_2$) or ammonium nitrate (NH_4NO_3). Phosphorus can be formulated as phosphoric acid (P_2O_5) or monocalcium phosphate (CaP_2O_5). Potassium can be included as potassium nitrate (KNO_3), potassium chloride (KCl), or potassium sulfate (K_2SO_4).

The salts of N, P, and K are used to make houseplant fertilizers dissolve readily in water, and in fact some are packaged predissolved as concentrated liquids. The dissolving ability makes it possible for them to be carried through the medium in water to reach the roots of plants. This ability is particularly valuable for phosphorus which will not move readily in a medium if applied to the surface.

Interpreting Fertilizer Labels

Because of the importance of the three macronutrients for healthy plant growth, they are contained in almost all houseplant fertilizers. By law, any product sold as a fertilizer or "plant food" must list the percentage of these elements it contains in bold type and separated by dashes. These numbers are known as the *fertilizer analysis* and are of more help in determining the value of the fertilizer than most claims made by the manufacturer.

For example, a fertilizer with an analysis of 5-10-5 contains 5 percent of its weight in nitrogen, 10 percent in a standardized phosphorus-containing compound, and 5 percent in a standardized potassium-containing compound, for a total of 20 percent nutrients by weight. The 5-10-5 formula is a relatively common analysis for houseplant fertilizers, but they range from 1.6 to 60 percent total nutrient content, depending on the manufacturer. Table 15-3 shows the analyses of a number of houseplant fertilizers which were being sold in a garden center in 1991.

When purchasing a fertilizer, you do not necessarily get what you pay for. Price appears to be governed more by packaging and volume than by the total amount of nutrients the fertilizer contains. A fertilizer with relatively few nutrients may be nearly the same price as a nutrient-rich one. When selecting a fertilizer, look for the largest numbers in the analysis at the least cost to get the best buy.

The comparison of three fertilizers in Table 15-4 illustrates this point. Brand A is the best buy, brand B the second, and brand C the least economical. Comparison shopping obviously can save considerable money.

Although it is not crucial, a *balanced* fertilizer with equal portions of N, P, and K compounds is widely used by commercial houseplant growers with excellent results. A 5-5-5 fertilizer would be an example of one balanced formula. Barring this, a formula should be chosen that has a nitrogen content equal to or greater than its phosphorus and potassium portions, such as 18-6-12.

Fertilizer Forms

Houseplant fertilizers are available in many forms, from tablets to granules and ready-to-use liquids. Each has

Table 15-3 *Examples of Houseplant Fertilizers*

Rapid Grow Houseplant Food	23-19-17	Dissolvable powder	8 oz. by wt.	$2.50
Ocean Bloom Fish Emulsion	5-2-2	Liquid concentrate	1 pt.	$3.50
Kelp Sea Life	0.3-0.25-0.15	Liquid concentrate	32 fl. oz.	$4.49
Liquinox "Bloom"	0-10-10	Liquid concentrate	32 fl. oz. (.95 l)	$3.39
Roger's Gardens Oxygen Plus Indoor Plant Food	1.3-2.5-1.2	Liquid concentrate	16 fl. oz.	$5.69
Osmocote Outdoor and Indoor Plant Food (270 day)	18-6-12	Encapsulated granules	16 oz. (454 g)	$5.69
Stern's Miracle-Gro All-Purpose Water Soluble Plant Food	15-30-15	Dissolvable powder	24 oz. (1.5 lb.)	$5.69
Professional Blend Fish Emulsion	5-1-1	Liquid concentrate	1 pt. (16 liq. oz.)	$3.69
Uni-gro Fizz-n-feed	5-2-9	Dissolving tablets	2.5 oz. (72 g)	$3.99
Schultz-Instant Liquid Plant Food	10-15-10	Liquid concentrate	12 oz. wt. (340 g)	$3.89
Stern's Liquid Miracle-Gro Plant Food Drops with Iron	8-7-9	Liquid concentrate	9.8 oz. wt.	$2.89
Ortho Fern and Ivy Food + minor elements	10-8-7	Liquid concentrate	8 fl. oz. (0.63 lb.)	$3.99
Peter's Professional Soluble Plant Food	15-30-15	Dissolvable powder	8 oz. wt. (226 g)	$2.99
Jobes Plant Food Spikes	13-4-5	Spikes	1.1 oz. (31 g)	$1.39
Jobes Sunsplash Houseplant Food	1-2-1	Liquid concentrate	8 fl. oz.	$2.49
VF 11	0.15-0.85-0.55	Liquid concentrate	16 fl. oz.	$4.69

Table 15-4 *Cost Comparison of Three Houseplant Fertilizers*

	Analysis	Nutrients	Weight	Total nutrition	Package price	Price per oz. of nutrient
Brand A	23-19-17	59%	8 oz.	4.72 oz.	$2.50	$ .52
Brand B	5-2-9	16%	2.5 oz.	0.4 oz.	$3.99	$9.98
Brand C	1-2-1	4%	8.5 oz.	0.34 oz.	$2.49	$7.32

advantages with regard to price and ease of application, but all are equally acceptable fertilization methods.

Dissolving Powder Concentrates

Powder fertilizers designed to be mixed with water are one fertilizer form. Economically they are usually the best buy since they are the most concentrated. The nutrients dissolve and are applied to plants in place of plain water, are carried down to the root zone in the water, and are immediately available for uptake by plant roots. A disadvantage with older brands was a poor dissolving ability due to filler materials that settled out, but this is uncommon now. Some brands contain colored dyes which can stain skin or clothes when they are used in concentrated form.

Tablets and Spikes

Dry fertilizers can be purchased as tablets designed to be laid on the medium or as small spikes to be pushed down into it (Figure 15-10). As a plant is watered, these solids break apart, dissolving their nutrients into the water as it flows down to the roots of the plant.

Solid fertilizers avoid the bother of mixing and are usually moderate in price.

Liquid Concentrates

Fertilizers in liquid concentrate form are actually dissolvable powders to which water has been added. They are further diluted before use and applied in place of plain water at watering time.

Figure 15-10 Houseplant fertilizer in tablet and spike form. (*Photo by Rick Smith*)

Liquid concentrates mix in water instantly but are more expensive than almost all other fertilizer forms based on the amount of nutrients.

Ready-to-Apply Liquids

Ready-to-apply liquids are poured directly on a plant in place of water and, accordingly, always have a very low nutrient concentration. They avoid mixing inconvenience but are the worst buy because of the substantial percentage of water they contain.

Encapsulated Slow Release

Encapsulated fertilizer (Figure 15-11) can be worked into the medium surface or mixed into the medium at planting with equal effectiveness. The material consists of pinhead-sized beads containing concentrated liquid or solid fertilizer. Covering each bead is a water-permeable plastic coating. The slow-release feature is made possible by moisture present in the medium seeping in and out of the beads. It carries in it dissolved nutrients which are moved to the roots of the plant at each watering. In the end only the plastic coating is left. The covering does not break down readily and may last for several years in the medium, but it is not harmful.

Slow-release encapsulated fertilizers require the least work and mess of any fertilizer type. The beads do not stain, and one application is effective for three to nine months. The expense is moderate when compared to that of other forms.

Figure 15-11 An encapsulated, slow-release fertilizer for houseplants. (*Photo by Rick Smith*)

Granular Slow Release

Granular slow-release fertilizers look very much like garden fertilizers. They do not possess the plastic coating of encapsulated slow-release types; instead they release their nutrients slowly because they are a blend of several chemical compounds. One compound releases nutrients immediately, whereas the other must be chemically broken down before it can be absorbed by the plant. Thus a plant receives nutrition over a long span of time from one application.

Granular slow-release fertilizers are moderately expensive. They have no particular advantage over other fertilizer types applied at proper rates and intervals.

Fertilizer Frequency and Rate

Because of the wide variety of concentrations of fertilizers available, it is impossible to prescribe a standard dilution rate or application frequency. The container mixing directions are the main source of mixing information.

However, many houseplant fertilizers advise weekly or monthly applications of fertilizer on a year-round basis, and it is seriously questionable whether this is advisable. In some cases adherence to container directions could result in overfertilization because the frequency recommendation is based on a rapidly growing plant.

The fertilizer needs of plants growing indoors vary greatly depending on the species and its growing conditions. A naturally slow-growing species like yew pine (*Podocarpus* spp.) would require much less fertilizer than a fast-growing Wandering Jew (*Zebrina* spp.), for example. Fertilizing both at the same rate is therefore not desirable.

Plant nutrient requirements are also greatly affected by the light intensity they receive. In the north, winter light intensities are very poor, and indoor plants experience a slowdown in growth. At this time fertilizer should be applied sparingly or not at all. However, increased light intensity in the spring causes an increased growth rate, boosting water and fertilizer requirements. Fertilizer frequency may need to be increased to keep pace with the rapidly growing plant and to replace those nutrients leached by frequent watering.

As a general rule, fertilization should be done only when a plant is actively growing. The dilution rate should be equal to or weaker than what the package directions stipulate. Application should be infrequent, with the fertilizer applied only during periods of growth and at intervals equal to or longer than those recommended. Few houseplants are harmed seriously by underfertilization, but all are susceptible to damage from overfertilization.

General Guidelines for Fertilizing Houseplants

One of the most important facts to remember when fertilizing a plant is not to increase the fertilization rate or frequency unless you are sure that the plant is suffering from nutrient deficiency. Even if nutrient deficiency is suspected, fertilizer should never be mixed stronger than package directions indicate. Instead, increase the fertilizer frequency slightly, and wait for signs of improvement.

A second precaution is never to apply fertilizer to a dry medium, with the exception of extremely weak ones designed to be used in place of water. A concentrated fertilizer solution can draw water from the roots, and fertilizer "burn" will result.

Third, fertilizing should not be considered a cure for plants that are declining from an unknown cause. It can do actual harm by creating stress on the plant because of excess fertilizer buildup around the roots. Instead, stop fertilizing entirely, resuming it only when the problem is corrected and the plant is growing well again.

Special Fertilizers

A number of special fertilizers for houseplants are sold today. Some are promoted as being strictly for African violets, some are derived from fish, and others claim that one drop in water satisfies all plant requirements.

Organic Fertilizers

These fertilizers are extracted from plant or animal sources instead of being obtained by mining or chemical reactions. Fish emulsions, seaweed extracts, and manure formulations are several types of organic fertilizers.

Organic fertilizers contain the same nitrogen, phosphorus, and potassium as other fertilizers, but only in naturally occurring chemical forms. In the process of nutrient uptake, plants absorb fertilizer only as simple inorganic molecules. It is therefore necessary that microorganisms break down organic fertilizer components before absorption can take place. The organisms work slowly, so organic fertilizers are available over a long period of time.

By analysis, organic fertilizers usually contain a lower percentage of nitrogen, phosphorus, and potassium than inorganic ones, but they sell for about the same price per package. They also contain small quantities of "micronutrients," and these may have a beneficial effect on houseplants. Although organic fertilizers

are no less beneficial for plants than inorganic types, they have no particular proven advantages over them. It should be noted that the microorganisms needed to break down organic fertilizers into usable chemical forms may be scarce in sterilized media. This can cause nutrients to become available extremely slowly and create nutrient deficiency problems in spite of regular fertilization.

African Violet and Flowering Plant Fertilizers

Fertilizers sold as specially formulated for blooming plants contain the same N, P, and K compounds as all-purpose houseplant fertilizers but usually in a different ratio. Less nitrogen, a lot of phosphorus, and a moderate to generous quantity of potassium typify a flowering plant formulation.

Although a deficiency of phosphorus can be a cause of poor flowering, using a phosphorus-rich fertilizer will not guarantee a blooming plant. Other requirements such as light intensity or duration, maturity of the plant, and temperature must be satisfied. Nor can a phosphorus-rich fertilizer guarantee a longer bloom life or be an insurance against bud and flower drop, for other factors are also involved in these processes.

In price, flowering plant fertilizers are about the same as all-purpose fertilizers. There is no research to support claims that these are better than general-purpose ones for flowering houseplants. An evenly balanced (N, P, and K equal) fertilizer will generally supply all the nutrients needed for flowering.

Micronutrient Fertilizers

Whereas macronutrients are used in moderate amounts by plants, the micronutrients are needed only in minute quantities. The micronutrients include such elements as iron, copper, zinc, manganese, boron, and molybdenum. Micronutrient fertilizers can be purchased either with micronutrients alone or as a part of a regular all-purpose fertilizer in dissolvable or encapsulated form.

Potting media usually contain enough micronutrients to supply a plant indefinitely due to the included soil and the decomposition of organic materials. But occasionally plants will develop nutrient deficiency symptoms that are not corrected by all-purpose fertilizers, and a micronutrient deficiency is the likely cause. In these cases a micronutrient fertilizer is advisable, and it can be applied either to the roots or as a spray on the affected foliage.

Many times, micronutrient fertilizers will state that their nutrients are in *chelated* form. Chelated nutrients are used when the pH of the potting medium is causing micronutrients to be "tied up." Even though the micronutrients are present, at an abnormally high or low pH they are held to the medium and are unavailable to a plant. However, chelated nutrients are less affected by pH and will remain available even at unfavorable pHs.

Acid-Forming Fertilizers

Acid-forming fertilizers react in the medium to make it more acidic or lower the pH. They are also sold under the name of azalea fertilizer, since azaleas are common plants that require a very acidic potting medium. But other plants listed in Table 15-2 are also "acid-loving" and will show deficiency symptoms for iron or manganese if the medium pH rises out of the acid range. For these plants, acid-forming fertilizers are useful to keep the pH in the 5.0–5.5 range.

Foliar Fertilizers

Some fertilizers are sold exclusively as foliar fertilizers, and others give instructions for mixing a solution for use as a foliar fertilizer. These fertilizers usually contain N, P, and K plus micronutrients or just the micronutrients alone.

Fertilizers applied to the foliage are used primarily for two purposes: to apply micronutrients and to fertilize epiphytic plants adapted to receiving nourishment chiefly in that manner. Foliar fertilizers are very weak in concentration and are applied with a mister, spraying until the solution runs off the foliage. Repeat applications are usually necessary before enough nutrient is absorbed to eliminate deficiency symptoms.

The ability of a plant to absorb nutrients foliarly is dependent on its species and the thickness of the waxy cuticle which covers the leaf surface. It is also dependent on environmental conditions when the fertilizer is sprayed. Best absorption takes place in bright light and warm temperatures, but not in direct sunlight as this dries the fertilizer on the leaves before it has adequate time to be absorbed. High humidity is also an aid to absorption because it increases fertilizer drying time.

When applying a foliar fertilizer, bear in mind that young leaves are more effective at absorbing nutrition foliarly than older ones, and the lower surfaces are more effective than the upper ones. Additionally a surfactant should be used (if one is not included in the fertilizer) to aid in spreading the liquid over the leaf surfaces to maximize contact and absorption. A drop of detergent in a quart (liter) of mixed foliar fertilizer liquid would be sufficient. Chelated fertilizer formulations are recommended because they increase mobility of the nutrients within the plant.

Oxygen-Generating Fertilizer

Plant roots need oxygen for respiration. Lack of oxygen around the roots causes a deleterious effect because it stops root growth. This in turn restricts growth of aboveground portions of the plant due to decreased root absorption. Normally adequate oxygen reaches the roots of plants from the soil surface, being conveyed downward through the macro pore spaces of the medium. However in some cases the oxygen content of a medium may be inadequate. This can happen when a plant sets in its drainage water continually, has been growing for too long in the same medium which, over time, has decomposed and compacted, or is potted in a medium that was formulated initially without sufficient pore space.

For these situations in which insufficient oxygen in the root zone is the limiting factor to healthy growth, there is a brand of fertilizer that will help alleviate the problem. Plant Research Laboratories has a houseplant fertilizer called Oxygen Plus® that, through a chemical reaction with the urea peroxide it contains, releases oxygen around plant roots. The actual release of oxygen into the growing medium and subsequent improvement in growth (and sometimes flowering) of a number of species[1] *growing under water-logged conditions* has been proven experimentally at both Plant Research Laboratories of Irvine, California, and the Department of Soil and Environmental Sciences of the University of California at Riverside.[2] Although not all species tested showed improvement, many did, and the product can therefore be used with some measure of confidence to aid in solving plant growth problems associated with water-logged media.

Vitamins and One-Drop-Does-All Chemicals

Such chemicals cannot properly be termed fertilizers because they do not contain the elements that have been proven essential for plant growth. Instead they contain vitamins and plant hormones (see Chapter 3). Although these chemicals are necessary for healthy plants, they are manufactured in the plant itself and generally do not need to be additionally supplied. Vitamins and plant hormones should not be considered as fertilizer replacements or substitutes.

[1]Species on which Oxygen Plus® proved effective were *Syngonium, Chamaedorea elegans, Chlorophytum comosum, Dieffenbachia* 'Exotica,' *Dizygotheca elegantissima, Chrysanthemum × hortorum* cv. Bright Golden Ann, *Monstera deliciosa,* and African violet.
[2]Plant Research Laboratories research conducted by P. D. Gaede and M. P. Boghosian. University of California research conducted by E. M. Herr and W. M. Jarrell.

Medium- and Fertilizer-Related Indoor Plant Problems

Medium Problems

Any problem present in the potting medium will affect the roots of the plant, and this in turn will produce symptoms on the top growth. Careful observation of the medium for any problems that might be developing is essential when growing plants.

Compaction

Compaction is the excessive packing or settling of potting medium. This packing can exclude air from the roots and results in rotting of roots or a decline in plant vigor.

Compaction is a characteristic of a medium or a result of decomposition of the medium over time, and is common in soil-containing media. The symptoms are yellowing foliage, weak new growth, and a decrease in growth rate. One cure is to remove the plant from its pot, gently break or soak off the majority of the medium from the roots, and repot in fresh medium. To make the new medium less prone to compaction, add a greater portion of peat moss and coarse aggregates and less or no soil. Switching to Oxygen Plus® houseplant fertilizer (see previous section) can also help with the problem.

Poor Drainage

Poor drainage (assuming the pot has a drainage hole) is again a symptom of an incorrectly formulated medium. Media that are prone to compaction are also prone to poor drainage.

A plant growing in a medium with poor drainage will exhibit the same symptoms as one growing with compaction, and for the same reason: air is excluded from the roots, causing them to decay and die. The cure is also the same; remove the old medium from the roots and repot.

Water Repellency

Water repellency is not common, but it can be a problem in media composed of high percentages of sphagnum moss. Although the moss is capable of holding up to 20 times its weight in water, when completely dried it will shrink and become water-repellent. Water applied to the medium will run through or down the sides of the shrunken soil mass with little being absorbed. Rewetting can be achieved by submersion or bottom watering with warm water, and the medium should not be allowed to become as dry between waterings again.

When the problem is recurring, apply a wetting agent (surfactant) mixed in the regular water.

Unsuitably High or Low pH

Although most media start at a slightly acid pH suitable for most plants, changes in pH can occur over time due to the pH of water used on plants. In such cases, the pH of the irrigation water can be raised or lowered by adding small amounts of common household chemicals: vinegar to lower pH and pure ammonia to raise it. To use these chemicals safely, buy a small, inexpensive pH test kit such as is sold for use on a pool. It should be used initially to test the natural pH of the water and at intervals while adding drops to a gallon or several liters of water to check the progress in raising or lowering the pH to the desired level. The amount of chemical necessary to change the pH can then be recorded and used as a reference when preparing more irrigation water.

If the pH is too high, the level should be adjusted to between 5 and 6 for a month or so and then raised to 6 or 7 to maintain the medium at the acidity level best for most tropicals. If the water is too acidic (a rare situation), ammonia should be used to alter it to 7 to 8, and then after a few months the level should be dropped back to 6 or 7 for maintenance. Acid-loving plants can be maintained on irrigation water in the 5 to 6 range. When using ammonia, be aware that ammonia contains nitrogen, which will probably not then need to be applied in the form of fertilizer.

Fertilizer Problems

Both over- and underfertilization can be injurious to the health of houseplants. The conditions are relatively easy to correct provided the problem can be identified.

Overfertilization (Toxicity)

Overfertilization results from an excess concentration around plant roots of compounds called *fertilizer salts*. This concentration pulls water out of the roots and creates, in effect, a drought condition. Wilting, browned leaf tips, and stunted growth are symptoms of both maladies. But unlike a drought-stricken plant, an overfertilized plant will not immediately recover when watered, since the cause will still be present. The only cure is to remove the excess salts as quickly as possible. This is done by leaching. Large amounts of water should be poured through the medium to flush the fertilizer salts out. (See Chapter 17 for a more complete description of the leaching process.) To prevent a salt accumulation in the soil always water with enough water so that about 10 percent drains through the bottom of the pot, and discard the drainage water.

Nitrogen Deficiency

Lack of nitrogen shows up first as yellowing of the older leaves of a plant, but it can eventually cause both older and younger leaves to discolor (Figure 15-12). Yellowing is also symptomatic of low light and overwatering, but in these instances the leaves soon drop, whereas in the case of nitrogen deficiency they do not. Additional deficiency symptoms are small new leaves and thin stems.

Phosphorus Deficiency

The first symptoms of phosphorus deficiency are a stunting of growth and the production of undersized new leaves.

Potassium Deficiency

Potassium deficiency appears on older leaves first, frequently as a browning of the leaf margins and tips. The areas between the veins may also be affected.

Iron Deficiency

A lack of iron is the most common micronutrient deficiency found in houseplants. It occurs frequently in acid-loving houseplants growing in media of an unsuitably high pH. The first symptom is yellowing between the veins of the young leaves (Figure 15-13). The yellowing will eventually include the entire leaf, and the symptoms are known as *iron chlorosis*.

Iron deficiency can be temporarily corrected by foliar sprays of micronutrients and permanently corrected by changing the pH of the irrigation water (see previous section).

Figure 15-13 Iron deficiency in a coffee plant (*Coffea arabica*). (*Photo by Rick Smith*)

Fertilizer Salt Deposits

Buildups of whitish fertilizer deposits on the medium surface or the outside of clay pots (Figure 15-14) are a common occurrence. They are formed when fertilizer salts dissolved in water move to the outside of the soil ball in response to evaporation of water. The water leaves as vapor, but the salts remain.

Salts on the pot or medium surface do not automatically indicate that there is excess fertilizer in the medium; however, a leaching of the medium is a worthwhile precaution.

Figure 15-12 Typical nitrogen-deficiency symptoms on two wandering jews. The shoot on the left is bleached looking, whereas the one on the right has a healthy, dark green color. (*Photo by Rick Smith*)

Figure 15-14 Fertilizer salt deposits on the outside of a clay pot. (*Photo by Rick Smith*)

Selected References for Additional Information

Bunt, A. C. *Modern Potting Composts, A Manual on the Preparation and Use of Growing Media for Pot Plants.* University Park: Pennsylvania State University Press, 1976.

Furuta, T. *Environmental Plant Production and Marketing.* Arcadia, CA: Cox, 1976.

Joiner, Jasper. *Foliage Plant Production.* Englewood Cliffs, NJ: Prentice-Hall, 1981.

Mastalerz, J. W. *The Greenhouse Environment: The Effect of Environmental Factors on Flower Crops,* 2nd ed. New York: Wiley, 1977.

16

Light and Indoor Plant Growth

Outdoors, sunlight is not frequently considered when planning gardens. It is simply assumed that unless an area is tree-shaded, sufficient light will be available to grow whatever is planted. When plants are placed indoors, ample light for growth is no longer automatically available. In separating themselves from nature with walls and ceilings, people restrict the available light and must properly manage and supplement the remaining light to grow plants successfully.

Effects of Light on Plant Growth and Development

Light is essential for photosynthesis, and without it, chlorophyll, carbohydrates, hormones, and many other plant-manufactured chemicals could not be made. Nor would energy be available for respiration and growth.

A logical conclusion would be that plants cannot grow without light. But this is not strictly true. Plants grown in insufficient light do not stop growing, but they do not produce healthy growth. Instead, they develop abnormally long stems with few leaves (Figure 16-1) and develop a stringy or lanky look. During this process, called *etiolation*, the plant respires its stored carbohydrates. When most of the carbohydrate has been respired, it will die.

Phototropism and Similar Responses

Phototropism is the leaning of a plant toward light (Figure 16-2). It is similar to etiolation in some ways, except that only the side of the stem away from the light

gains auxin and lengthens. Because the sides of the stem are growing at different rates, the stem curves in the direction of light. Thus the seeking of light by plants is actually a hormonal response.

In other plants, differences in light level change the leaf form. A weeping fig (*Ficus benjamina*) grown in bright light will have large numbers of thick, leathery leaves. When moved to a shadier location, it will drop many of these and refoliate with "shade leaves." The shade leaves will be larger, thinner, and fewer in number than the previous "sun leaves."

Cutleaf philodendron (*Monstera deliciosa*) shows a different reaction to varying light levels (Figure 16-3). The number of splits that new leaves contain decreases at lower light levels. A plant growing in poor light may have no splits at all, whereas one growing in moderate light may have several in each leaf.

Light and Flowering

In addition to affecting photosynthesis and hormone manufacture, light also plays a key role in the flowering of many plants.

Intensity

Insufficient light intensity can inhibit the flowering of many plants grown indoors. Relatively bright light intensities are necessary to trigger flowering in most cultivated plants, and those grown indoors are no exception. Unfortunately, bright light intensity is not often available indoors.

Figure 16-1 Etiolation of a plant due to insufficient light. Both plants have the same number of leaves and were propagated at the same time, but the plant on the right received less light. (*Photo by Kirk Zirion*)

Photoperiod

Correct photoperiod combined with sufficient light intensity causes or hastens flowering in some indoor plants. The word *photoperiod* (discussed in Chapter 3) refers to the day/night ratio in each 24-hour period and can affect not only flowering but also tuber and bulb formation.

Plants can be classified with regard to photoperiod as *short day, long day,* or *day neutral.* Many indoor plants are from tropical regions near the equator where day length varies little throughout the year. Accordingly, most are day neutral and flower due to combinations of age, light intensity, and temperature. But a few indoor plants such as chrysanthemum and poinsettia require specific photoperiods to bloom (see Table 16-1 for additional listings).

The number of weeks a photoperiodic plant must be kept under correct day and night lengths to trigger flowering varies among species. If the light/dark cycle is not maintained long enough, the plant may revert to its vegetative state, ceasing bud development in the middle of the flowering process. If kept too long under light/dark control, no harm will usually be done. How-

ever, a short-day plant on a long-night schedule too long will lose valuable photosynthetic time and may lose vigor. As a general rule, a photoperiodic schedule should be maintained until the flower buds are completely developed.

Techniques for Controlling Photoperiod. Most houseplants in which photoperiod controls flowering are short-day plants. They require daily periods of uninterrupted darkness before flowering will take place. Plant species vary in their sensitivity to light, which accidentally "breaks" the necessary dark period. Poinsettias are very sensitive, and bloom will be delayed if the dark period is ever broken by light or if the darkness is not absolute. Most plants, however, are not so sensitive.

One common home procedure for supplying a short-day plant with a long dark period is placing the plant in a closet each day at 5 P.M. and taking it out the next morning at 8 A.M. The schedule can be varied to accommodate personal schedules provided at least 14½ continuous hours per day are spent in darkness. The darkness should be checked by standing inside with the door closed. If light can be seen around the

Figure 16-2 Phototropism caused by light reaching the plant from only the left side. (*Photo by Rick Smith*)

Figure 16-3 Splitting of philodendron (*Monstera deliciosa*) leaves is a function of light intensity. The unsplit leaf developed in dim light, the split leaves in brighter light. (*Photo by Rick Smith*)

door, the leak should be covered or another closet should be used.

Flowering long-day plants is less exacting. The plant can simply be left in a room that is used in the evenings. The artificial lights will supplement the natural day length with enough light to total 14 or more hours, and blooms will be initiated.

Light and Plant Health

The minimum amount of light a plant must receive to remain in a healthy condition is dependent on its species. However, the following generalizations can be made regarding light requirements.

Foliage Color

In many instances the darker the foliage of a plant, the less light it requires. The most common plants that grow well in dim light—snake plant, rubber plant, cast iron plant, Chinese evergreen, and philodendron—all have deep green foliage. When grown without much light, these plants develop a dense chlorophyll layer near the leaf surfaces. This makes them efficient users of light.

In the opposite fashion, many variegated plants such as variegated spider plant, tricolor dracaena, and variegated Wandering Jew require more light than their nonvariegated relatives. They lack chlorophyll in the white areas of the leaves and require more or brighter light to compensate.

Many plants that have red coloring pigments in addition to their chlorophyll are also bright light requirers. Coleus, ti plant, and croton have their chlorophyll masked by red pigment and require strong light to penetrate through to the chlorophyll below.

Variegated plants frequently show their most attractive coloration only in moderate to bright light. Under less light they often revert to solid green for more efficient photosynthesis.

Fleshiness and Succulence

Plants with thickened leaves and/or stems such as jade plant, kalanchoe, and cacti require more light than those with thin leaves.[1] Their fleshy structure minimizes the surface area vulnerable to transpiration water loss and enables them to survive in dry climates. However, at the same time the photosynthesizing area is reduced. Consequently, these plants can be grown indoors more successfully in bright-light locations.

---

[1]Exceptions include Christmas, Easter, and orchid cacti.

Table 16-1   *Photoperiodic Houseplants*

Common name	Botanical name	Photoperiodic classification	Remarks
Gardenia	*Gardenia jasminoides*	Short day	Causes flowering
Strawberry begonia	*Saxifraga stolonifera*	Long day	Runners and new plantlets form
Spider plant	*Chlorophytum comosum*	Long day	Earlier production of runners and plantlets
Variegated spider plant	*Chlorophytum comosum 'Vittatum'*	Short day	Earlier production of runners and plantlets
Hardy begonia	*Begonia grandis*	Short day	Forms aerial stem tubers
Rex begonia	*Begonia rex*	Long day	Increases leaf area and stem elongation only
Wax begonia	*Begonia semperflorens*	Long day	Flowers under long days at warm temperatures but day neutral at cool temperatures
	Begonia socatrana	Long day/ short day	Flowers under long days; forms aerial stem tubers under short days
Tuberous begonia	*Begonia* × *tuberhybrida*	Long day/ short day	Forms blooms under long days and underground stem tubers under short days
Pocketbook plant	*Calceolaria crenatiflora*	Long day	Accelerates flowering but is not essential
Cattleya orchid	*Cattleya* spp.	Short day	Causes flowering
Chrysanthemum	*Chrysanthemum* × *morifolium*	Short day	Flowering may also be hastened by cool temperatures
Poinsettia	*Euphorbia pulcherrima*	Short day	Causes flowering
Fuchsia	*Fuchsia* × hybrida	Long day	Causes flowering in some cultivars
Kalanchoe	*Kalanchoe blossfeldiana*	Short day	Causes flowering and increased leaf succulence
Evening primrose	*Oenothera* spp.	Long day	Flowering accelerated by cool temperatures also
Shamrock	*Oxalis* spp.	Short day	Forms underground tubers
Sedum	*Sedum spectabile*	Long day	Causes flowering
Cineraria	*Senecio* × hybridus	Short day	Short days hasten flowering but are not essential
Christmas cactus	*Schlumbergera bridgeslii*	Day neutral or short day	Day neutral at night temperatures between 50–55°F (10–13°C). At night temperatures of between 55–60°F (13–16°C), it requires 13 hours or more of darkness to flower. Seldom flowers if night temperatures are greater than 70°F (21°C)

Native Growing Conditions

A plant that grows naturally in the shade of other plants is usually better adapted to indoor growing than a plant accustomed to unobstructed sunlight. Ferns are dim light plants that illustrate this well. Because they grow naturally in the shade of trees, they also grow well in a dim area indoors. This generalization holds for many tropical plants including devil's ivy, peperomia, and arrowhead.

Conclusions

Ultimately, the amount of light required to maintain a plant species or cultivar indoors is best determined from a reliable reference book. The generalizations stated would hold true for the majority of indoor plants, but exceptions are not uncommon. A plant might decline due to an improper light environment if these guidelines were followed in lieu of a specific reference.

Light Intensity

Light intensity is one factor that contributes to total light a plant receives. The other is light duration. A lack of total light is frequently the cause of poor condition in houseplants, particularly in northern areas and in winter.

Natural light intensity is affected by many factors. Some are the result of geographical or altitudinal variation, and some are climatic. Understanding these factors and how they interrelate to affect light intensity will help the indoor gardener understand his growing handicaps or assets with respect to light.

Latitude

Because of the way the Earth is oriented in space, the sun rises exactly in the east and sets exactly in the west only at the equator. For people outside the tropics living in the Northern Hemisphere, the sun always ap-

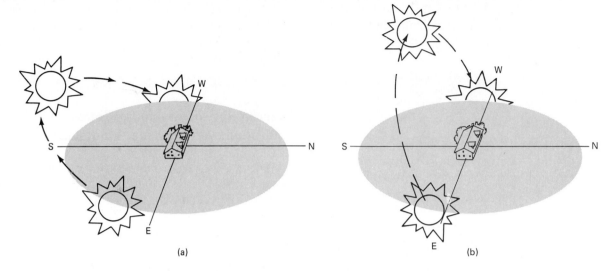

Figure 16-4 Typical positions of the sun in winter (a) and summer (b) in the Northern Hemisphere. Note that the sun remains in the South and that the positions of rising and setting differ with the season.

pears slightly in the southern sky (Figure 16-4). How far off the southern horizon it appears depends on the distance from the equator (latitude), with the sun dropping lower toward the southern horizon the further north one proceeds. The opposite is true in the Southern Hemisphere.

This orientation of the sun in the southern sky affects its intensity substantially. By hitting the Earth's atmosphere at an angle instead of directly above, the distance the light must travel through the atmosphere to reach the Earth is increased. Light is lost in the atmosphere, and therefore its intensity when it ultimately reaches Earth is much less. For this reason, the further north one lives, the less the light intensity.

Time of Year

Time of year also affects light intensity because of the sun's angle. Although the sun stays in the southern sky throughout the year, it drops lowest toward the southern horizon in winter and approaches closest to directly overhead at midday in summer (Figure 16-4). How much of an angle it changes over again depends on the distance from the equator. In the southern United States it will change only slightly from summer to winter, and intensity will vary only moderately. But in the far North the change will be very pronounced. The sun will appear low on the horizon and be relatively dim in intensity in winter, while rising to nearly overhead with a bright intensity in summer. In spring and fall it will be midway between and moderate in intensity.

Time of Day

Light intensity is also affected by time of day. Midday, when the sun is at its highest point, provides the greatest intensity of light. In either morning or evening the sun is lower toward the horizon, and its intensity will decrease accordingly because of the distance it is penetrating through the atmosphere.

Local daily weather variation can change this however. In the West, some areas have routinely foggy mornings which lessen light intensity early in the day. Accordingly, the afternoon sun is much brighter in these areas.

Altitude

People living at high altitudes (greater than 1000 feet [300 meters] above sea level) frequently receive greater light intensity than would be expected at their latitude. The effect of the atmosphere on decreasing sunlight intensity is the reason. Air is less dense and the atmospheric layer is thinner at high altitudes, and less intensity is lost passing through it. Denver, Colorado, illustrates this effect of altitude perfectly. Located at over 5000 feet (1524 meters) in altitude, its light intensity year-round is substantially higher than that of Boston, Massachusetts [elevation 21 feet (7 meters)], even though the two are less than 4° apart in latitude.

Clouds, Fog, and Smog

In cloud-, fog-, or smog-covered areas, light intensity is habitually lower than in clear areas at the same latitude. The water vapor or pollution haze acts as a reflector, sending much of the sunlight back toward space. In addition, clouds, fog, and smog diffuse sun rays, giving the earth shadowless illumination rather than direct sunlight.

Snow Cover

Persistent snow cover affects light intensity by reflecting sunlight indoors. Cold areas that have snow cover most of the winter receive a slightly increased amount of light through all windows because of this reflection.

This principle also applies to houses surrounded by light-colored sand, white rock mulch, or a similar material. While dark soil will absorb light, light sand reflects it and will increase the light entering a house.

Light Duration

Light duration (day length) is a second component of total light. To some extent an increased duration can compensate for a low intensity in providing plants with adequate light for photosynthesis. However, below a certain minimum light intensity, an increased duration will not compensate.

Days are longer in summer than in winter, and again this is a result of latitude. Days and nights at the equator are equal in length all year round. But in the North it is not uncommon for winter days to be as short as 9 hours and summer days as long as 15 hours. The briefness of winter days and the increased length of summer days become more pronounced the further north one travels. In Alaska, for example, days are 20 hours or more long in the summer.

Unfortunately, the shortest winter days correspond with the period when the sun is lowest on the horizon, reaching a climax on December 21, the winter solstice. At this point the sun again begins moving toward the northern sky and days begin to lengthen.

Summary

Latitude, day length, altitude, cloud and snow cover, and the other topics discussed all combine to give total light. Because of these many variables it is almost impossible for a person to determine whether an area is generally bright, moderate, or dim except by observation. However, the Environmental Sciences Service Administration has recorded light intensity and duration throughout the United States for the months of July and December. Their findings give a rating of total light for the United States in midwinter and midsummer.

Figure 16–5 codes the continental United States into bright-, medium-, and dim-light areas for winter and summer. Bear in mind that for most areas, summer light is three to four times as great as winter light.

Determining the Amount of Total Solar Radiation Entering Windows

The factors affecting the intensity and duration of light that reaches the Earth are important to understand. But it is the total light entering windows that is of the greatest importance to an indoor gardener.

Several types of light meters can be used to measure light, but for nonprofessional home use the only practical one would be a thermopile or bolometer which has a flat response to all wavelengths, meaning it does not favor reading of any particular wavelength. It is an *incident* light meter because it measures the light intensity falling on an object, in this case a plant. To obtain a representative reading, measurements would need to be taken at several parts of the plant while aiming the meter at the light source.

Readings would also need to be taken several times a day, and under cloudy and sunny conditions, to determine the average light a plant is receiving. After determining this, the light requirement for the plant must be found. Although this has been determined for many plants and is published in Cooperative Extension Service bulletins, the list is far from complete.

In some cases the light requirement of a species will be listed in a range of footcandles. The footcandle is an old measurement [2] denoting the illumination of a surface 1 foot from the light of a standard candle. For reference purposes, a comfortable light for reading is 50 footcandles, and a bright summer day yields about 10,000 footcandles. The United States Department of Agriculture classified over 100 houseplants by their light requirements, with the range of 75–200 footcandles (767–2153 luxes) deemed low, 200–500 footcandles (2153–5382 luxes) moderate, and 1000 + footcandles (10,764 luxes) high. *Compensation-point* (light intensity below which respiration exceeds photosynthesis) intensities were evaluated as 25 footcandles (229 luxes) for low-light plants, 75–100 footcandles (767–1076 luxes) for medium-light species, and 1000 footcandles (10,764 luxes) for high-light types.

[2]The footcandle measurement of light intensity is outdated and has been replaced by the metric lux. However, some publications dealing with the light requirements of plants still use this measurement.

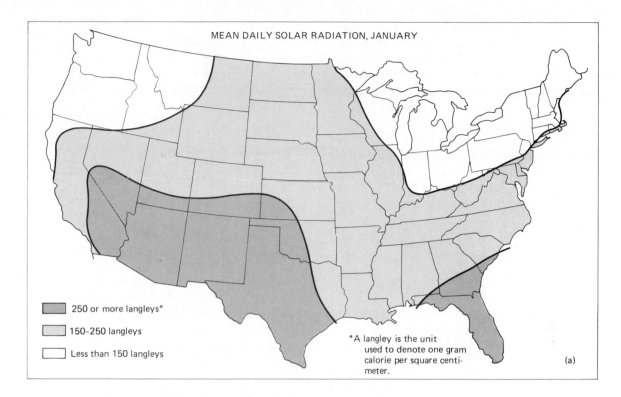

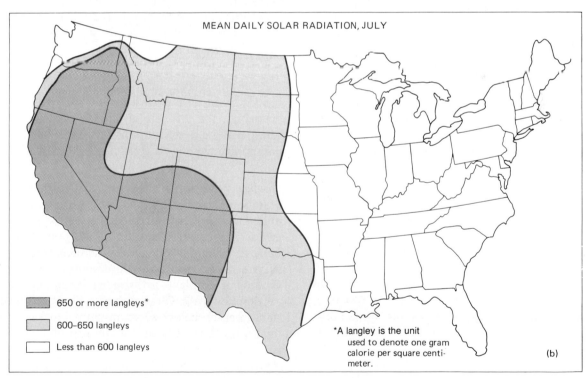

Figure 16-5 Mean daily solar radiation throughout the United States in January and July.

Interactions occur between intensity and temperature, watering frequency, and fertilization frequency. Temperatures less than 63°F (17°C) conserve stored carbohydrate by reducing respiration rate, while frequent watering and fertilization exacerbate and accelerate senescence in the forms of stem lengthening and aging of old foliage.

For most gardeners, simple bright, medium, or dim designations of light based on the direction a window faces and distance from the light are more workable than footcandles. Exact light measurements are not necessary to grow houseplants successfully. The use of a few reference books, the ability to quickly spot the signs of light deficiency, and trial and error are the tools of most successful indoor gardeners.

Window Direction

Several factors make south-facing windows the brightest and best for growing many indoor plants. First, since the sun shines from the southern sky all year, unobstructed south-facing windows receive the longest duration of daily direct sunlight all year. Only south-facing windows receive sun from midmorning through late afternoon, and an indoor gardener with southern windows is fortunate. He has enough light for almost any plant and can modify the light by moving plants further from the window.

East- and west-facing windows provide an intermediate amount of light, less than southern but more than northern. However in the western states west-facing windows may provide substantially more total solar radiation than in eastern states. Each receives direct sunlight for about half of the day and only reflected light for the remainder. The total light entering from east- and west-facing windows is sufficient for growing the majority of houseplants, provided they are not placed too far from the glass.

North-facing windows provide the least light but are still acceptable for growing a number of species in most areas. Their dimness results from the fact that they receive no direct sun, only light reflected from the ground or nearby buildings.

Other Factors Decreasing Sunlight Entry through Windows

Several factors, such as the presence of overhanging eaves or tall buildings, can decrease the sunlight entering through windows. These factors should be taken

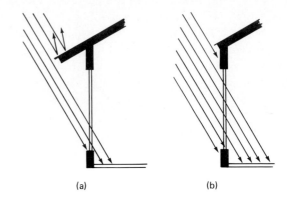

(a)   (b)

Figure 16-6 The effect of eaves on the amount of sunlight entering through a window. Note the depth of penetration into the room with (a) and without (b) eaves.

into account in determining the light level of windows. They are discussed in the following sections.

Overhanging Eaves. Wide eaves (Figure 16-6) can decrease the time or distance direct sunlight enters a room appreciably. They shade the window when the sun is high in the sky, at midday, and several hours before and after. This effect is more pronounced in the summer than in the winter because the sun is higher off the southern horizon.

Buildings or Fences. Tall buildings or fences close to windows can also decrease direct sunlight. As with overhangs, the problem is one of shading from the south, east, or west, but unlike blockage due to eaves, the problem will be more pronounced in winter.

Shrub or Tree Plantings. Shrubs or trees planted close to a window can seriously limit light entry, sometimes reducing it to only reflected light. Trees and shrubs that are deciduous will defoliate in fall and cause little blockage of light during the crucial winter months, but evergreens exclude light continuously. Removal or pruning of the plants is the only solution. Limbing up or thinning will work on trees. Shrubs should be thinned and headed back to minimize blockage of sunlight. Chapter 11 describes all these pruning styles.

Curtains and Shades. Most homeowners open their curtains or shades in the morning and close them after dark, but valuable light can be lost this way during the early morning hours. East-facing windows, in particular, can lose a substantial portion of their direct sunlight when curtains remain closed after sunrise.

Even when curtains are open, they can decrease the light area around the outside edge of the window.

To maximize light, hang curtains so that they expose the entire window area when opened. Avoid overhanging valances, which block overhead light.

Tinted or Stained-Glass Windows. These windows can affect light entry only slightly or block it almost completely, depending on the shade of glass used. Although little can be done short of removing them, their effects should be taken into account when positioning plants for proper lighting.

Placement of Plants for Optimum Light

Just as windows are classified as providing bright, medium, or dim light, plants can be classified as bright-, medium-, or dim-light-requiring and positioned within a room to satisfy their requirements. Generally the dilemma revolves around how far from a window a plant can successfully be grown. This minimum light level at which a plant can maintain itself is called the *compensation point*. The light it receives is sufficient to photosynthesize carbohydrate for its respiration but leaves no surplus carbohydrate for growth. With less light the respiration use of carbohydrate will exceed its rate of manufacture, and the plant will slowly die. At brighter light levels excess carbohydrate will be used for growth.

The concept of compensation point is very important in indoor plant growing. Plants being used for decoration can be bought at the desired size and maintained in good condition with the correct amount of light. Because the plant is growing very slowly if at all, the time devoted to maintenance is minimal. Also, plants grown at compensation-point light levels can be placed further back from windows than is optimum for growth, creating space for additional plants that require more light to survive.

The following descriptions and figures of bright, moderate, and low light are intended only as rough estimates of available window light. However, they include the range from a compensation-point amount of light through the highest light advisable or available for plants indoors.

Bright Light

The designation "bright"- or "high"-light-requiring is given to a number of plants including most flowering species, cacti, and succulents (Table 16-2). Plants that

Table 16-2 Bright-Light Plants

Common name	Botanical name
Flowering maple, Chinese lantern	*Abutilon* spp.*
Chenille plant	*Acalypha hispida**
Copperleaf	*Acalypha wilkesiana* cultivars*
Lipstick vine	*Aeschynanthus* spp.
Agave	*Agave* spp.*
Bamboo	*Bambusa* spp.*
Ponytail palm	*Beaucarnea recurvata**
Wax begonia	*Begonia* × *semperflorens-cultorum**
Cacti	Xerophytic spp.*
Camellia	*Camellia* spp.
Natal plum	*Carissa grandiflora**
Cattleya orchid	*Cattleya* spp.
Rosary vine	*Ceropegia woodii*
Chrysanthemum	*Chrysanthemum* × *morifolium**
Citrus	*Citrus* spp.*
Croton	*Codiaeum variegatum**
Coleus	*Coleus* × *hybridus**
Goldfish plant	*Columnea, Nematanthus*
Cigar plant	*Cuphea* spp.*
Cymbidium orchid	*Cymbidium* spp.
Tricolor dracaena	*Dracaena marginata* 'Tricolor'
Poinsettia	*Euphorbia pulcherrima**
Fatshedera	*Fatshedera lizei**
Purple velvet plant	*Gynura aurantiaca* cultivars
Variegated ivy	*Hedera helix* cultivars
Amaryllis	*Hippeastrum* spp
Wax plant	*Hoya carnosa* cultivars
Polka-dot plant	*Hypoestes phyllostachya*
Bloodleaf	*Iresine herbstii*
Shrimp plant	*Justicia brandegeana*
Gold hop	*Justicia brandegeana* 'Yellow Queen'
Kalanchoe	*Kalanchoe blossfeldiana**
Ice plant	*Lampranthus, Mesembranthemum,* and other genera*
Lantana	*Lantana* spp.*
Wax leaf privet	*Ligustrum japonicum, Ligustrum lucidum**
Oxalis	*Oxalis* spp.*
Geranium	*Pelargonium* spp.*
Avocado	*Persea americana**
Yew pine	*Podocarpus* spp.*
Rose	*Rosa* spp.*
Stephanotis	*Stephanotis floribunda*
Succulents	Many species*

*Indicates no maximum light level indoors.

require bright light are, as a whole, the hardest to grow successfully indoors. Difficulty in providing them with sufficient sunlight during the darker winter months is the main problem.

Although light conditions vary enormously, depending on the area, as a general rule, bright-light

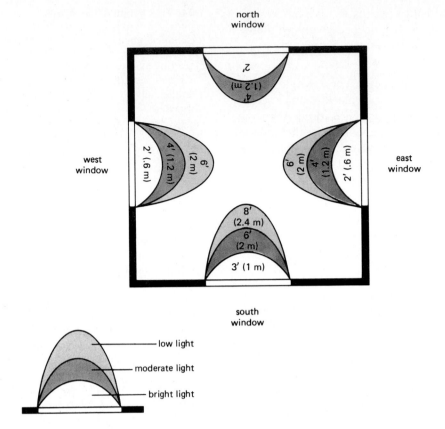

Figure 16-7 Suggested distances from windows for growing bright-, medium-, and dim-light plants.

plants should be placed no further than 3 feet (1 meter) from a southern window or 2 feet (0.6 meter) from an east or west window (see Figure 16-7). It is nearly impossible to give bright-light-requiring plants too much light. The more light they receive, the healthier they will be. Most adjust readily to full sunshine and will grow vigorously outdoors during warm weather after they have been acclimated.

Moderate Light

Most foliage houseplants fall into the moderate or medium light requirement category (Table 16-3). They can be provided with adequate light if placed up to 6 feet (2 meters) from a southern window, up to 4 feet (1.2 meters) from east or west windows, or up to 2 feet (0.6 meter) from a north window. As with bright-light-requiring plants, moderate-light users grow better when placed close to the glass, especially in winter. However, as summer approaches, those in south or west windows will receive increasing amounts of light. They should be watched for symptoms of excess light and moved back from the window if symptoms appear.

Low Light

Only a small percentage of indoor plants grow satisfactorily in low light or what is considered "dim" from a plant's perspective (Table 16-4), and most of these are better in a medium-light location. However, a number of species classified as moderate-light requirers can survive low-light conditions for a period of up to several months without harm.

Low-light plants can be grown 4 to 6 feet (1.2–2 meters) from east and west windows and anywhere up to 4 feet (1.2 meters) from north windows (Figure 16-7). They will also grow 6 to 8 feet (1.8-2.4 meters) from a south window. Unlike bright-light plants, some low-light plants can be injured by excess light. They will live well in moderate light but generally should not be grown in bright light.

Maximizing Natural Light

In general, an indoor gardener must work with whatever natural light is available. He cannot alter his climate, and short of making structural modifications for larger windows or cutting skylights, not much can be

Table 16-3 *Moderate-Light Plants*

Common name	Botanical name
Maidenhair fern	*Adiantum* spp.
Anthurium	*Anthurium* spp.
Aucuba	*Aucuba japonica*
Begonia	*Begonia* spp.
Umbrella plant	*Brassaia* or *Tupidanthus* spp.
Fishtail palm	*Caryota* spp.
Spider plant	*Chlorophytum comosum* cultivars
Kangaroo and grape ivies	*Cissus* spp.
Coffee	*Coffea arabica*
Ti plant	*Cordyline terminalis*
Earth star	*Cryptanthus* spp.
Dumbcane	*Dieffenbachia* spp.
False aralia	*Dizygotheca elegantissima*
Dracaena	*Dracaena* spp.
Euonymus	*Euonymus* spp.
Fatsia	*Fatsia japonica*
Fig, weeping	*Ficus benjamina*
Fig, fiddle-leaf	*Ficus lyrata*
Fig, creeping	*Ficus pumila*
Ox tongue	*Gasteria* spp.
Prayer plant	*Maranta leuconeura* cultivars
Boston fern	*Nephrolepis* spp.
Lady slipper orchid	*Paphiopedilum* spp.
Peperomia	*Peperomia* spp.
Dwarf date palm	*Phoenix roebelenii*
Pileas	*Pilea* spp.
Staghorn fern	*Platycerium bifurcatum*
Swedish ivy	*Plectranthus* spp.
Hare's foot fern	*Polypodium aureum*
Aralia	*Polyscias* spp.
Moses in the cradle	*Rhoeo spathacea*
African violet	*Saintpaulia ionantha* cultivars
Strawberry begonia	*Saxifraga solonifera*
Epiphytic cacti	*Schlumbergera, Rhipsalis, Epiphyllum,* and other genera
Jerusalem cherry	*Solanum pseudocapsicum*
Baby tears	*Soleirolia soleirolii*
Piggyback plant	*Tolmiea menziesii*
Wandering Jew	*Tradescantia, Zebrina*
Ginger	*Zingiber officinale*

Table 16-4 *Dim-Light Plants*

Common name	Botanical name
Miniature flag plant	*Acorus gramineus*
Chinese evergreen	*Aglaonema* spp.
Norfolk island pine	*Araucaria heterophylla*
Asparagus fern	*Asparagus* spp.
Cast iron plant	*Aspidistra elatior*
Bird's nest fern	*Asplenium nidus*
Parlor palm	*Chamaedorea elegans*
Holly fern	*Cyrotomium falcatum*
Corn plant	*Dracaena fragrans* cultivars
Pleomele	*Dracaena reflexa*
Devil's ivy	*Epipremnum aureum*
Rubber plant	*Ficus elastica*
Fittonia	*Fittonia verschaffeltii* cultivars
Kentia palm	*Howea fosteriana*
Philodendron	*Philodendron* spp.
Hart's tongue fern	*Phyllitis scolopendrium*
Snake plant	*Sansevieria* spp.
Spathe flower	*Spathiphyllum* spp.
Arrowhead vine	*Syngonium podophyllum*
Bromeliads	*Vriesea, Guzmania, Nidularium*

taken back to the greenhouse for a recuperation period. During this rotation in the greenhouse they accumulate carbohydrate which will be used for respiration when they are again in the display area.

The home gardener without ample light for all his plants can practice a modified type of rotation, moving some plants closer to windows and others further away every two to three weeks. This technique works best on fairly hardy indoor plants and on moderate- or low-light-requiring species. The requirements of high-light plants are less flexible. They can be placed in moderate or dim areas to a limited extent but should not be kept there more than a few days to avoid leaf drop. In addition, plants that are in bud should not be moved because the sudden change in environment can cause the buds to drop. However, after the buds open, the plant can be moved to be more effectively displayed.

Reflective Colors

Light is reflected by white surfaces, completely absorbed by black, and absorbed to some degree by all colors between. Once light enters through a window, its contact with plants can be maximized by light-colored walls, carpets, and drapes, which bounce light back rather than absorbing it. White walls reflect up to 90 percent of the light they receive. They can contribute significantly to the amount of light plants receive. An increase in reflected light makes it possible to

done to increase light entry. However, several "light management" practices can make the most efficient use of available light.

Rotation

Plant rotation has proven its workability commercially and is also practical for the home gardener. Many of the large foliage plants found in lobbies and public shopping malls are rented from firms specializing in interior landscaping and plant maintenance. Every few months the plants in the low-light display area are replaced and

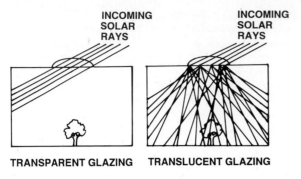

INCOMING SOLAR RAYS

INCOMING SOLAR RAYS

TRANSPARENT GLAZING   TRANSLUCENT GLAZING

Figure 16-8 Light penetration patterns in transparent and translucent skylights. (*Figure based on one by Leonard A. Kersch of Garden Milieu of Ann Arbor, MI, and published in* Interior Landscaping Industry *vol. 5 [10]*)

grow moderate- and especially low-light-requiring plants further back from windows.

Skylights

Skylights are probably the ideal solution to the problem of insufficient light indoors for growing plants. Once very expensive because they were always a custom construction item, they are now available prefabricated in a variety of sizes and materials. The original skylights in buildings in the late 19th and early 20th centuries were glazed with wire glass, the safety glass of the day. New skylights must be made of plastic, tempered glass, and/or laminated glass for safety.

One point of importance to a person who is purchasing a skylight for growing plants is that translucent rather than transparent glazing should be selected. Just as humidity in the air diffuses sunlight and makes everything bright regardless of the direction from which the sunlight is coming, translucent glass or plastic breaks an incoming light ray at the point at which it passes through the material and enables it to illuminate a larger area over a longer time than clear glazing (Figure 16-8). Computer models of commercial-sized atriums hypothetically installed in Detroit (42° latitude, a very dim area) by Leonard A. Kersch (see "Selected References for Additional Information") showed that when the sun is unobstructed by clouds the atrium will be four to five times brighter (Figure 16-9).[4] The scattering effect of the translucent glazing is indepen-

[4]When there is no direct sunlight, the illumination under the skylight would be the same regardless of the type of glazing, but even in the cloudiest areas of North America the diffusive effect of translucent glazing would be a significantly positive factor to improved plant growth.

dent of the angle at which the sunlight strikes the glass except for extreme latitudes in northern Canada and Alaska, and this is particularly relevant because at very northerly latitudes the sun can drop very far toward the horizon (21° in Detroit on December 21, for example) and in such situations frequently illuminates the walls instead of the floor and plant-growing beds in an atrium.

Translucent glazing also eliminates the problem of "hot spot," an area of direct bright sun that moves across the atrium as the sun moves across the sky. With clear glazing this hot spot is the main bright-light plant-growing area, when what would be preferred would be slightly less intensity over a larger area.

Artificial Lighting for Plant Growth

Another solution to the problem of insufficient light is the use of artificial lighting. The indoor gardener who branches out into artificial lighting has a practically unlimited choice of plants. He or she can grow almost any flowering plant successfully, as well as herbs, vegetables, and transplants for outdoor use. "Light gardening," as it is called, can be a fascinating hobby.

The Makeup of Light

Sunlight is naturally suited for plant growth, but some artificial light is not. An understanding of the makeup of light and the kinds of light necessary for plant growth is necessary for a gardener to purchase lighting equipment wisely.

Light energy is divided into wavelengths; Figure 16-10 shows the relationship between wavelengths of visible light and other types of radiant energy. Of the energy types listed, only three—ultraviolet, visible, and infrared rays (heat)—are contained in solar radiation and are significant to plant growth.

Ultraviolet light causes tanning, and plants growing outdoors are exposed to it all day. However, it is not necessary for plant growth. The fact that ultraviolet light does not pass into glass greenhouses further proves that plants can be grown very successfully without it.

Infrared waves transmit heat by traveling through air or space and warming whatever objects they contact. Consequently the infrared rays in sunlight warm plants outdoors during the day, but at night infrared will be given off from previously warmed surfaces like

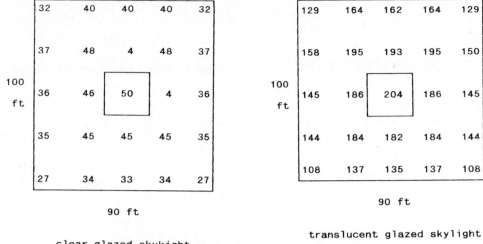

32	40	40	40	32
37	48	4	48	37
36	46	50	4	36
35	45	45	45	35
27	34	33	34	27

100 ft

90 ft

clear glazed skykight

129	164	162	164	129
158	195	193	195	150
145	186	204	186	145
144	184	182	184	144
108	137	135	137	108

100 ft

90 ft

translucent glazed skylight

Figure 16-9 Light intensity variation between clear and translucent skylights. The dotted square represents a skylight 50 feet (15 meters) above the floor. (*Figure based on one by Leonard A. Kersch of Garden Milieu, Ann Arbor, MI, and published in* Interior Landscaping Industry *vol. 5 [10])*

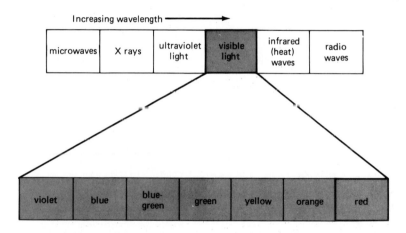

Increasing wavelength ⟶

| microwaves | X rays | ultraviolet light | visible light | infrared (heat) waves | radio waves |

| violet | blue | blue-green | green | yellow | orange | red |

Figure 16-10 The radiant energy spectrum.

the soil. Heat is needed for plant growth, but need not be supplied by a light fixture.

This leaves only visible light necessary for plants, and even all the colors that make up that range of light have not been proved necessary. Light in the orange-red and blue-violet portions of the spectrum have the most significant influence on plant growth. The green and yellow wavelengths are commonly reflected back by plants, giving them their characteristic green appearance.

Both orange-red light and blue-violet light are used in photosynthesis. Red light has been proved to be the wavelength range that triggers flowering in photoperiodic plants, whereas blue-violet light is responsible for phototropic responses.

Types of Artificial Lights

Two basic types of lights are used for artificial lighting, incandescent and fluorescent.

Incandescent Lights

Incandescent lights were the first lights invented. They produce light by channeling electricity through a wire, which then heats up and glows. Compared to fluorescent lights, incandescent lights use more electricity to produce the same amount of light, the extra energy being given off as heat. In wavelengths of light emitted, they are high in orange-red but low in blue-violet, giving objects they light a warm appearance. Although the

Figure 16-11 A simple fluorescent light setup for flowering plants.

red-orange is useful for flowering plants and makes an attractive spotlight, the lack of blue-violet makes normal incandescent bulbs unsatisfactory for plants when used alone.

Plants grown exclusively under incandescent lighting are paler, thinner, and more etiolated than plants grown under the other light sources listed. In addition, they will not branch, and flowering and subsequent senescence are very rapid.

Household-Type Fluorescent Lights

Fluorescent lights (Figure 16-11) are long tubes that can only be used in specially designed light fixtures. The inside of the tube is coated with phosphorescent material and has an electrode at each end. Electricity flows from each electrode and, by a chemical reaction with mercury, stimulates the phosphorescent coating to give off light.

Although initially more expensive, fluorescent lights are more economical to operate than incandescent lights, producing up to five times as much light from the same wattage of electricity. They give off almost no heat from the tubes, and only a small amount from the ballast, the part of the fixture that regulates the electricity flow to the electrodes. Finally, fluorescent tubes have a much longer life than incandescent bulbs.

The wavelengths of light emitted by fluorescent tubes vary depending on the blend of phosphorescent materials used to coat the tube. Table 16-5 lists the most common types of fluorescent tubes and their

Table 16-5 *Wavelength Emissions of Household-Type Fluorescent Tubes*

Tube type	Blue-violet	Orange-red
Cool white	Good	Poor
Cool white deluxe	Good	Fair
Warm white	Fair	Poor
Warm white deluxe	Good	Fair
Daylight	Excellent	Poor
Natural white	Good	Fair
Soft white	Fair	Fair

emissions in the orange-red and blue-violet wavelengths. Of the normal fluorescent tubes, cool white has been used most often for growing foliage plants. However, it is poor in the orange-red area of the spectrum and should be supplemented by an incandescent light for flowering species. Characteristics of plants grown under cool white lights are richly colored foliage, slow stem elongation, and proliferous branching, the latter two of which create a sturdy, bushy, and consequently, attractive plant.

Plant Lights

Plant lights in both incandescent and fluorescent styles are now commonly available. They are designed to provide light in both the blue-violet and the orange-red wavelengths in the best combination for plant growth. Because fluorescent lights are economical to operate and provide a higher intensity, most plant lights are of the fluorescent type.

Plant lights are sold in fewer numbers than normal fluorescent tubes, so they are more expensive. For a small light garden the cost may not be important. But for larger setups, combinations of household tubes from Table 16-5 can be substituted to provide both blue-violet and orange-red light. There will be little difference in plant growth.

Setting Up a Fluorescent Light Garden

Even with the efficiency of fluorescent lighting, more than one tube is generally needed to grow bright-light-requiring plants such as flowering species and annual or vegetable transplants. A minimum of two (and preferably four) tubes placed side by side is commonly used by experienced light gardeners. There is no set formula regarding the wattage of bulbs or how close to place them from one another. In general, tubes of 40 watts or more are best, placed not more than 3 inches (10 cm) apart.

A homemade light garden is not difficult to construct. One widely used setup is a shop light containing two tubes and a built-in reflector. The fixture is mounted on a metal stand or shelving unit, and a lamp timer regulates the hours of operation.

A more attractive light garden can be built using a reflectorless two-tube unit mounted in a bookcase stereo cabinet, or room divider. The only stipulation is that the area to which the fixture is attached be painted white for maximum light reflection. Detailed diagrams for homemade light gardens are plentiful in light gardening and home building project books.

Growing Plants under Lights

The distance from the tubes at which plants are grown and the number of hours per day the lights are left on largely determine the success of a light garden.

Distance from Tubes

For 40-watt household tubes, seedlings and bright-light-requiring plants should be about 2 inches (5 centimeters) from the bulbs. With higher-wattage bulbs, the distance can be increased but should not be greater than about 6 inches (15 centimeters). Since fluorescent lights do not give off heat, there is no problem of heat damage to plants placed too close to the bulbs. Some plants grow so rapidly that they touch the bulbs, but they are seldom injured.

Foliage plants can be grown further from the tubes since their light requirements are lower. Many foliage houseplants will grow well placed up to 2 feet (0.6 meter) from the light source.

Hours per Day

The length of time lights should be on is not an absolute figure. In general, 12 to 16 hours per day is considered suitable if there is no additional source of light. Although foliage plants can grow with less, there is no advantage to running the lights a shorter time except for controlling the photoperiod. The amount of electricity used to operate the tubes is minimal.

Maintenance of Fluorescent Lights

Although fluorescent light bulbs may appear to burn out spontaneously, they actually lose intensity slowly over their entire lifetime of 12,000 or more hours. For this reason it is wise to replace the bulbs after about 75 percent of the stated hour-life has passed because the intensity will have decreased by 15 to 20 percent. New bulbs should be dated with indelible marker, and replacement time calculated by estimating the hours the bulb burns per day multiplied by the number of days of use. Replacing one bulb every three or four months will avoid excess light symptoms from the sudden intensity increase.

Light-Related Houseplant Disorders

Insufficient Light

Lack of light is one of the most common causes of houseplant problems; it is also among the easiest to diagnose. Etiolation of new growth is one symptom. Yellowing and dropping of the older leaves usually also occur. Once the leaves have begun to yellow, they will fall regardless of any improvement then made in the light.

Plants in the advanced stages of light deficiency have thin weak stems and few leaves. The leaves remaining may be widely spaced, pale, and undersized. Such plants should be placed under increased light to recuperate.

Excess Light

Excess light is seldom a problem for indoor plants. However, a plant will occasionally develop symptoms of *chronic* excess light exposure when, for example, a low-light plant is grown in a south-facing window. The new leaves will be small and a pale yellowish-green. The internodes will be very short, giving the plant a compact but unhealthy appearance.

Figure 16-12   Bleached foliage caused by sunburn on a cast iron plant (*Aspidistra elatior*).

A more common case is *acute* excess light damage which occurs when houseplants are placed outdoors after growing indoors all winter. The sudden increase in intensity destroys the top layers of cells on the leaf surfaces, and the leaves acquire a silvery cast (Figure 16-12). The silvery cast turns brown or bleached yellow in a few days. Unfortunately, sunburned leaves do not heal, and the only way to improve the appearance of the plant is by removing these leaves.

Plants "summered" outdoors must have a lengthy acclimatization period to avoid sunburn. Bright-light plants will benefit from summer sunlight, but medium- and dim-light species should be placed in shaded or filtered sun areas. They will thrive under outdoor conditions without risking sunburn.

To acclimate houseplants to outdoor conditions, they should first be placed in the brightest indoor area for about a week. They can then be moved to a sheltered outdoor location that receives early morning or late afternoon sun but is shaded the rest of the day. After another week the bright-light plants can be moved to a brighter outdoor area but should still receive protection from intense midday sun.

When bringing the plants indoors again, the acclimatization procedure should be reversed. Gradually adjusting the plants to decreased indoor light will prevent leaf drop and other signs of light deficiency. The plants should also be checked during this period for insects and diseases, since these problems can spread rapidly indoors.

Selected References for Additional Information

Cathey, H. M., and Lowell Campbell. "Light and Lighting Systems for Horticultural Plants." pp. 491–538 in *Horticultural Reviews* Vol. 2. Jules Janick, ed. Westport, CT: AVI Publishing, 1980.

Elbert, G. A. *The Indoor Light Gardening Book*. New York: Crown, 1975.

Gaines, R. L. *Interior Plantscaping: Building Design for Interior Foliage Plants*. New York: McGraw-Hill, 1977.

Kersch, Leonard A. "Computers Shed New Light on Indoor Illumination Levels." pp. 21–32 in *Interior Landscaping Industry* Vol. 4(3), March 1987.

Kersch, Leonard A. "Something for Nothing." pp. 51–55 in *Interior Landscaping Industry* Vol. 5 (10), October 1988.

Kramer, J. *Plants under Lights*. New York: Simon and Schuster, 1974.

Ortho Book Division. *The Facts of Light about Indoor Gardening*. San Francisco: Chevron Chemical Company, 1975.

Sunset Books Staff. *Windows and Skylights*. Menlo Park, CA: Sunset Publishing, 1982.

17

Indoor Plant Watering and Humidity

Watering and Indoor Plant Growth

Watering is the most frequent maintenance activity associated with growing houseplants. It is also (along with improper lighting) one of the main causes of houseplant failure, probably because absolute directions are not available. Proper watering depends on an understanding of the water requirements of the species, how water is held in and lost from the potting medium, and how it is used in plants.

Pathways of Water Loss in Houseplants

The water needs for the majority of indoor plants are met by water absorbed from the potting medium surrounding the roots. Water held in the medium can be thought of as a reservoir of moisture for the roots that needs to be refilled periodically to assure that adequate water is always available. When the reservoir is full, water can be lost in one of two ways: through transpiration or evaporation (Figure 17-1).

Transpiration

After water enters the roots, it moves upward through the stems and out to the leaf blades to be used in photosynthesis and the manufacture of other chemical compounds (hormones, pigments, etc.) in the plant. However, relatively little of the total absorbed water is used for these synthesis operations; the rest is lost as vapor through the leaves in a process called *transpiration*.

Transpiration occurs almost continuously in houseplants, but its rate is greatly affected by environmental conditions. First, transpiration speeds up as the temperature rises. This is true of almost all processes involved in plant growth. Second, it increases in rate as humidity decreases. A plant growing in a dry room would lose water faster than one growing in a humid greenhouse, for example.

Transpiration rate is also affected by the plant species and form. Large-leaved or densely foliated tropicals transpire considerable water because of the large area of leaf surface exposed to the air. Cacti are more compact and have a limited amount of exposed surface. Consequently, they transpire less water.

Evaporation

Like transpiration, evaporation occurs continually. Water leaves the area where it is concentrated (the wet potting medium) and moves as vapor into the drier adjacent air. Humidity and temperature cause changes in the evaporation rate of water as they do with transpiration, speeding it up as the temperature increases or the humidity drops.

With some types of pots, evaporation of water can occur through the sides of the container. Pots made of materials that are porous to water (unglazed clay or pressed wood fiber, for example) absorb water from the medium and release it into the air.

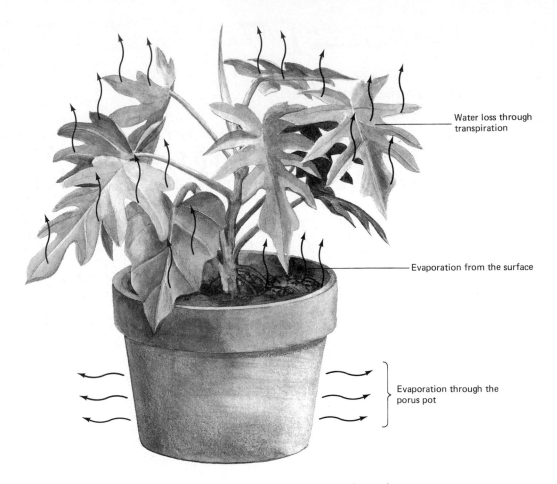

Water loss through transpiration

Evaporation from the surface

Evaporation through the porus pot

Figure 17-1 The pathways of water loss in a plant potted in a clay pot.

Bottom Drainage

Drainage of excess water out the bottom of a pot is the third pathway of water loss. The amount of water retained by a potting medium after water is applied will depend on the ingredients of the medium (including their respective water-holding capacities, particle sizes, and subsequently created pore spaces), the depth of the medium in the pot, and the presence or absence of an *interface* in the pot. In Chapter 15 the roles of moisture-retaining organic ingredients and non-moisture-retaining coarse aggregates were explained. Thus it is clear that the former would promote poor drainage (excess water retention) and the latter, "good" drainage.

Depth of the medium and presence of an *interface* also play roles. Basically, the shallower the pot, the greater the volume of water retained after irrigation. A simple demonstration, which can be done in class, will illustrate the superior drainage of a taller pot (Figure 17-2). A rectangular sponge (representing a pot of medium) should be saturated with water and held hor-

izontally (to represent a shallow pot) until the water stops draining from it. A layer at the bottom will be completely saturated. One would naturally assume that the sponge would hold the same amount of water no matter which way it were turned. But such is not the case. When the sponge is turned vertically, water will pour out the bottom.

The forces at work in this situation are *cohesion* of water molecules to themselves and *adhesion* of the water molecules to a solid (the sponge) *against gravity* pulling out the water. (See Chapter 5 for a fuller description of adhesion and cohesion.) More water pours out when the sponge is turned vertically because the weight of the water increases as it is distributed over a smaller area, the narrow end of the sponge. Therefore, a shallow pot is not the ideal choice for plants sensitive to wet media as a greater proportion of the total medium remains saturated after watering than in a deeper pot.

From this demonstration one might assume that the old tradition of putting a layer of gravel in the bottom of a pot even though it has a drainage hole is a wise

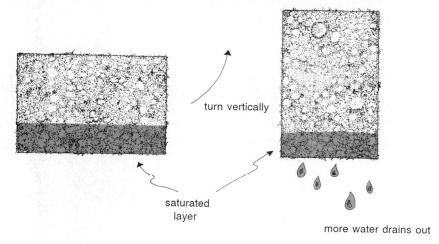

turn vertically

saturated layer

more water drains out

Figure 17-2 The saturated sponge demonstration of water-holding capacity of potting medium.

idea, and that the gravel would become the saturated layer and keep the potting medium unsaturated, but again this is not true. By putting a layer of gravel in the bottom, one is creating an *interface,* a change in pore size due to a change in particle size. The forces that hold the water in the potting medium will not be affected by the gravel layer; the gravel layer cannot influence the medium above it. Its only effect is essentially to shorten the depth of the pot, thus worsening the drainage of the medium.

Watering Frequency

The ways in which temperature and humidity affect the rate of water loss demonstrate one reason why an exact watering schedule cannot be prescribed for a particular plant: the environmental conditions under which a plant is grown can vary radically. Water loss by a plant growing in a cool, humid bathroom will be much slower than that from an identical plant in a warm, dry living room, and the frequency at which each requires water will differ accordingly. Added to that is the change in rate of water loss depending on whether the plants are in porous or nonporous pots.

The environment and type of pot partially determine how frequently a plant must be watered, but there are other factors on which the watering frequency should be based.

Pot Size and Material. Almost without exception, the smaller the pot in which a plant is growing, the more frequently it will have to be watered. This is because a small amount of potting medium holds less moisture than a large volume. Even at the same evaporation and transpiration rates, the water will be depleted sooner from the small pot.

Also, as discussed earlier, a porous pot will lose water through the sides and deplete the water reservoir in the potting medium sooner than a metal, plastic, or glazed clay pot would.

If the plant is in a pot that is both small *and* porous, the water will be depleted at a magnified rate because the surface area of the pot through which water is evaporating will be great in comparison to the volume of potting medium. The depletion rate will be much greater than in the case of a large porous pot.

Potting Medium. As discussed in Chapter 15, potting media differ in the amount of water they can hold because of differing ingredients. A potting medium largely composed of peat moss will be able to hold several times more water than one heavily fortified by perlite. Such a medium will take longer to dry out completely, and watering will not be required as frequently.

Growth Rate. Some plants naturally grow faster than others, and they require more water to photosynthesize and provide energy for that growth. These vigorous growers absorb water from the potting medium very rapidly, necessitating that it be replenished more frequently.

Even plants that are not vigorous growers may pass through changes in growth rate. A rapid spring and summer growth rate is common in houseplants. The watering frequency may have to be increased to keep up with the water demand brought on by the plant's size increase.

Size of the Plant in Relation to the Pot. The larger a plant is in relation to its pot, the more frequently it will require watering. As the top of the plant increases in size, more leaves require water for photosynthesis and transpiration. The rot mass will also have increased, giving a much larger absorption area. However, the plant is still restricted to the same volume of medium from which to absorb water. It will deplete the available water quickly, and the water will need to be replenished sooner. Therefore when a plant is grown in the same pot for a long time and continues increasing in size, its watering frequency will also need to be increased.

Species. The many plant species grown for indoor decoration differ widely in their needs for water. Cacti and succulents grow best when the potting medium is kept fairly dry through most of the year, whereas bog plants like umbrella sedge (*Cyperus alternifolius*) grow with their roots completely submerged in water. Most other plants tend to be somewhere in between. They grow best when there is a slight to moderate amount of water in the potting medium instead of when it is either fully dry or completely saturated.

With the exception of ferns and a few other species, most plants can survive up to a week with dry potting medium and not sustain permanent damage. Many experienced indoor gardeners grow plants very successfully on a "wet-dry" cycle, watering only when the medium is dry but before the plant wilts. However, a wet-dry cycle may minimize new growth because of water stress. Watering more frequently is advisable when rapid growth is desired.

Soil Moisture Testing

Since differing environmental conditions rule out a prescribed watering schedule, the question of timing the watering frequency remains. The easier answer is to evaluate the moisture content of the potting medium by touch.

Touch Method

Feeling the potting medium (Figure 17-3) has proved to be a reliable method of testing the moisture and determining the watering frequency. The fingertips touch the surface of the potting medium or are pushed about ¼ inch (6 millimeters) deep to feel for moisture. A cool damp feeling indicates that moisture is still present near the surface and the medium surrounding the roots still contains adequate water. A dry feeling indicates that moisture surrounding the roots needs to be replenished and the plant should be watered.

Figure 17–3 Testing moisture in the potting medium by feel. (*Photo by Kirk Zirion*)

Color Method

The color method is less reliable than the touch method, but it is used successfully by some gardeners. The method works on the principle that wet potting medium is darker in color than dry, and the moisture content is estimated visually. A drawback of this method is that the surface of the medium may appear dry several days after watering, but the medium underneath may still be damp.

Weight Method

The weight method is reasonably reliable but requires more skill than the touch method. Immediately after watering, a potted plant will be heaviest because of the weight of water contained in the potting medium. As the water is transpired or evaporates, the medium loses water weight, and the potted plant will weigh considerably less. With practice, a gardener can estimate water need by picking up the pot and comparing its present weight to the weight immediately after watering.

Figure 17-4 A moisture meter. (*Photo courtesy of Luster Leaf Products, Crystal Lake, IL*)

Commercial Moisture Sensors

Several types of chemical and electrical sensors are now available. The first type has a piece of blotter paper impregnated with a chemical that changes color in the presence of moisture. The paper is enclosed in plastic and mounted on a probe that is pushed into the medium. The changing color of the paper shows whether moisture is present in the root zone.

The electrical water sensor (Figure 17-4) is an adaptation of a laboratory instrument called a *solu-bridge*, which is used to measure soil nutrient content. It measures the ease of conducting a minute electrical current through the medium between two electrodes (conductors of electricity). The electrodes are on different sections of a long probe, which is inserted into the medium, and are separated on the probe by a section of insulator to force the electrical current to pass through the medium and take the measurement of the ions (dissolved salts) present in the soil water. The more water present, the more ions will be dissolved, and the greater the ease of conductivity.

This type of water sensor is accurate at a normal nutrient range. However, it will read falsely (giving a wet reading in a dry medium) when the nutrient level is extremely high. Conversely, it will read "dry" when

placed in distilled water because of the complete absence of ions in the water.

Accuracy is also affected by the depth to which the probe is inserted. If the majority of the second electrode (actually a long metal sleeve) is left exposed to obtain a shallow moisture measurement, the readings will vary radically between media with the same moisture contents. To solve this problem, the upper electrode can be wrapped with Mylar-type Scotch tape leaving exposed only an area equal to the size of the first electrode.

Commercial experience with these probes has shown them to be most accurate when the reading is taken about 30 seconds after insertion of the probe (see Snyder reference).

Applying the Correct Volume of Water

The goal when watering most houseplants is to wet *all* the potting medium thoroughly at each watering, allow it to dry to a point of slight dampness, and then rewet the medium again.

Pots with Drainage Holes

Assuming that the plant is watered from the top, the volume of water applied to a plant in a pot with a drainage hole is not critical. The makeup of the medium will allow it to hold only a limited amount of water, and the potting medium should be formulated so this capacity is not too much for healthy root growth. Excess water will then drain out the bottom and should be discarded.

With many commercial media it is practically impossible to overwater because the water-holding capacity of the media is so low. Damage from overwatering could result, however, if the pot remained standing in drainage water. With other media (particularly those containing soil) overwatering is possible. These media have a great water-holding capacity, and watering too frequently could cause rotting of the plant roots.

Pots without Drainage Holes

The volume of water applied to plants in pots without drainage holes is critical. Because there is no way for excess water to leave the pot, any amount greater than that required to wet the potting medium will collect at the bottom. For this reason, it is advisable to provide a reservoir of pebbles or charcoal in pots without holes to serve as a collection area for excess water. Standing water in the pebbles will normally be below the root

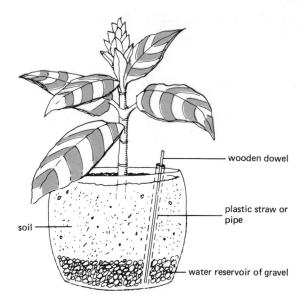

Figure 17-5 A dipstick system for checking the water level in the reservoir of a pot.

growth level. The water there will not damage the plant and will slowly be absorbed back into the potting medium.

To decide how much water should be applied, the volume of the potting medium should be estimated. A volume of water one-fourth as great should then be applied. After about five minutes the water will have been absorbed, and the pot can be gently tipped sideways to allow the excess to run out. If more than a couple of teaspoons (10 milliliters) runs out, the water volume should be reduced next time. If none runs out, enough water may not have been applied to wet all the medium. More should be added and the pot tested again in a few minutes. Ideally just a few drops should flow out to indicate that the entire medium is wet.

Another way of determining the volume of water to apply is by using a dipstick. When the plant is potted, a length of plastic straw or pipe is buried with one end in the gravel reservoir (Figure 17-5). The other end remains above the surface. By inserting a wooden skewer or dowel into the straw, the presence of water in the reservoir can be determined. This technique is valuable for large specimens that are too heavy or cumbersome to be checked by tipping.

Watering Techniques

Top Watering

Most people water from the top. It is easy and completely satisfactory for almost all indoor plants. Top watering is so common that it requires little explanation, but two general rules should be followed. First, enough water should be applied to wet all the potting medium. For pots with holes this means applying enough so that some runs out the bottom. Second, a few minutes should be allowed for water to drain out, and then the drainage water should be discarded. Many plants develop root rot and die if they are left in drainage water.

Bottom Watering

Bottom watering (Figure 17-6) can be used only on plants in pots with drainage holes since the water is taken up through the bottom of the pot. It is very useful for plants whose leaves are susceptible to spotting from water and for flat rosette plants whose leaves cover the soil surface, making it difficult to pour in water. Bottom watering is also good for watering large cacti whose bodies nearly fill the top of the pot. Water applied from the top frequently does not reach the roots of such cacti because it can be poured in only around the rim of the pot. The water flows down the outside of the root ball and out, never adequately wetting the roots. Bottom watering is also useful for rewetting a medium that has dried excessively and retracted from the pot.

For bottom watering, the pot must sit in a plant saucer or bowl. The saucer should be filled with water, and in five to ten minutes the medium will have absorbed the water by capillary action. The saucer should be filled again and, if necessary, a third time until the medium stops absorbing. The excess should then be discarded.

One precaution must be taken with bottom watering. Because water does not flow through the medium and out the bottom, the fertilizer salts contained in the medium are never flushed out. There is a possibility that they will accumulate to toxic levels. As a precaution, a plant which is constantly bottom watered should be submersion-watered or leached every 6 to 12 months to rinse through excess fertilizer.

Submersion Watering

Submersion watering (Figure 17-7) is also used only for pots with drainage holes and for the same purposes as bottom watering. It can also be useful for large, unwieldy hanging baskets.

Submersion watering is similar to bottom watering, in the sense that water enters through the drainage holes, but it is quicker. In submersion watering, the plant is set in water deep enough to come up to the

Figure 17-6   Bottom watering of a dumbcane. (*Photo by Rick Smith*)

rim of the pot. In 5 to 15 minutes the water will have penetrated to the surface of the medium. The pot is then taken out of the water, allowed to drain, and replaced in its growing area.

In submersion watering there is no likelihood of excess fertilizer buildup. During the wetting, the medium becomes saturated, and the fertilizer dissolves in the water. It is carried out with the excess water during the draining.

Wick Watering

Wick watering (Figure 17-8) is not the ideal solution for watering plants because it keeps the medium too moist for most species. However, it is successful with plants that must have constant moisture (African violets and most ferns, for example) and for those in which complete drying could cause death.

Wick watering follows the same principle of capillary action which pulls water up through the medium in bottom watering. In this case capillary action occurs in the wick. Water is absorbed into the wick from the water reservoir, an area of water concentration. The water is pulled through the wick by capillary action to an area of lesser concentration (the potting medium). The

wick action continues until the potting soil is saturated with water.

Shower Watering

Showering, as discussed in Chapter 14, is principally a method of leaf cleansing which waters the plant at the same time. In some regions the mineral content of the water is excessive, and watering by showering will leave unsightly mineral deposits. In these cases avoid showering or gently towel dry the leaves to prevent the deposits.

Special Watering Techniques

Bulb- and Tuber-Bearing Houseplants

During their growth and flowering periods, bulb- and tuber-bearing houseplants such as amaryllis and gloxinia require no special watering techniques. They can be treated like most other houseplants. But as their blooming period ends, these plants may enter a yearly dormancy phase. The top growth dies, putting the bulb in a rest period for several months before growth begins again. The dormant period for

Figure 17-7 Submersion watering of a *Spathiphyllum* in the kitchen sink. (*Photo by Rick Smith*)

bulb- and tuber-bearing plants coincides in nature with climate changes such as decreased rainfall. This change should be simulated to meet the plant's growth requirements.

After the foliage begins yellowing, watering frequency and volume should be slowly decreased over about two months. During this period the foliage will slowly die back to the bulb. When all the foliage is dead, watering should completely stop for several months while the plant rests. It can then be divided or repotted while dormant and regular watering resumed to encourage fresh growth.

The key to correctly watering bulb- and tuber-bearing houseplants lies in careful observation of the plant during all its growth phases. Some bulbous plants never enter dormancy indoors and may be severely injured by underwatering. Others may be unintentionally overwatered during dormancy and rot. Bulbous plants show signs of approaching dormancy by yellowing and death of older leaves. These are signals to reduce watering. Plants emerging from dormancy often begin new growth before water is applied. As soon as this new growth begins, watering should be started. If watering is delayed, the bulb will respire excessive amounts of carbohydrate from the storage organ and may be stunted or die.

Leaching

Leaching is a watering technique used to remedy the problem of excess fertilizer salts in potting media. It works on the principle that the salts are soluble in water and consists of watering a plant four or five times in succession at 5- or 10-minute intervals. Each watering dissolves some of the salts, and the following one flushes them out in the drainage water.

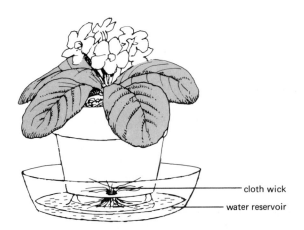

cloth wick
water reservoir

Figure 17-8 A wick watering system. Note that the pot is positioned above water level.

A succession of waterings is not the only form of leaching, although it is probably the most common. Plants can also be submerged and drained several times or left under a slowly running tap for 10 minutes for the same result.

Water Quality and Indoor Plant Growth

For most plants, the amount of water they receive is more important than the quality of that water. In most parts of the country, plants grow well given normal tap water, with only a few being adversely affected by its temperature, pH, or chemical content. One exception to this generalization can be found in the southwestern states, where high-pH and high-salt-content water can cause plant health problems.

Temperature

Correct water temperature is not critical for most houseplants provided it is within a 60 to 90°F (15–32°C) range. Letting water become room temperature is probably ideal for houseplants, but it is not practical if there are many houseplants to be watered.

African violets, though, have been proved sensitive to water temperature. Water more than 10°F (5°C) colder than air temperature will cause permanent yellow spotting on the leaves due to chlorophyll destruction (Figure 17-9). Bottom watering or warming the water to room temperature will avoid the problem.

pH

Acidity or alkalinity (pH) is a characteristic of water as well as potting media. It varies depending on the water source. Continual watering with very acidic or alkaline water will eventually alter the original pH of the medium to an undesirable range.

Although not highly accurate, an inexpensive pH test kit can determine if water is exceptionally acidic or basic. The gardener will then be aware of media pH problems that could occur because of its use. (See Chapter 15 for information on modifying the pH of water for houseplants.)

Chemical Content

Tap water, depending on its source, can contain any number of "impurities" including (but not limited to) sodium, fluoride, chlorine, calcium, ammonia, and iron.

Figure 17-9 Spotting of African violet leaves caused by cold water. (*Photo by Rick Smith*)

Sodium. Sodium often enters water through the softening process. It is used to replace the calcium in calcium carbonate, the mineral that often causes hard water. In areas of the southwest it may be naturally present in the groundwater.

The sodium found in tap water can occasionally build up to concentrations harmful to houseplants with normal watering. To avoid this problem, do not use water from the hot-water tap. In home water softeners, hot water alone is treated with sodium, and cold water remains sodium-free. If the sodium is present in the natural groundwater, or if all household water is softened, rainwater or distilled water should be used. Leaching every two to three months will also help flush the sodium from the potting medium.

Fluoride. Fluoride is put into most city water supplies as a preventative against tooth decay. The concentration used is very small—a few parts per million. However, some untreated water supplies contain more than that concentration naturally.

Relatively small amounts of fluoride are capable of causing damage to plants, but only members of the lily family are known to be particularly sensitive. Common houseplants in this family include dracaenas, spider plants, and asparagus ferns. Excess fluoride can also be found as a result of amendments like perlite. The injury appears as browning of the leaf tips.

Chlorine. Chlorine is added to city water for sanitation purposes; it kills bacteria, fungi, algae, and other potentially harmful organisms. It is generally added as a salt that dissolves in the water or is occasionally bubbled in as a gas under pressure.

Chlorine is essential to plant growth in small quantities and is normally supplied to plants outdoors in the soil and sometimes by rainfall. Tap water does not contain sufficient chlorine to damage houseplants.

Contrary to popular belief, allowing tap water to stand overnight will not eliminate the chlorine. Instead, it escapes extremely slowly as a gas, although the process can be speeded up by exposing the water to ultraviolet light.

Alternatives to Tap Water

Rainwater

The belief that rainwater is pure is not true. It contains a surprising amount of chlorine and some nitrogen, sulfur, and other substances as well. It can also contain chemicals from air pollution and be affected by whatever roof surface it runs off before it is collected. Although rainwater is certainly no worse than tap water for plants, it has no special plant-growing properties because it came from the atmosphere instead of a well or reservoir. However, it is a valuable alternative in areas with very salt-laden water.

Distilled Water

Distilled water has been treated to remove all substances but the hydrogen and oxygen which form water. It is truly "pure" water that does not contain minerals or other foreign chemicals.

If the natural water supply has plentiful sodium, fluoride, or another chemical injurious to houseplants, distilled water is a good alternative water source. It is also useful for leaching plants because it can carry away more salts than the same volume of tap water.

Symptoms of Improper Watering

Underwatering. Plants that are receiving too little water or are not being watered frequently enough may incur root injury. They may show different symptoms, depending on the plant species. Usually they wilt but can be revived provided the leaves have not lost moisture to the point of becoming leathery or crisp, a stage called *permanent wilting point.* Even if the top portions die completely, many plants will regenerate from the roots if the medium is kept moist.

Other symptoms of underwatering are not as easy to diagnose. Slow growth, browned leaf tips and margins, and bud and flower drop can be due to insufficient water as well as several other causes. The symptoms of under- and overwatering can even be the same since both result in root injury.

Overwatering. Excess water excludes oxygen from the potting medium. Eventually root death will occur because the roots are unable to respire.

Overwatering can be caused by watering a poorly draining medium too frequently, by applying too much water to a pot without a drainage hole, or by a medium that holds too much moisture, often due to packing down and loss of pore spaces with age. Symptoms of overwatering include weak and slow growth, lower leaf drop, and discoloring of the roots. In the advanced stages plants wilt because the dead roots can no longer absorb water.

Vacation Watering of Indoor Plants

Keeping houseplants watered during vacation can be inconvenient. However, there are ways to keep them sufficiently moist without interim watering.

One approach is to slow down plant metabolism. This will decrease the water requirement and make the water in the medium last the longest possible time. Photosynthesis, transpiration, and respiration must all be slowed, and the following suggestions can be used separately or in combination to accomplish this.

Lowering the Temperature. As temperature decreases, biological processes slow down. A drop from 75°F (24°C) to 55°F (13°C) will decrease water use appreciably.

Moving Plants out of Direct Sunlight. Direct sunlight warms the leaves, increasing transpiration. It also increases the photosynthesis rate and the water needed for that process. Most bright-light plants can be maintained without full sunlight for up to two weeks without deteriorating, but they should be given bright diffused light to prevent leaf drop. A sheer curtain will provide this type of light.

Grouping Plants Together in the Bathroom. A small bathroom can be turned into a holding room for plants fairly easily. The windows should be closed and the tub and sink filled with water to maximize the humidity. All the plants should then be watered and set in the bathroom with the lights on (not sun or heat lamps) and the door closed. The high humidity will slow transpiration, and the lights will permit limited photosynthesis. Most plants can be left up to two weeks under these conditions. The main drawback is the production of weak growth on vigorous plants because of insufficient light. This undesirable growth should be pruned off when the plants are returned to their normal growing locations.

Enclosing Plants in Plastic Bags. Dry-cleaning bags work well for this purpose, creating a closed system from which no water is lost. Plastic drop cloths can be used for large numbers of plants. One is spread on the floor and the recently watered plants are grouped together on top of it. A second cloth is used to cover them, and the edges are sealed with tape or rolled and stapled to make an airtight envelope. One precaution must be followed: bagged plants should not be placed in strong sunlight because the sun can heat the air inside the plastic and cause heat damage.

For specimens too large for bagging, cover just the pot, sealing around the crown as snugly as possible with tape or string. The plastic will eliminate water evaporation through the medium and thus conserve part of the moisture.

Overall, bagging is the most reliable of all vacation watering ideas. Plants completely sealed in plastic can be left for a month or more without care (Figure 17-10).

Mulching. If bagging is not practical, houseplants can be mulched an inch or two deep (3–5 cm). A bag of shredded bark mulch would not introduce pests and would be sufficient to mulch all the houseplants in an average home. It has a pleasant smell as well.

Figure 17-10 Enclosing a plant in a plastic bag will enable it to survive for a long period without watering. (*Photo by Rick Smith*)

Humidity and Indoor Plant Growth

Humidity is the amount of water vapor in the air, and sufficient humidity is essential for healthy plant growth. Indoor plant species vary in their humidity needs. Cacti and succulents grow satisfactorily in relatively low humidities, but leafy tropicals grow best at higher humidities.

The effect of humidity on plant growth is due mainly to its influence on transpiration rate. Generally, the higher the humidity, the slower the loss of water through transpiration. Without constant and rapid water loss, plant roots do not expend as much energy absorbing water, and that energy can be used for growth and development.

Relative Humidity

The humidity content of the air is generally recorded in a percentage figure called the *relative humidity*. This figure takes into account the fact that the ability of air to hold water vapor varies with its temperature. Air can hold increasingly greater amounts of water as its temperature rises. The relative humidity is the percentage of water vapor the air is presently holding in relation to the total water it is capable of holding at that temperature. For example, 50°F (10°C) air at 80 percent relative humidity means the air contains 80 percent of the water vapor it is able to hold at that temperature. If the air were at 60° and contained the same amount of moisture, its relative humidity would be lower, say 70 or 75 percent.

The outdoor humidity of an area is influenced by the amount of precipitation (generally rain) the area receives and by its proximity to large bodies of water. Accordingly, the East, Northeast, and Northwest have high natural humidities, and the dry inland portions of the West have less. But outdoor humidity is not necessarily equal to indoor humidity. It will be equal only if there is free air circulation between the two. Once windows are closed and heating or air conditioning is used, the humidity inside will drop.

Low indoor humidity in winter is blamed for many houseplant problems. Actually indoor air contains the same amount of water as outside air, and the relative humidity of the outside air may be 60 to 70 percent. The temperature difference creates the problem. Cold outdoor air at a moderate humidity is brought indoors, warmed, and circulated to freshen the indoor air. But the water it contained at 20°F (−6°C) outside is much less than it is capable of holding at 65°F (18.3°C) indoors, and its relative humidity drops from a typical 60 percent down to 20 percent.

Measuring Humidity

Several instruments are available for measuring relative humidity in the indoor environment. A human hair hygrometer is one. It shows the effect of rising or falling humidity on the length of a single human hair and indicates the humidity on a simple dial. It is a relatively expensive instrument, but it is the only dial-type device that is reasonably accurate.[1]

A sling psychrometer is the second instrument for measuring humidity. It consists of two identical thermometers mounted side by side on a small plaque. The mercury bulb of one measures ordinary temperature. The bulb of the second is covered with a small piece of wick (Figure 17-11). When measuring the relative humidity, the wick covering the one thermometer bulb is first wet. The two thermometers are then whirled in the air. When both thermometers show constant readings, the difference between the two temperatures is noted. The "wet-bulb thermometer" with the wick will always show a lower temperature because of the evaporation of the water around the bulb. Using the temperature difference and special tables, the relative humidity can be quickly calculated.

In general, a reading of 35 percent or less can be considered low humidity, 35 to 70 percent moderate, and 70 percent or greater high humidity. The majority of tropical plants grows best at humidities of at least 50 percent, although most will grow with less.

Raising the Relative Humidity Indoors

A number of techniques are used to raise humidity indoors, with differing degrees of success.

Misting

Although misting is widely advocated as a means of raising the relative humidity around plants, it actually does not do much good. In dry air, water will evaporate very rapidly. A thin film of mist applied to leaves can dissipate into the air in less than half an hour. It is unlikely that such a short period of improved humidity would have much effect.

Misting is, however, beneficial to bromeliads and orchids, which absorb water directly through cells on their leaves.

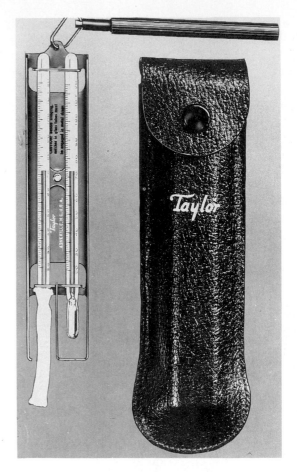

Figure 17-11   A sling psychrometer for measuring humidity. (*Photo courtesy of Nasco International, Ft. Atkinson, WI*)

Pebble- and Water-Filled Trays

This method works because of the constant evaporation of water into dry air. Trays or saucers are filled with a layer of pebbles or marble chips, which are partially covered with water (Figure 17-12). Plants are then grouped on the tray to benefit from the vapor rising up from the water. The water reservoir is refilled as needed, but never to a depth where the pots would be standing in water.

Humidifiers

The most reliable way to increase room humidity is by means of a room humidifier. Different models range upward in price from $10.00. Among the most economical are cool vapor humidifiers. In addition to being inexpensive, they use very little electricity.

Symptoms of Improper Humidity

Insufficient Humidity

Browning of the tips and margins of leaves on tropical plants is the most common low-humidity symptom.

---

[1]It should be noted that a hygrometer requires frequent recalibration to maintain accuracy.

Figure 17-12 Keeping houseplants on a layer of wet pebbles will help increase the humidity. (*Photo by Rick Smith*)

Rapid transpiration causes water to be lost from the leaves before it is able to reach the furthest portions, and browning results. Although unsightly, the browning does little harm to the plant.

Although *tip burn* (Figure 17-13) is the most easily diagnosed symptom of insufficient humidity, others may occasionally appear. Dropping of both new and old leaves may be a symptom, although it is usually associated with other growing problems. Undersized new leaves, a general lack of vigor, and bud drop are others.

Excess Humidity

Most indoor plants grow well at humidities of up to 90 percent. However, the disease powdery mildew can develop under humid conditions. It affects roses and begonias among other plants. Another fungus disease, botrytis, is also common at high humidities.

Selected References for Additional Information

American Water Works Association. *Water Quality and Treatment: A Handbook of Public Water Supplies.* 4th ed. New York: Osborne-McGraw-Hill, 1990.

Snyder, Stuart D. *Building Interiors, Plants and Automation.* Englewood Cliffs, NJ: Prentice-Hall, 1990.

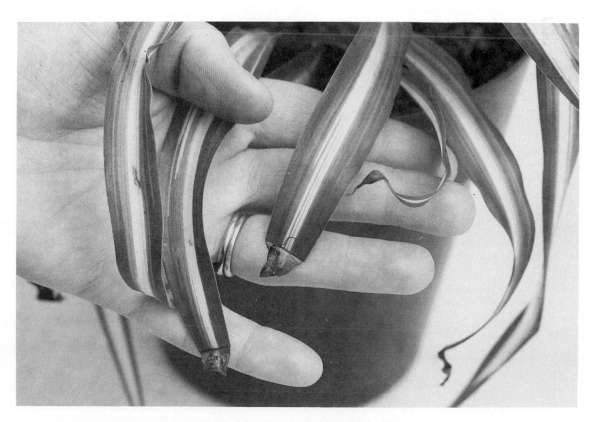

Figure 17-13 Tip burn on a spider plant. (*Photo by Rick Smith*)

18

Controlling Houseplant Pests and Diseases

Houseplant problems are almost always attributable to one of two sources: the environment or organisms contained in it. Those caused by improper environmental conditions (light, water, temperature, and the like) are called *cultural problems* or *physiological diseases*. (These problems are not discussed in this chapter, since the effects of environmental factors were dealt with in the chapters on light, watering and humidity, and potting media.) The organisms harmful to plants can be anything from submicroscopic viruses to snails. The only characteristic they share is that all are detrimental to plant health, whether by eating foliage, sucking sap, or causing spots on leaves.

Houseplant Insects and Related Pests

There are insects and also *insectlike* pests that feed on houseplants. Although it might appear overly technical to separate the two, the distinction can be important if chemicals are used for their control.

Insects, which are six-legged, can be controlled by *insecticides*. Pests like mites, nematodes, and sowbugs belong to other classifications and may require different chemicals for control. Mites, for example, are susceptible to miticides, and nematodes to nematicides. Other than for control purposes these distinctions are unnecessary; all will be discussed under the broad heading of pests (Table 18-1).

Foliage Pests

The easiest pests to detect are often those that damage the aboveground portions of plants—the leaves, stems, buds, and so on. They can be collectively called *foliage pests*. Some foliage pests are more frequently encountered than others; their listing here is ranked from most to least common.

Mites (spider mites, red spider mites, two-spotted spider mites, cyclamen mites)

As mentioned, mites (Figure 18-1) are not insects. They have eight rather than six legs and are related to spiders; hence the name "spider mites." At first encounter, a mite problem can be among the hardest to diagnose. Mites are so small—about the size of dust specks—they are almost invisible to the naked eye.

Several kinds of mites attack indoor plants. Red or two-spotted spider mites can be red, brown, or creamy white. Their damage is done when they insert their mouthparts into a leaf repeatedly to feed. Each feeding site shows as a pinpoint-sized white dot, and eventually leaves become highly speckled with feeding marks, shrivel, and die. Spider mites attack almost any houseplant having thin leaves, as well as cacti and succulents. They are by far the most common type of mite and thrive in low humidity.

The surest diagnosis for spider mites is by examination of the undersides of the damaged leaves in strong light. The mites will be barely visible, but

325

Table 18-1 *Insects and Related Pests of Houseplants and Their Common Hosts*

Pest	Remarks	Common plant hosts
Spider mite	Common on almost all houseplants	African violet (*Saintpaulia ionantha*) Asparagus fern (*Asparagus densiflorus* 'Sprengeri') Azalea (*Rhododendron hybrid*) Cacti Citrus (*Citrus* spp.) English ivy (*Hedera helix* cultivars) Prayer plant (*Calathea, Maranta*) Umbrella plant (*Brassaia, Tupidanthus*) Croton (*Codiaeum variegatum*)
Cyclamen mite	Less common than spider mite and attacks fewer species	Begonia (*Begonia* spp.) *Crassula* spp. Cyclamen (*Cyclamen persicum giganteum*) English ivy (*Hedera helix* cultivars) Gesneriads (*Saintpaulia, Gloxinia, Episcia,* and other genera) Peperomia (*Peperomia* spp.) Purple velvet plant (*Gynura* spp.) Zebra plant (*Aphelandra squarrosa*)
Scale	Attacks primarily woody plants with some exceptions	Cacti Cast iron plant (*Aspidistra elatior*) Citrus (*Citrus* spp.) Coffee (*Coffea arabica*) Croton (*Codiaeum variegatum pictum* varieties) Cycad (*Cycas* spp.) Dracaena (*Dracaena* spp.) English ivy (*Hedera helix* cultivars) Fatsia (*Fatsia japonica*) Ferns Figs (*Ficus* spp.) Palms Gardenia (*Gardenia jasminoides*) Kalanchoe (*Kalanchoe* spp.) Orchids Poinsettia (*Euphorbia pulcherrima*) Zebra plant (*Aphelandra squarrosa*)
Whitefly	Generally uncommon	Begonia (*Begonia* spp.) Citrus (*Citrus* spp.) Coleus (*Coleus blumei* cultivars) Ferns Fuchsia (*Fuchsia* × *hybrida*) Gardenia (*Gardenia jasminoides*) Pepper (*Capsicum annuum*) Poinsettia (*Euphorbia pulcherrima*)
Mealybug	Common and relatively nonselective	Cacti Coleus (*Coleus blumei* varieties) Croton (*Codiaeum variegatum pictum* cultivars) Cycad (*Cycas* spp.) Dracaena (*Dracaena* spp.) Ferns Fuchsia (*Fuchsia* × **hybrida**) Gardenia (*Gardenia jasminoides*) Jade plant (*Crassula argentea*) Palms Poinsettia (*Euphorbia pulcherrima*) Rubber plant (*Ficus elastica*)

(continued)

Table 18-1 *Insects and Related Pests of Houseplants and Their Common Hosts (continued)*

Pest	Remarks	Common plant hosts
Aphid	Occasionally found on the new growth of these and other plants	Cineraria (*Senecio cruentus*) Chrysanthemum (*Chrysanthemum* × *morifolium*) English ivy (*Hedera helix*) Ferns Fuchsia (*Fuchsia* × hybrida) Purple velvet plant (*Gynura* spp.)
Thrip	Relatively uncommon	Arrowhead (*Syngonium podophyllum* cultivars) Azalea (*Rhododendron hybrid*) Citrus (*Citrus* spp.) Dracaena (*Dracaena* spp.) Fuchsia (*Fuchsia* × hybrida) Gardenia (*Gardenia jasminoides*) Gloxinia (*Sinningia speciosa*) Palms *Philodendron* spp. Rubber plant (*Ficus elastica* cultivars) Umbrella plant (*Brassaia actinophylla*)
Nematode	Relatively uncommon	African violet (*Saintpaulia ionantha* varieties) Begonia (*Begonia* spp.) Bloodleaf (*Iresine herbstii* varieties, *Calathea* spp.) Cyclamen (*Cyclamen persicum giganteum*) Ferns Fuchsia (*Fuchsia* × hybrida) Gardenia (*Gardenia jasminoides*) Rubber plant (*Ficus elastica*)
Root mealybug	Occasionally found in unsterilized media	Arrowhead (*Syngonium podophyllum* cultivars) Boston fern (*Nephrolepis exaltata*) Bromeliads Cacti Dumbcane (*Dieffenbachia* spp.) Palms

Note: Plants listed frequently in this table are not necessarily prone to disease and insect problems but are simply the more thoroughly researched indoor plants.

they can be seen clearly with a magnifying glass. In advanced cases of spider mite infestation, webbing will cover leaves. It is to this webbing that the female attaches her eggs.

Cyclamen mites are the second main type of mite affecting houseplants. It is from a main host plant, the florist cyclamen, that their name is derived. Unlike spider mites, which attack old and young leaves indiscriminately, cyclamen mites cluster on very young leaves and buds for feeding.

Cyclamen mites are nearly transparent and feed by puncturing leaves, but their feeding damage does not show speckling or webbing. Instead it causes curling and cupping of the leaves, giving the young leaves a dwarfed appearance.

Cyclamen mites are not as universal as spider mites. They are an unusual problem affecting house-plants, as contrasted to the persistent problem spider mites pose.

Mealybugs

Appearing like specks of cotton, the many types of mealybugs (citrus, long-tailed, and Mexican, for example) are destructive pests. They cause stunting, leaf drop, and eventually the death of a plant.

The adult insects are pink and shaped like sowbugs (Figure 18-2). They appear white because of the white waxy fibers they secrete and in which they lay their eggs. They are very slow moving and frequently found in the axils of leaves or on the undersides along the veins.

Adult mealybugs are often difficult to kill with sprays because of the protection afforded them by

Figure 18-1 Spider mite damage on a palm leaf. (*Photo courtesy of Dr. A. W. Ali*)

their waxy coats. However, newly emerged young lack this covering and are susceptible to most all-purpose insecticides.

Scale Insects (hard scale, soft scale)

Scales (Figure 18-3) are relatively immobile and are frequently not recognized as being insects. Like mealybugs, scales secrete a protective covering. The type of covering determines whether they are called *armored* or *soft scales.*

Armored scales are most common on houseplants. They are covered with hard, blisterlike coats and range up to ⅛ inch (3 millimeters) in diameter. Their color can be clear to white, yellow, or brown, depending on the species, age, and sex of the insects.

Soft scales are very much like mealybugs, secreting a soft waxy covering attached to their bodies. They are nearly round in shape and generally larger than armored scales.

Scales of both types do their damage by puncturing plant tissues and extracting sap through their needlelike mouthparts. As a result of this feeding, plants become weakened, and parts or the whole plant will eventually die.

Scales are found on both sides of leaves, on twigs, and on branches, frequently concealing themselves in the leaf axils. When they attach to leaves, a yellow spot will sometimes be found on the reverse side.

In the life cycle of scales, the most visible stage is the adult, who is permanently anchored in one place

Figure 18-2 Mealybugs. A young adult without protective covering can be seen on the right. Mature adults covered with cottony fibers are left of it. (*Photo by Rick Smith*)

Figure 18-3   Scale insects at the base of a palm frond. (*Photo by Rick Smith*)

and grows to a large size. In the immature crawler stage, scale insects are mobile, and it is at this period when they are most vulnerable to spray insecticides.

Whiteflies

Whiteflies (Figure 18-4) resemble tiny moths. They cluster on the undersides of leaves, flying up when disturbed and resettling almost immediately. Their damage is done by piercing the undersides of leaves to extract plant sap, which causes pale or spotted leaves and weakens the plant.

In a generalized life cycle of a whitefly, the eggs are laid under leaves and hatch in four to twelve days to a crawler stage. The *crawlers* move over the leaf for several hours, then puncture the leaf and feed in one spot. They shed their skin three times, then enter a resting stage and emerge as the familiar white adult. The process takes almost six weeks.

When whiteflies are detected, it is almost always in the adult stage. At this point they are bothersome and difficult to spray because of their flying. It is advisable at that point to begin spraying under the leaves to control the crawlers, with the expectation that control will not be complete for several weeks.

Aphids (plant lice or green flies)

These soft-bodied insects (Figure 18-5) are easily visible and are about 1/32 inch (1 millimeter) across. They can be green, red, black, or any other number of colors and tend to cluster in large numbers on the youngest leaves and buds of plants. Aphids suck plant juices, causing deformed young leaves.

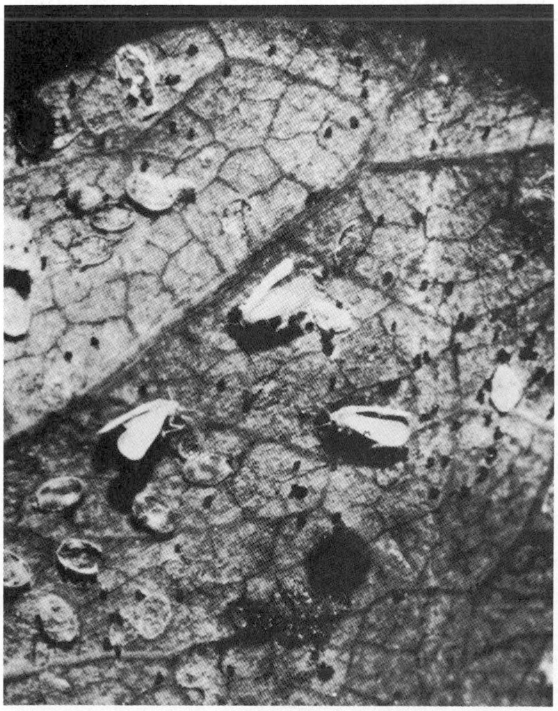

Figure 18-4   Whiteflies. (*USDA photo*)

Figure 18-5   Aphids. (*USDA photo*)

Figure 18-6 Thrip damage on Ardisia. (*Photo courtesy of Dr. Dick Henley, Central Florida Research and Education Center, Apopka, FL*)

Figure 18-7 The larva of a fungus gnat. (*Photo courtesy of Marge Coon*)

In addition, they secrete a syrupy substance called *honeydew* from their backs. The honeydew makes plant leaves sticky and promotes the growth of sooty black mold, a nonparasitic mold which is nonetheless unsightly. Aphids also transmit virus diseases.

Aphids are easily killed with chemicals during any stage in their life cycle. Unlike other insects, young aphids are born live instead of hatching from eggs. The adult female does not require mating, and she can give birth to up to 100 self-fertile females during her 20- to 30-day lifetime.

Thrips

These insects (Figure 18-6) are just slightly larger than spider mites. They are winged in the adult stage and wingless in earlier immature stages. Foliage damaged from their feeding is silvery-colored, later turning brown. The flowers they attack become streaked, and the buds fail to open.

It is very difficult to see thrips on plants as they are extremely small and move fast. If thrips are suspected, the damaged plant part should be tapped over a white sheet of paper. This will dislodge the thrips, which can then be seen briefly before they crawl away.

Slugs and Snails

These invade from outdoors, often entering hidden in the drainage holes of pots. They chew large irregular holes, primarily at night, and hide during the day. Baits and sprays designed for outdoor use are both effective.

Root and Media Pests

Gnats (fungus gnats, manure flies, mushroom flies)

Flying gnats and their immature worm stages (Figure 18-7) are annoying to people and harmful to houseplants. Potting media rich in organic matter provide ideal breeding areas, and a few gnats can multiply to several hundred in a short time.

The adults are brown to black and about ⅛ inch (3 millimeters) long. They are usually seen crawling along the soil surface, where they lay eggs that hatch into tiny white maggots. The burrowing and feeding of the maggots on plant roots cause stunting and root diseases. As the larvae increase in number, the damage intensifies.

Adult gnats can be killed with spray insecticides, but the maggots that do the actual damage must be controlled by drenching.

Springtails

Springtail insects derive their name from the sudden jumping motion they go through when they are disturbed, as by watering. The insects spend their entire life cycle in the soil, growing to an adult size of about ¹⁄₁₆ inch (1.5 millimeters).

Unlike parasitic gnat larvae, springtails in all stages eat only dead vegetation, causing no harm to houseplants. They are, however, bothersome and can be eliminated by an insecticidal soil drench.

Nematodes (Eelworms)

Nematodes are microscopic or near-microscopic worms which live in the medium and feed on the roots of plants. Certain species live on plant leaves, entering through the stomates to feed, but they are rare on houseplants.

Different species of nematodes inflict different types of damage. The feeding of the root-knot nematode causes irritation of the root and leads to the formation of a knot of tissue at the wound site. Numerous such knots interfere with the root's ability to transport water and nutrients to the foliage, and the top of the plant becomes stunted and pale.

Root-lesion nematode does not cause swellings on the roots but does destroy root cells and leaves an entryway for disease. The effects on the top portions of the plant are the same as with root-knot nematode.

Because they are usually microscopic, soil infestations on houseplants are difficult to diagnose. But nematodes can be suspected if the medium was not sterilized before use or if it contains soil. Unpotting the plant and checking for knots will diagnose root-knot nematode. Brown decayed roots could indicate the

presence of other types. An exact diagnosis can be made only through a medium and root sample submitted to a soil-testing laboratory. Nematodes are most effectively controlled on growing plants with chemical nematicide soil drenches.

Root Mealybugs

These root-parasitizing insects look like regular mealybugs but are smaller. Since their feeding weakens plants, they should be controlled with an insecticidal soil drench. (See Figure 18-8)

Sowbugs (pillbugs)

These small pests are not insects but are related to the armor-covered crayfish. They live only in damp places and will often be found hiding under pots or living inside the drainage hole. Although they eat mainly dead organic matter, they also feed on roots and small seedlings and should be eliminated by a chemical drench.

Earthworms

Earthworms do not eat live plants and are seldom seen at the surface. During their burrowing they digest potting medium and excrete it behind, leaving a network of tunnels. Unless present in large numbers, they do no damage.

Organisms Causing Diseases of Indoor Plants

All houseplant diseases are caused by one of four types of microscopic organisms: fungi, bacteria, viruses, or mycoplasmas.

Fungi

Fungi (sometimes called molds) are technically plants, but they do not contain chlorophyll and they are incapable of manufacturing their own carbohydrates. They depend instead on absorbing it from dead or living organisms.

The fungi that cause diseases in plants are parasitic, invading the individual cells of plants with small projections called *hyphae* and extracting nutrition (Figure 18-9). The cells die, and the progression of the disease is begun.

Fungi that attack houseplants usually prey on the roots; foliar fungi-caused diseases are less common. The former are called *soil-borne* because plant roots are ex-

Figure 18-8   Root mealybugs. (*Photo courtesy of Dr. Dick Henley, Central Florida Research and Education Center, Apopka, FL*)

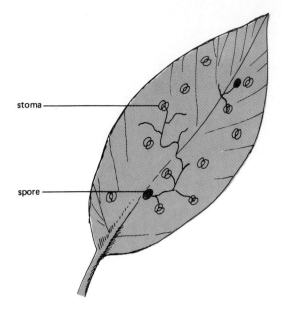

Figure 18-9 Fungus spores germinating and invading a leaf through stomata.

posed to the organisms through fungi-contaminated media. Treatment is through modifying environmental conditions (moisture in particular) and chemical fungicides.

Bacteria

Bacteria are single-celled plants also incapable of manufacturing their food. Some use living plant cells for nourishment and will reproduce rapidly through division under proper environmental conditions. A few bacteria attack indoor plants, causing such disorders as leaf spots and root rots. They can be partially controlled by chemicals.

Viruses and Mycoplasmas

Submicroscopic viruses and mycoplasmas are occasionally encountered in houseplants. When they are, it can be assumed that they were present in the plant at the time it was purchased.

Viruses and mycoplasmas are often transmitted between plants through the mouthparts of contaminated insects. Contaminated tools such as pruners can also pass the organisms from plant to plant.

Specific Diseases of Houseplants

Root Rot

This general term is used for both fungal and bacterial diseases that cause the decay of root tissues. Root rot is

Figure 18-10 Crown rot on a nerve plant (*Fittonia argyroneura*). (*Photo by Kirk Zirion*)

usually associated with poor growing conditions such as a heavy medium that excludes air or lacks an outlet for excess drainage water. Initial symptoms are a slowdown in growth and a yellowing and dropping of foliage, signs of the inability of the roots to take up water and nutrients. In extreme cases, permanent wilting occurs, and the plants die. An inspection of rot-infected roots will show them to be brown and mushy.

Treatment consists of chemical soil drenches combined with improvements in culture. Repotting in a better-drained medium in a pot bearing a drain hole and less frequent watering are recommended.

Crown or Stem Rot

This disease attacks the stem or crown of a plant at the point where it enters the medium. Numerous bacteria and fungi found in unsterilized potting media often cause the disease, but it is also possible that the organisms may be airborne.

The main symptom of crown rot is brown discoloration of the stem (Figure 18-10). The rotting prevents water passage up through the plant, and it can wilt and break off at the weakened area.

Cacti show a form of stem rot called *dry rot* (Figure 18-11). This disease causes the stem base to discolor to yellow or brown and become sunken. The end effect on the plant is the same as with stem and crown rot.

Once infected, there is little that can be done to cure a plant with this disease. The recommended procedure is to take cuttings of the uninfected portions and root them in sterile medium for new plants. The procedure is the same for cacti; that is, the entire top of the cactus is cut off above the rot area and rerooted.

Figure 18-11   Dry rot of cactus. (*Photo by Andrew Schweitzer*)

Mildew

Mildew (Figure 18-12) is a foliar fungus disease that attacks a limited number of susceptible houseplants (see Table 18-2). The first symptom is a gray powdery residue on the leaves. Later the leaves yellow and drop, and the plant eventually dies.

Mildew is controlled by reducing the humidity and applying a fungicide spray. After treatment, keeping the humidity low and all water off the foliage will prevent further problems.

Figure 18-12   The fungus disease mildew. (*Photo courtesy of Marge Coon*)

Leaf Spots

A number of bacteria and fungi occasionally cause leaf spot diseases on foliage plants grown indoors. The symptoms vary with the disease and host, but in general, small pinpoint-sized spots will appear, enlarging and spreading slowly from leaf to leaf. The spots will usually show a pattern such as a center black spot surrounded by yellow or an irregular yellow spot ringed in brown. Figure 18-13 shows a typical fungal leaf spot.

Fungicidal sprays may prevent spread, but often there is no cure. Removing infected leaves as soon as they are detected will slow or stop the spread, and a plant can sometimes be saved in this manner.

Virus and Mycoplasma Infections

The symptoms of a virus infection range widely, depending on the organism and its host plant, but often include yellowing, curled foliage, and stunting.

Although viruses are unlikely to spread between different species of houseplants, they do occasionally. Bacterial leaf spot diseases of *Pilea* and *Pellionia* can move to *Anthurium* spp. and Queensland umbrella tree (*Brassaia actinophylla*). No cures are known and diseased plants should be discarded.

Pest and Disease Prevention

Prevention is better than any cure that can be administered. It involves avoidance of both the initial contact with the insect or disease and its spread.

Plants that are healthy and under good growing conditions have been proved to be less susceptible to attacks by insects and diseases than those under "stress." Stress conditions include any factors that slow the normal growth and development of the plant such as poorly drained medium, overly warm temperature, insufficient light, and the like. In addition, plants in a weakened state are less able to survive such disorders.

Potting Medium

All potting media used for houseplants should be sterilized prior to use.[1] Unless this is done, fungi, bacteria, and soil pests can be introduced and spread rapidly.

[1]See Chapter 15 for directions on sterilizing potting media.

Table 18-2 Diseases of Houseplants and Their Common Hosts

Disease	Common hosts	Disease	Common hosts
Leaf spot	African violet (Saintpaulia ionantha cultivars)	Root rot	Cacti
	Aluminum plant (Pilea cadierei)		Succulents
	Anthurium (Anthurium spp.)		Many foliage plants
	Arrowhead (Syngonium podophyllum)	Stem and	African violet (Saintpaulia ionantha)
	Baby rubber plant (Peperomia obtusifolia)	crown rot	Aralias (Polyscias spp.)
	Balfour aralia (Polyscias balfouriana)		Arrowhead (Syngonium podophyllum)
	Begonia (Begonia spp.)		Begonia (Begonia spp.)
	Cast iron plant (Aspidistra elatior)		Cacti
	Century plant (Agave spp.)		Dumbcane (Dieffenbachia spp.)
	Chinese evergreen (Aglaonema spp.)		Dracaenas (Dracaena spp.)
	Croton (Codiaeum variegatum pictum cultivars)		English ivy (Hedera helix cultivars)
	Dragon lilies (Dracaena spp.)		Figs (Ficus spp.)
	Dumbcane (Dieffenbachia spp.)		Geranium (Pelargonium × hortorum)
	English ivy (Hedera helix)		Gloxinia (Sinningia speciosa)
	Fern		Japanese aralia (Fatsia japonica)
	Gardenia (Gardenia jasminoides)		Kalanchoe (Kalanchoe spp.)
	Geranium (Pelargonium × hortorum cultivars)		Nerve plant (Fittonia spp.)
	German ivy (Senecio macroglossus)		Orchids
	Goldfish plants (Columnea spp.)		Peperomias (Peperomia spp.)
	Impatiens (Impatiens spp.)		Philodendrons (Philodendron spp.)
	Ivy tree (× Fatshedera lizei)		Pileas (Pilea spp.)
	Indian laurel (Ficus retusa nitida)		Pothos (Epipremnum aureum cultivars)
	Jerusalem cherry (Solanum pseudocapsicum)		Sinningias (Sinningia spp.)
	Lollipops (Justicia spp.)		Snake plant (Sansevieria trifasciata cultivars)
	Nerve plant (Fittonia verschaffeltii argyroneura)		Succulents
	Orchids		Ti plant (Cordyline terminalis)
	Philodendron (Philodendron spp.)		Umbrella tree (Schefflera and Brassaia genera)
	Piggyback plant (Tolmeia menziesii)		Zebra plant (Aphelandra squarrosa)
	Poinsettia (Euphorbia pulcherrima)	Virus	Anthurium (Anthurium spp.)
	Ponytail palm (Beaucarnea recurvata)		Begonia (Begonia spp.)
	Prayer plants (Maranta spp.)		Cacti
	Queensland umbrella tree (Brassaia actinophylla)		Calathea (Calathea spp.)
	Royal red bugler (Aeschynanthus pulcher)		Chinese evergreen (Aglaonema spp.)
	Rubber plant (Ficus elastica)		Christmas cactus (Zygocactus truncatus)
	Satin pellonia (Pellonia pulchra)		Dumbcane (Dieffenbachia spp.)
	Silver vase (Aechmea fasciata cultivars)		Figs (Ficus spp.)
	Snake plant (Sansevieria trifasciata cultivars)		Geranium (Pelargonium × hortorum)
	Stone crops (Sedum spp.)		Goldfish plants (Columnea spp.)
	Thread flower (Nematanthus spp.)		Moses in the Cradle (Rhoeo spathaceae)
	Ti plant (Cordyline terminalis)		Nerve plant (Fittonia verschaffeltii cultivars)
	Watermelon begonia (Pilea cadierei)		Orchids
	Weeping fig (Ficus benjamina)		Peace lily (Spathiphyllum spp.)
	Zebra basket vine (Aeschynanthus marmoratus)		Peperomia (Peperomia spp.)
	Zebra basket vine (Aeschynanthus pulcher)		Philodendron (Philodendron spp.)
Mildew	African violet (Saintpaulia ionantha)		Prayer plant (Maranta spp.)
	Begonia (Begonia spp.)		Wandering Jew (Zebrina spp.)
	Cissus ivies (Cissus spp.)		Zebra plant (Aphelandra squarrosa)
	Kalanchoe (Kalanchoe blossfeldiana cultivars)		

Second, the medium should be composed of ingredients in the proper proportions. This will assure that it drains quickly and thoroughly when watered. A drainage hole or other means for the escape of excess water should be provided in the pot. Media that retain an excessive volume of water or from which water is prevented from escaping have conditions leading to poor root and healthy microorganism growth. Water-logged media can frequently be traced as the initial cause of root rot diseases.

Figure 18-13 A fungal leaf spot disease on German ivy (*Senecio mikanioides*). (*Photo by Marge Coon*)

Finally, a medium should contain adequate nutrients. Poorly nourished plants, like those suffering from any other kind of stress, are commonly attacked by pests and diseases.

Humidity

The humid conditions optimal for the growth of most indoor tropicals can be both desirable and undesirable from the standpoint of pest and disease control. Perhaps the worst pests of houseplants, spider mites, are adversely affected by high humidity. Raising the humidity and misting can be used to slow their breeding and control their damage. But humidity can cause a foliar fungus problem because the spores of the fungus require high humidity to germinate. Overall, unless a plant is commonly susceptible to leaf spots or mildew and has been attacked previously (see Table 18-2), high humidity is advisable.

Water

Stagnant water, whether in the medium or on the crown, is detrimental to plant health. Crown rot disease, which attacks such houseplants as the African violet, has often been attributed to water left standing in the center of the plant after watering. Barrel-shaped cacti, in which the growing point is in a depressed area on top of the plant, are also susceptible to rot if water is left standing there.

In summation, although misting foliage and showering are recommended on most plants, they should be avoided under certain circumstances.

Isolation and Inspection

Isolation of newly acquired plants and older ones showing symptoms is a standard practice for controlling both diseases and insects. Isolation outdoors or in a separate room is preferred since air currents carry disease spores and insects readily.

Weekly inspection, mentioned in Chapter 14, will help catch disease problems before they become epidemic. Showering (except for disease-susceptible species) is also helpful in washing off insect pests before they become established.

Pest and Disease Control Methods

Modification of the environment, home remedies, and commercial chemicals all have use in controlling diseases and pests of houseplants. Modification of the environment for prevention has already been discussed, and the same instructions can be applied to plants that are already infested.

Home Remedies

Home remedies are generally used for pests and not diseases, and used with persistence, they can be effective. In addition, they are completely safe to humans and avoid any danger of accidental pesticide poisoning.

Wiping or washing with soapy water made from a few drops of detergent in a sink of water is a home remedy useful for reducing populations of mites or aphids. The pests can be sponged off with a dilute solution, or the entire plant can be held upside down and swished in the soapy water. Soap will not harm the plant, and the agitation will dislodge most of the pests. As a precaution against toxicity and to avoid soap film, rinsing of plants afterward is advisable.

Hand picking of large insects like sowbugs, scales, and mealybugs is another home remedy that works. Toothpicks or cotton swabs dipped in rubbing alcohol and touched to the insects (Figure 18-14) can be used . These remedies are most effective if only a few pests are present.

The key to success with both these methods lies in repetition. Probably not all of the pests will be killed, but the population will be reduced.

Figure 18-14 Touching mealybugs or aphids with a cotton swab dipped in alcohol is a reliable control method for small infestations. (*Photo by Kirk Zirion*)

Few home remedies are useful for soil insects. An atomizer filled with alcohol will kill many surface pests on contact, but the alcohol should not touch the plant.

Commercial Pesticides

Because of the constantly changing laws regarding pesticide use and the many new chemicals developed each year, it is useless to attempt to prescribe a chemical cure for every houseplant ailment discussed here.

Instead, chemicals currently approved by the Environmental Protection Agency to control diseases will be covered. All of these chemicals can be dangerous to the health of people, pets, and plants if not used in accordance with package directions and in observance of safety rules. Plant-derived insecticides should not be considered safer than synthesized ones.

The term "pesticide" is often used to refer only to "bug killers." But it actually includes all chemicals used to kill organisms that adversely affect the growth of plants including fungicides, bactericides, insecticides, miticides, nematicides, and herbicides. Insecticides are then further broken down by the source of the *active ingredient* (the chemical in the product that actually kills the pest, as opposed to the materials used to dilute, chemically "carry", or scent it).

Organic or Botanical Pesticides. These are derived from materials that occur in nature, primarily plants. Pyrethrum, rotenone, and nicotine are the most common.

Pyrethrum is a nerve poison, which was originally marketed in 1947. Its killing properties were excellent, but the chemical was not stable in sunlight.

This and the development of pest resistance to the original led to the development of a number of *synthetic pyrethroids.* The chemistry of these is based on the original, but varies sufficiently to overcome insect *resistance* (see p. 337). Resmethrin is an example of a synthetic pyrethroid. So many derivatives of the original pyrethrum exist that they are grouped into *generations* based on when they originated. The current generation is called the fifth.

Rotenone was first used in gardens in 1848 to kill caterpillars, but before that it was used in South America as a fish poison. It paralyzed the fish, causing them to float to the top where they could be gathered. It is derived from two genera of legumes and works by inhibiting a respiratory enzyme and causing the respiratory system to fail.

Nicotine is derived from tobacco and was used as early as 1890 to kill sucking insects on garden plants. It is a nerve poison.

Insecticidal Soaps. The insecticidal soaps are the fastest-growing branch of chemicals for houseplant pest control. They kill insects and insectlike pests such as mites by disruption of the cell and tissue structures by the action of potassium salts of fatty acids. They are of very low toxicity to humans.

Synthetic Pesticides. Among the synthetic pesticides only those of extremely low toxicity are approved for indoor use. Currently approved synthetics are mainly *organophosphates.* These insecticides were discovered in Germany in World War II when Germany was seeking alternatives to the insecticide nicotine, which was in short supply. They are nerve poisons and some have a systemic action, meaning they enter the plant, move within it, and kill whatever insects eat the plant or suck the sap. Currently used houseplant pesticides in this category are acephate (Orthene®), malathion, and disulfoton.

Implants. These are incapsulated "bullets" of insecticide which are implanted into trees for systemic control of insect pests or for alleviation of micronutrient deficiencies. Their advantage is that they apply the insecticide systemically in a "closed system," which means there is no pollution of the air with insecticidal spray. A hole is drilled in the trunk of the tree and the capsule inserted (Figure 18-15). The wound will heal, and the capsule will dispense insecticide into the phloem for 12 to 18 weeks.

Indoor trees such as *Ficus* with trunk diameters of 1½ (3–4 cm) or greater can be treated by this method for infestations of thrips, aphids, scales, mites, and white flies using acephate insecticide.

Figure 18-15 A medication-containing implant in place in a tree trunk. (*Photo courtesy of Creative Sales, Inc., Fremont, NE*)

Safety Precautions for Using Pesticides

Following general safety precautions for pesticide use is the first prerequisite for effectively controlling pests without harm to either the gardener or the plant being treated.

Personal Safety

The methods of applying pesticides vary. Sprays are most common for foliage insects, mites, and diseases. But drenches, in which the chemical is mixed in water and applied in place of plain water, are used for soil insects, nematodes, and root diseases. For personal safety, all contact with the pesticide should be avoided. Rubber gloves should be used and any skin spills should be washed off immediately with soap and water.

Pesticide sprays, whether aerosol or hand pump type, should be applied outdoors. During cold weather an unoccupied room that can be aired out afterward can be used. The plant to be treated should be placed on newspapers to prevent accidental spraying of carpets or furniture. Particular care should be taken when applying aerosol pesticides. They disperse small droplets into the air, and the airborne pesticide can be inadvertently inhaled.

After spraying is completed, safety precautions should include washing the spray bottle, measuring equipment, and hands with soap. The plant should not be touched by people or animals.

Some houseplant insecticides are applied to the soil as a liquid drench in place of plain water. Others may be applied to the soil surface as granules, and when the plant is watered, the insecticidal chemical is released. Sometimes these chemicals are *systemic*. These chemicals are taken up into the vascular system of the plant and poison insects and disease organisms which prey on that plant. Because a plant to which a systemic has been applied is toxic for a prolonged period, systemics are not often available for home use. Table 18-3 lists some of the pesticides used on houseplants and the pests they control.

Plant Safety and Pesticide Effectiveness

The first rule for plant safety is to follow the label directions. A stronger concentration than recommended may burn foliage severely, whereas a weaker one may fail to kill the pests. Many species of houseplants, including ivies, ferns, and succulents, are easily injured by pesticides; these susceptible plants should be listed on the pesticide label. Test spraying several leaves is always advisable when treating a new species. The plant should be observed for several days for signs of reaction to the pesticide before the remainder of the foliage is treated.

In the application of water-diluted spray pesticides, "spray to runoff" is usually advised. This means that the chemical should be applied heavily enough to drip from the leaves. Coverage should include all aboveground plant parts and the undersides of leaves, even those not presently infested. Aerosol pesticides can present dual hazards to plant health. The propellant used to carry the pesticide may be toxic, and the heat pulled from the leaves when the propellant vaporizes may cause cold injury. To prevent the latter, spray no closer than the directions advise. At a moderate spray distance most of the propellant will vaporize before touching the leaf.

When applying soil drenches, the chemical should be mixed according to directions given for a drench. The pesticide should be applied in sufficient quantity to wet all the potting medium, and any that drains out should be discarded.

Repeat sprays as recommended on package directions are the key to successfully controlling most houseplant pests. Many generations and life stages, from egg to adult, occur at one time, and not all will be chemically susceptible. Repeat sprayings are necessary to do more than temporarily slow reproduction rates.

Resistance of insects and mites to specific pesticides is occasionally encountered in houseplants.

Table 18-3   *Selected Houseplant Pesticides*[a]

Chemical name	Uses
Rotenone	Plant-derived contact and stomach poison for insects; used for aphids, thrips, whiteflies, mites, and immature scales and mealybugs
Resmethrin	Synthetic pesticide similar chemically to rotenone; used for the same pests
Pyrethrum	Plant-derived insecticide with contact action; effective against the same pests as rotenone
Copper oleate and other copper compounds	Fungicide for powdery mildew and other fungus diseases and some bacteria; registered for use on fuchsia, ferns, ivy, gardenias, philodendrons, geraniums, begonias, and succulents
Malathion	Insecticide for aphids, red spider and other mites, mealybugs, thrips, whiteflies, and root mealybugs; not recommended for use on crassulas, *Pteris* ferns, *Adiantum* ferns, anthuriums, or kalanchoes
Kelthane®	For red spider and other mites; registered for use on gardenias, rubber plants, bracken ferns, Boston ferns, ivy, African violets, and philodendrons
Nicotine sulfate	For aphids, thrips, spider mites, orchid scales, and fungus gnats; registered on philodendrons, fuchsia, ferns, and African violets
Di-Syston®	Used for aphids, thrips, mites, and whiteflies; granular systemic applied to the medium and watered in
Orthene®	For aphids, thrips, leaf miners, mealybugs, scales, spider mites, and whiteflies; use with caution on gloxinia, hibiscus, and philodendron
Oxydemeton-methyl (Metasystox-R®)	Sold in granular form for use on a number of species
Karathane	Miticide and fungicide for powdery mildew, spider mites, and two-spotted mites; registered for use on begonias and cineraria
Methoprene (Altosid®)	Insect growth regulator which prevents maturation of larvae of insects
Pentac®	Contact poison for mites; long residual action

[a]All pesticides must, by law, be approved and registered with the Environmental Protection Agency and state agencies for use on any plant species. The above table is not intended as authorized recommendations for pesticide use since the registrations are constantly changing.

Generally all but a few pests will be killed by a chemical, but those remaining continue to breed and eventually build a *resistant population* that cannot be killed by the insecticide. Resistance can be overcome in part by spraying alternately with different pesticides.

Finally, never save leftover mixed insecticides. Mixed pesticides are often inactivated in storage and are a common cause of accidental poisonings.

Selected References for Additional Information

Alfieri, S. A., Jr., K. R. Langdon, and C. Wehlburg. *Index of Plant Diseases in Florida.* Bulletin 11. Gainesville, FL: Division of Plant Industry, Florida Department of Agricultural and Consumer Services, 1984.

Ali, A. D. "A Back to Basics Look at Pests." pp. 90–95 in *Interior Landscape Industry,* Vol. 6(1), January 1989.

Chase, A. R. *Compendium of Ornamental Foliage Plant Diseases.* St. Paul, MN: ARS Press, 1987.

Lindquist, Richard. "Pesticide Resistance: Why It Occurs, How to Deal with It." pp. 56–59 in *Interior Landscape Industry,* Vol. 6(3), March 1989.

Manaker, G. W. *Interior Plantscapes.* 2nd ed. Englewood Cliffs, NJ: Prentice-Hall, 1987.

Pirone, Pascal P. *Diseases and Pests of Ornamental Plants.* 5th ed. New York: Wiley, 1978.

Simone, Gary. "New Disease Problems in the Foliage Industry." pp. 36–42 in *Interior Landscape Industry,* Vol. 6(4), April 1989.

Simone, Gary. "Virus Diseases Affecting Foliage." pp. 70–84 in *Interior Landscape Industry,* Vol. 5(6), May 1988.

Simone, Gary W., and A. R. Chase. *Disease Control Pesticides for Foliage Production,* p. 30. Gainesville, FL: University of Florida Institute of Food and Agricultural Science, 1989.

Simone, Gary W., T. A. Kucharek, and R. S. Mullin. *Plant Disease Control Guide.* Gainesville, FL: University of Florida Institute of Food and Agricultural Sciences, 1989.

19

Decorating with Growing Plants and Fresh Flowers

Plants are beautiful creations in themselves. But as with other works of beauty, a little artistic treatment will enhance and focus attention on their most attractive features. A new field of "interior landscaping" or "interior plantscaping" has evolved to deal with the selection and arrangement of indoor plants. An interior planting specialist must have not only a knowledge of artistic design but also an understanding of the care requirements associated with particular species. This person must be an artist *and* a horticulturist, arranging plants in a manner that is pleasing visually while satisfying their growth requirements.

Most experienced houseplant growers have already tried their hands at interior landscaping, although they may not realize it. For example, when the host or hostess moves an attractive plant to the coffee table before company comes, he or she is practicing interior landscaping. Similarly, selecting attractive pots for houseplants is part of the design process. With a bit of artistic sense, knowledge of plant growth requirements, and sample ideas given in this chapter, anyone can create an interior landscape.

A Historical Perspective

Using plants for indoor decoration is not a new idea. Indoor plants were much in vogue during the 1800s, and plant explorers undertook lengthy expeditions to secure exotics for wealthy patrons. Unfortunately, the preference for dark interiors and the lack of central heating made it impossible to grow all but the hardiest species indoors. Cast iron plants, potted palms, and rubber plants were the standard indoor species. Boston ferns also adapted to the low light and cool temperatures and were proudly displayed on ornate fern stands (Figure 19-1).

Partly because of the inadequate indoor growing conditions, the terrarium and solarium gained favor as plant-growing areas. Terrariums were originally designed as a method of holding plants during sea voyages. From this basically utilitarian structure, they developed into elaborate "parlor cases," highly embellished in the style of the Victorian era. It was not unusual for a terrarium of that period to be made of hand-blown glass, with its supporting structure of filigreed iron or brass or enameled metal.

The solariums or conservatories of that period were the garden rooms of today. They were used to grow and display many exotics, a hobby of the rich during that period. With the turn of the century and the death of Queen Victoria, the fashion in interior decoration changed to a lighter and airier look. The heavily draped floor-to-ceiling windows typical of Victorian homes were uncovered, and light lace curtains or simple valances allowed light to enter homes freely. However, central heating still did not exist, and this precluded growing many of the cold-sensitive species that are so popular today.

After World War I, middle-class bungalow-style houses were built with large window areas and, frequently, with sun porches in the front. By this time more low-light-requiring houseplants had been discovered, and central heating was becoming common. This facilitated indoor plant growing and removed it from being the exclusive province of the rich. But the

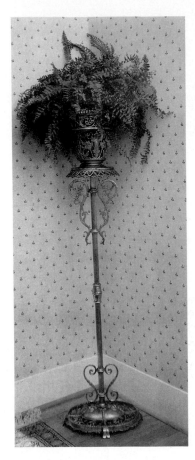

Figure 19-1   A bronze Victorian fern stand. (*Photo by Andrew Schweitzer*)

original plant fervor of the 1800s never returned, and there was a slump of interest until the late 1960s. Architecture reflected this lack of enthusiasm for indoor plants. Ranch-style houses became popular. Although many had large picture windows, these were designed for viewing and were seldom used for plants. The remaining windows were small or, at most, medium-sized, with emphasis on sleek, horizontal lines.

In the 1970s, the pendulum swung back, and indoor plants are very much in demand for homes, offices, and shopping malls. Houses are being designed with large expanses of thermal pane glass that not only provide enough light for growing almost any houseplant but are also energy efficient. Indoor gardeners are no longer restricted to pots on a windowsill, and large interior plants are flourishing.

From demand comes supply, and the horticulture industry has risen to the occasion. Large specimens of many species are sold today that years ago would have been impossible to find. Species that were previously

obscure are now common, and innumerable new cultivars have been and are being developed.

All of this presents today's indoor gardeners with a wide variety of plants and accessories from which to choose. Their only problem will be determining how to fit plants into the living area attractively—the subject of this chapter.

Functional Uses of Plants

Specimen Plants

Specimen plants (Figure 19-2) are those displayed purely for their aesthetic appeal. They are often large trees or floor plants, although they may also be hanging baskets, flowering species, terrariums, or bonsai. Specimen plants are generally used alone and can be thought of as living sculpture. In some instances a specimen plant will become the focal point of the room, attracting more attention than any of the other furnishings.

Specimen plants are usually expensive and should be considered a major plant purchase. Before you invest in one, try to grow a small plant of the species you intend to buy so that there will be less chance of your losing the plant from improper care. Table 19-1 lists a number of plants available in large sizes suitable as indoor specimens.

Plants as Wall Art

A bright wall near a window can be an ideal spot for a wall grouping of plants (Figure 19-3). The grouping takes the place of a picture or other wall decoration.

Pots can be attached to the wall with the type of clips used to hang plants on fences in outdoor gardens, and a structured design or random pattern can be used. Interest can be added by using unusual potting containers such as antique tins and handmade pottery, although this will make hanging more difficult.

Among the plants that lend themselves most readily to wall grouping are epiphytes. Certain orchids, bromeliads, philodendrons, and ferns can be grown mounted on osmunda or bark plaques. These are lightweight and easily hung.

Plants That Accent Other Objects

Another function of plants can be to draw attention to some other element of the room design, to accent it. An example of this is shown in Figure 19-4

Figure 19-2 The specimen bamboo in the corner is the focal point of the room and takes advantage of the light entering through the high windows near the cathedral ceiling. (*Photo courtesy of Landscape Images, a Division of British Indoor Gardens, Inc., El Toro, CA*)

Table 19-1 *Selected Large-Specimen Plants for Indoor Culture*

Common name	Botanical name
Norfolk island pine	*Araucaria heterophylla*
Bamboo	*Bambusa* spp.
Umbrella plant	*Brassaia actinophylla*
Cacti	*Cereus, Echinocactus, Ferocactus*
Palms	*Chamaedorea, Howea, Rhapis*
Jade plant	*Crassula argentea*
Cyperus	*Cyperus alternifolius*
Dumbcane	*Dieffenbachia maculata*
Thread-leaf aralia	*Dizygotheca elegantissima*
Corn plant	*Dracaena fragrans* 'Massangeana'
Red-margin dracaena	*Dracaena marginata*
Euphorbia	*Euphorbia millii, Euphorbia lactea, Euphorbia trigona*
Fatsia	*Fatsia japonica*
Weeping fig	*Ficus benjamina*
Rubber tree	*Ficus elastica*
Fiddle-leaf fig	*Ficus lyrata*
Wax-leaf privet	*Ligustrum lucidum*
Avocado	*Persea americana*
Philodendron	*Philodendron, Monstera*
Yew pine	*Podocarpus macrophyllus* or *Podocarpus gracilior*
Ming tree	*Polyscias* spp.

with flanking palms. Other examples of plants used as accents would be a small fern setting on an antique desk or a collection of small potted plants on a table as background for one or more family photographs.

Plants as Traffic Channelers

People generally walk the shortest distance from room to room by the route that offers the least resistance. Once these traffic patterns are established, they can be difficult to change, but strategically placed indoor plants can be helpful in this regard. By placing several plants between a chair and sofa, for example, people will naturally be directed to enter the sitting area by another way. Figure 19-5 shows a dramatic and obvious use of plants as traffic channelers.

Plants as Screening

Visual screening is another functional use of plants. A strategically placed group of plants can separate two living areas (Figure 19-6).

Figure 19-3 A wall arrangement of plants.

Figure 19-4 Palms on either side of the entryway are used as accents to focus attention onto the living area and fireplace beyond. (*Photo courtesy of Landscape Images, a Division of British Indoor Gardens, Inc., El Toro, CA*)

Figure 19-5 Built-in free-form planters filled with palms and peace lilies channel guests from the living room to the dining room in this residence. (*Photo courtesy of Landscape Images, a Division of British Indoor Gardens, Inc., El Toro, CA*)

If the screening problem is an objectionable window view, plants are the ideal solution. They thrive in the light, yet will not exclude the sunlight as a curtain would. Figure 19-7 shows how plants can be used for screening in a bathroom.

Plants as Space Fillers

There are often areas of a room in which no furniture can be conveniently placed because of space or design limitations. These are "dead" spaces which need to be filled to avoid an unbalanced look in the decor. Plants are the ideal solution since they can be bought in any size and shape, and can be chosen to live in the amount of light provided in the area. They are also very economical compared to art or furniture. Figure 19-8 shows a wonderful use of plants to fill the area under a spiral staircase.

Elements of Design

After approaching the functional use of plants and the restricting environmental requirements (primarily available light) that determine placement, the design elements are considered. These include

1. Color
2. Size
3. Texture (leaf size)
4. Decor association of the species

Figure 19-6 Fishtail palms in the center are both specimens and provide screening between two living areas in this large home. Their height bridges the gap between the cathedral ceiling and the low seating areas thus establishing human scale. (*Photo courtesy of Landscape Images, a Division of British Indoor Gardens, Inc., El Toro, CA*)

Figure 19-7   A built-in planter provides the screening for privacy necessary for this tub. (*Photo courtesy of Landscape Images, a Division of British Indoor Gardens, Inc., El Toro, CA*)

Color

Suitability of color to decor becomes an issue whenever any but a plain green plant is selected. If a plant has white, yellow, pink, or red on its foliage, this accent must be acceptable in the room color scheme. Flowering plants often have large expanses of color displayed as flowers (already attention getters) and must be used with careful forethought.

The range of variegated and color-accented foliage plants has expanded greatly in the past 15 years and hundreds of new cultivars are now available (Table 19-2). Likewise short-term and long-season flowering plants of many new species and cultivars are now commonly available (see Table 14-1), and using them effectively in a design requires mainly knowing their light requirements and functions in the room design (normally as a specimen or specimen group) after which the color can be selected.

One general rule is to limit brightly colored foliage and flowering specimens and use them against "neutral" green plant backgrounds. Exceptions are bromeliads and focusing groupings of massed flowering plants. However, if a room's furnishings are green,

an interior landscaper should be wary of using more green in the plants. Some contrast, in the form of lighter tones of green or variegation, is advisable to relieve the monotony.

Size

Size, both at time of purchase and ultimately when the plant is fully grown, is the second consideration. Plants may feel close, but should never be of a size that people feed crowded by them. For this reason, palms, hanging baskets, and the like should be chosen and positioned with their subsequent growth in mind. Variation in size (trees in back progressing to ground covers or creeping plants in front) is a major design issue in commercial interior landscapes such as those in shopping malls, but seldom is there adequate room in a private home for such a large plant area. Still a small "plant area" is an effective visual showcase for plants of varying sizes.

Houses or apartments with sliding glass doors or floor- to-ceiling windows are particularly adaptable to this idea. In its simplest form an area of roughly 2 to 10 square yards or meters of floor space is selected. The

Figure 19-8 Bromeliads and other species fill the otherwise "dead" area under the stairs, and a bamboo in the corner is a specimen focal point. (*Photo courtesy of Landscape Images, a Division of British Indoor Gardens, Inc., El Toro, CA*)

plants are then arranged within the area, with taller specimens in the back and shorter ones in the front. Plant stands, sections of logs, and small display stands can all be used to create height differences between plants and move the viewer's eye through the display. Hanging baskets can also be incorporated and add to the variety of form and texture.

A plant area can be placed directly on the floor, and plant saucers can be used to keep moisture from the carpet. An even more attractive display can be created by laying down a sheet of heavy plastic and covering it with a single layer of used bricks or bark chunks. This defines the area and attracts attention.

Texture

Texture refers primarily to leaf size although design professionals also take into account leaf surface character, leaf thickness, leaf density, and branch and petiole arrangements in assessing this quality. The three common designations are fine, medium, and coarse. The key to an interesting interiorscape is to vary texture.

Texture also needs to coordinate with room size. A small room will be overwhelmed by coarse-textured plants because they are viewed too closely, although one such plant could be acceptable as a specimen since its texture would be its specimen-making feature.

Decor Association

Decor association is the feeling people have regarding which indoor species are traditionally associated with which type of decor. While many plants are decor neutral and have no association, others can be selected to intensify the effect of other elements of interior decoration. Some examples of plants for antique, country, tropical, modern, oriental, masculine, feminine, and southwestern decors are given in Table 19-3.

Obviously pots selected for both material and form will enhance the decor association of a plant, although the standard unglazed clay pot in all its forms remains a classic.

The Design Process

A scale drawing of the room or rooms to be landscaped is the first step in the design process. It should include furniture placement, paths normally taken from room to room, and light levels throughout the area, designated as bright, moderate, or dim (Figure 19-9). Next the need of plants for functional uses discussed earlier should be assessed and then general areas where plants are to be placed shaded in on the drawing. The final step is to select the plants for each area based on the elements of design discussed in the previous section and on the light requirements of the species. Favorite species to be included or existing plants that will need to be incorporated can be selected and positioned first, after which a shopping list of all others can be prepared.

The final touch on the project is the selection of accessories, mainly pots, but sometimes plant stands, pot covers, or plant furniture.

Clear plastic tables (Figure 19-10) function as terrariums and are available through furniture stores and plastics manufacturers. They provide the high humidity needed by many ferns and orchids and can be a handsome addition to a modern decor.

Hanging baskets can also be equipped with artificial lights. They can be used for supplemental illumination in low-light areas or strictly for decorative night lighting.

Plant stands popular during the Victorian era are in style again, made from such materials as wood, wicker, and wrought iron. These pedestal-type stands can greatly enhance plants and give an illusion of

348 Growing Plants Indoors

Table 19-2   Foliage Plants That Give Color in an Interiorscape[a]

Variegated white:	Alocasia 'Frydek'	Elephant's ear
	Bromeliads	*Neoregelia carolinae meyendorfii* and *N. c. tricolor*
		Nidularium innocentii lineatum
	Chinese evergreens	*Aglaonema* 'White Raja'
		A. 'Superba'
		A. 'King of Siam'
		A. 'Fransher'
	Dumbcane	*Dieffenbachia amoena* 'Tropic Snow'
		D. maculata 'Camille'
		D. 'Perfection Compacta'
		D. 'Hilo'
		D. 'Rudolph Roehrs'
		D. 'Paradise'
	Dracaenas	*Dracaena fragrans* 'Warnecki'
		D. sanderiana
		D. surcolosa 'Florida Beauty'
	English ivy	*Hedera helix* 'Glacier'
	Nerve plant	*Fittonia verschaffeltii argyroneura*
	Panda plant	*Kalanchoe tomentosa*
	Baby rubber plant	*Peperomia obtusifolia variegata*
	Pineapple	*Ananas comosus* 'Variegatus'
	Pothos	*Epipremnum aureum* 'Marble Queen'
	Spider plant	*Chlorophytum comosum variegatum*
	Snake plant	*Sansevieria trifasciata* 'Bantel's Sensation'
	Wax ivy	*Senecio macroglossus* 'Variegatum'
	Wax plant	*Hoya carnosa* cultivars
	Wandering Jew	*Tradescantia albiflora* 'Albo-vittata'
	no common name	*Xanthosoma lindenii*
	Zebra plant	*Aphelandra squarrosa*
Variegated silver:	Aluminum plant	*Pilea cadierei*
	Begonia	*Begonia* × *argenteo-guttata*
		B. 'Rex' cultivars
	Bromeliads	*Tillandsia edithae*
		T. xerographica
		T. tectorum
	Chinese evergreen	*Aglaonema* 'April in Paris'
		A. 'Mona Lisa'
		A. 'Silver King'
		A. 'Silver Queen'
		A. 'Rhapsody in Green'
	Flame violet	*Episcia cupreata*
	Peace lily	*Spathiphyllum cannifolium* 'Silver Streak'
	Rosary vine	*Ceropegia woodii*
	Strawberry begonia	*Saxifraga stolonifera*
	Watermelon begonia	*Peperomia argyreia*
Variegated yellow or pale green:	Corn plant	*Dracaena fragrans* 'Massangeana'
	Dumbcane	*Dieffenbachia* 'Triumph'
		D. 'Bali Hai'
	English ivy	*Hedera helix* 'Golddust'
	no common name	*Holalomena wallisii* 'Camouflage'
	Japanese acuba	*Acuba japonica*
	Pothos	*Scindapsus aureus* 'Golden Pothos'

(continued)

Table 19-2 *Foliage Plants That Give Color in an Interiorscape*[a] (continued)

	Vase plants (bromeliads)	*Aechmea chantinii* 'Vista'
		Billbergia pyrimidalis striata
		B. 'Fantasia'
		Cryptanthus beuckeri
		Guzmania vittata
	Wax ivy	*Senecio macroglossus* 'Variegatum'
	Bromeliads	*Billbergia vittata*
Purple-foliaged or -accented:	no common name	*Billbergia* 'Muriel Waterman'
		Neoregelia melanodonta
		Hemigraphis 'Exotica'
	Moses-in-the-Cradle	*Rhoeo spathaceae*
	Persian shield	*Strobilanthes dyeranus*
	Prayer plant	*Calathea concinna*
	Purple passion plant	*Gynura aurantiaca*
	Tahitian bridal veil	*Gibasis geniculata*
	Wandering Jew	*Zebrina pendula*
Pink-Accented:	Bromeliads (vase plants)	*Aechmea luddemanniana* 'Mend'
		Cryptanthus 'Starlight'
		C. 'Pink Starlight'
		C. 'Carnival de Rio'
	Chinese evergreen	*Aglaonema* 'Rembrandt'
	Episcia	*Episcia* 'Pink Brocade'
	Jade plant	*Crassula argentea* 'Tricolor'
	Polka dots	*Hypoestes sanguinolenta* 'Pink Splash'
	Prayer plant	*Maranta leuconeura erythroneura* and other cultivars
	Strawberry begonia	*Saxifraga stolonifera* 'Tricolor'
	Wax plant	*Hoya carnosa* rubra
		H. c. compacta regalis
Red-accented:	Abyssinian banana	*Ensete ventricosum* 'Maurelii'
	Beefsteak plant	*Iresine herbstii*
	Bromeliads	*Aechmea* 'Burgundy'
		A. maculata
		A. 'Royal Wine'
		A. 'Foster's Favorite'
		Cryptbergia 'Rubra'
		Cryptanthus zonatus
		C. 'Feuerzauber'
		Neoregelia marmorata × *spectabilis*
		N. 'Fireball'
		Nidularium innocentii 'Innocentii'
		N. billbergioides
		Vriesia 'Kitteliana'
	Coleus	*Coleus blumeii* cultivars
	Nerve plant	*Fittonia verschaffeltii*
	Prayer plant	*Calathea* 'Red Star'
	Rex begonia	*Begonia* 'Rex' cultivars
	Red-margined dracaena	*Dracaena marginata*
	Ti plant	*Cordyline terminalis*

[a]The variegation on these plants may become yellow in insufficient light.

Table 19-3 *Decor-Associated Plants*

Decor	Common name	Botanic name
Antique/ Victorian:	Asparagus ferns	*Asparagus* spp.
	African violet	*Saintpaulia ionantha*
	Begonias	*Begonia* spp.
	Cast iron plant	*Aspidistra elatior*
	Fern	*Nephrolepis exaltata* cvs. (Boston fern)
		Adiantum spp. (Maidenhair fern)
	Gloxinia	*Sinningia speciosa*
	Palms	*Howeia fosteriana* (Kentia palm)
		Chamaedorea elegans (Parlor palm)
		Rhapis excelsa (Lady palm)
		Chrysalidocarpus lutescens (Areca palm)
	Peperomia	*Peperomia caperata* 'Emerald Ripple'
	Rosary vine	*Ceropegia woodii*
Country:	Baby tears	*Helxine soleirolii*
	Black-eyed Susan vine	*Thunbergia alata*
	English ivy	*Hedera helix*
	Geranium	*Geranium hortorum*
	Grape ivy	*Cissus rhombifolia*
	Mock strawberry	*Duchesnea indica*
	Piggyback plant	*Tolmiea menziesii*
	Primrose	*Primula obconica* or *malacoides*
	Strawberry begonia	*Saxifraga stolonifera*
	Swedish ivy	*Plectranthus australis*
Oriental:	Balfour aralia	*Polyscias balfouriana*
	Bamboo	*Bambusa, Phyllostachys* and *Pseudosasa*
	Bamboo palm	*Chamaedorea erumpens*
	Ming aralia	*Polyscias fruticosa*
	Mother fern	*Asplenium viviparum*
	Norfolk island pine	*Araucaria heterophylla*
	Threadleaf aralia	*Dizygotheca elegantissima*
Masculine:	Corn plant	*Dracaena fragrans* 'massangeana'
	Cycads	*Zamia* and *Cycas* spp.
	Dumbcane	*Dieffenbachia* species and cultivars
	Fiddleleaf fig	*Ficus lyrata*
	Fish tail palm	*Caryota mitis*
	Green cliffbrake	*Pellaea viridis*
	Hare's foot fern	*Polypodium aureum*
	Iron cross begonia	*Begonia masoniana*
	Japanese aralia	*Fatsia japonica*
	Moses-in-the-cradle	*Rhoeo spathaceae*
	Pony tail palm	*Beaucarnea recurvata*
	Rex begonia	*Begonia* 'Rex' cultivars
	Snake plant	*Sansevieria trifasciata* cultivars
	Staghorn fern	*Platycerium bifurcatum*
Feminine:	Boston fern	*Nephrolepis exaltata* (cultivars that are highly cutleaf, such as 'Fluffy Ruffles', 'Verona', and 'Shadow Lace')
	Cyclamen	*Cyclamen persicum giganteum*
	Dauphin violet	*Streptocarpus saxorum*
	Episcia	*Episcia* cultivars
	Gardenia	*Gardenia jasminoides*
	Maidenhair ferns	*Adiantum* spp.
	Miniature rose	*Rosa chinensis hybrida*
	Orchids	See list in Chapter 14

(continued)

Table 19-3 *Decor-Associated Plants* (continued)

Decor	Common name	Botanic name
Tropical:	Plumosa fern	*Asparagus plumosus*
	Tahitian bridal veil	*Gibasis geniculata*
	Bird's nest fern	*Asplenium nidus*
	Bromeliads	See list in Chapter 14
	Croton	*Codiaeum variegatum* cultivars
	Orchids	See list in Chapter 14
	Palms	*Howiea, Chamaedorea, Chrysalidocarpus,* etc.
	Philodendron	*Philodendron* spp.
	Prayer plants	*Maranta* and *Calathea*
	Split-leaf philodendron	*Monstera deliciosa*
	Staghorn fern	*Platycerium bifurcatum*
	Umbrella plant	*Brassaia* and *Tupidanthus*
Modern:	Bromeliads	See list in Chapter 14
	Cacti	See list in Chapter 14
	Moses-in-the-cradle	*Rhoeo spathaceae*
	Peace lily	*Spathiphyllum* cultivars
	Purple passion plant	*Gynura aurantiaca*
	Pyrrosia fern	*Pyrrosia macrocarpa*
	Red-margined dracaena	*Dracaena marginata*
	Snake plant	*Sansevieria trifasciata* cultivars
Southwestern:	Cacti and succulents	See list in Chapter 14
	Christmas cherry	*Solanum pseudo-capsicum*
	Peppers	*Capsicum annuum* cultivars with very small fruits, such as 'Fiesta' and 'conoides'

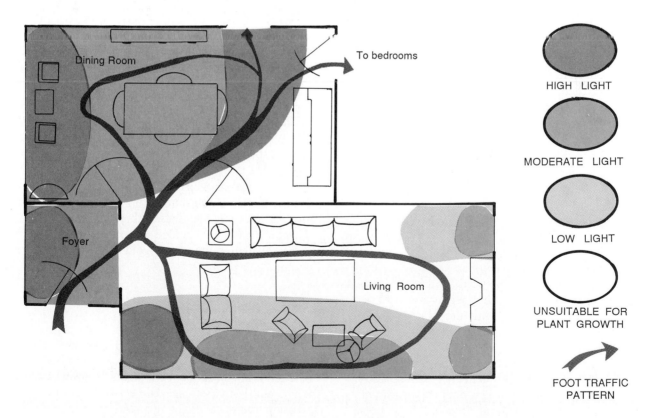

Figure 19-9 Basic room drawing for interiorscaping the main areas of a home.

Figure 19-10   A lucite coffee table used as a terrarium. (*Photo courtesy of Earthway Products*)

height thus substituting for more expensive large plants. Instructions for homemade plant stands, utilizing materials such as plywood and sections of clay drainage pipe, are also common in home decorating magazines.

Plant accessories (pots, hangers, stands) should be chosen carefully to enhance the plant's beauty. First, the size of the container should be in proportion to the size of the plant to avoid a top-heavy or bottom-heavy look. Double potting frequently creates a bottom-heavy appearance and emphasizes the need for proportion.

Second, select containers and accessories in design-coordinated colors, patterns, and styles. Simple unadorned clay pots or wicker pot covers are two of the most suitable. They blend with almost any decor and are classic in their appeal.

A bold container can visually overpower a plant, making the pot rather than the plant the focus of attention. However, this is a general guideline and exceptions exist. For example, a room in a Mexican motif might work nicely with bright red or yellow pots. A large specimen might look outstanding in a very elaborate antique container.

Use of Night Lighting

The decorative effects of lighting are well understood by interior designers. Night lighting, for example, is very striking when used with plants. One or two inexpensive spotlights can dramatize houseplants and provide supplemental lighting as well.

The lights can be positioned in many ways for accenting plants. Lights shining up from the floor create interesting ceiling shadows as they pass through plant foliage. Track lights aimed downward from several points on the ceiling can highlight individual plants within a collection.

Arranging Flowers

Using fresh, dried, or silk flowers in the home in place of, or to supplement, growing plants is another aspect of decorating indoors with plants. Flower arrangements add festivity to special occasions and bring the pleasing presence of living plants into a room that would otherwise be bare of plants due to an unsuitable growing environment. Creating custom floral designs is easy with simple instruction and a little practice. The focus of this chapter will be the care and arrangement of natural flowers with emphasis on fresh flowers, although the principles of arrangement can be applied to silk flowers as well.

Flower Selection

Flowers for arrangement can come from many sources: one's own garden, the wild, or by purchase.

Garden and Forced Flowers.

A perennial garden that has long-stemmed flowers—the most versatile for arranging—is one flower source. Table 9-6 lists flowers that can be grown in most climates for cutting and for drying for permanent arrangements. However, landscape shrubs and even trees should not be neglected as a source of floral materials for arranging. In early spring, flowering branches can be *forced* into bloom indoors to give a bright spot of color out of season (Figure 19-11). *Forcing* is bringing a plant into bloom ahead of its normal

Table 19-4 *Spring-Flowering Shrubs and Trees for Forcing*

Common Name	Botanical Name
flowering quince	*Chaenomeles japonica*
forsythia	*Forsythia* spp.
redbud	*Cercis* spp.
dogwood	*Cornus florida*
magnolia	*Magnolia* spp.
fruit trees	
peach, plum, pear, apple,	
apricot, cherry, quince	
witch hazel	*Hamamelis japonica*
white forsythia	*Abeliophyllum distichum*
buttercup winterhazel	*Corylopsis pauciflora*
shadbush	*Amelanchior* × *grandiflora*
Japanese apricot	*Prunus muma*
wisteria	*Wisteria sinensis*
	Wisteria japonica
crabapple	*Malus* spp.
flowering almond	*Prunus triloba*
pussywillow	*Salix discolor*

blooming season. Table 19-4 lists some shrubs and trees that set and mature their flower buds in fall. Because the buds are already mature and ready to open, the branches can be forced in early spring ahead of their normal blooming season. The earlier the shrub or tree blooms naturally, the earlier it can be forced. Branches that are picked before the buds are fully mature are incapable of continuing to mature those buds when severed from the parent plant and will simply die when brought in the house.

Examine the branches you are considering forcing to be sure that the flower buds are sufficiently mature. Mature flower buds are distinctly formed and easily distinguished from leaf buds by their larger size. In addition, they are usually round in shape, and should ideally be showing some color.

The forcing process requires very little in the way of special techniques. Simply cut the branches and place them in preservative solution (see section on Water and Preservative), then place them out of direct sun and wait for the buds to open. At normal household temperatures, this should take a week or less.

Collecting Wild Plant Materials

Wild plant materials, including many grasses found along the road (Figure 19-12), dried teasel in fall, cattails, native shrubs, goldenrod, sunflowers, and the like, can also be used to create striking floral arrangements. Be alert to sources of your own local wild

Figure 19-11 An arrangement of springtime and other flowers in a basket including forced forsythia branches on the right. (*Photo courtesy of Syndicate Sales, Design by Denise Lee, AIFD, PFCI*)

Figure 19-12 Wild roadside grasses are effective in natural-type arrangements. (*Photo courtesy of Sharp Brothers Seeds*)

materials as you drive by natural areas. They can be used to supplement purchased flowers or used alone. All florist flowers were originally wild, and nature is still an excellent source of arrangement materials.

Purchasing Flowers

Flowers from florists, a supermarket, or a roadside stand should be bought with care to be sure that they are fresh and will last in an arrangement. One key point to look for to assure the flowers were harvested within the last few days and are fresh is by checking to make sure they are not wilted. Check petal tips and especially the youngest buds, which will show aging first.

Lift the flowers out of the water and check the condition of the stems and leaves that have been submerged. If the leaves are beginning to soften or are very dark green, they have begun to decompose. The flowers have been picked too long ago and will not last well.

Occasionally, roses are "dry stored" before they are sold by the grower. This means that they are picked and then stored under very cool conditions without water. Although this practice is perfectly legitimate, if stored too long the roses will not open when put in water. The buds will remain closed and the stem under the bud (peduncle) will wilt, causing the flower to die prematurely. When roses are purchased and this occurs, return them to where you bought them, as the seller will certainly want to know that he or she has been sold roses that were stored too long.

Selecting Greens

Selecting *greens* means choosing the foliage that will be used to fill out the arrangement. With some arrange-

ment styles and species, the flowers have enough natural foliage that additional greens are not required. But if they are, they can be purchased from a florist or cut from landscape plants. Even in winter, landscape evergreens such as spruce, pine, and yew can provide foliage for arrangements.

Assembling Arranging Supplies

A flower arrangement normally consists of a vase or other decorative container, water and preservative, and sometimes, for more elaborate arrangements, an arrangement aid to hold the stems in position. In addition, shears to cut the flower stems are very useful.

Choosing a Vase or Other Arrangement Container

Vases that are neutral in color and simple in design, made of glass, ceramic, porcelain, or metal (Figures 19-13 and 19-14), are easiest for an amateur to work with. When choosing a vase, remember that the viewer's attention should focus on the flowers instead of on the vase.

The size of a vase can greatly affect how easy it is to create an arrangement. One useful size for an all-purpose mixed flower arrangement is about 8 inches high with a mouth about 3 inches across. This size is small enough that flowers of adequate height are easy to buy; most purchased flowers measure 18 inches to 2 feet tall.

A larger vase about 10 inches tall with a mouth 3–4 inches across and a smaller vase about 6 inches tall with a mouth about 2–3 inches across are also very useful. Vases with narrow mouths of less than 2 inches in diameter can be difficult to arrange flowers in without crowding.

If a flower arrangement is being made for a dinner party and will stay on the table during dinner, it is important that it not obstruct the guests' view of each other. For this reason, the container chosen should be shallow. A good rule of thumb is to select a container a maximum of 4 inches tall. When flowers are added, the arrangement will be 8 to 10 inches tall. A basket (Figure 19-15) can be a good choice, lined with a waterproof container that will not show, such as a plastic storage bowl.

Many everyday objects can function as containers for floral arrangements, if you use your imagination. A hollowed-out pumpkin can be used for

Figure 19-13 An assortment of containers suitable for flower arrangements. (*Photo courtesy of Syndicate Sales*)

Figure 19-14 An assortment of glass containers for flower arrangements. (*Photo courtesy of LBK Marketing Corporation*)

Figure 19-15 Baskets such as these are inexpensive containers for many types of floral arrangements. (*Photo courtesy of LBK Marketing Corporation*)

Thanksgiving or Halloween designs, and clay pots for natural designs. Decorative glass bricks, wine glasses, bottles, and so forth all make interesting and imaginative containers.

Water and Preservative

A vase should be washed thoroughly with detergent and rinsed well before use to remove all debris and dirt film. Unwashed, previously used vases harbor bacteria which will begin to multiply in water. The bacteria clog the xylem of the stem and block water flow to the blossoms, resulting in premature flower death.

The container need not be completely filled with water; it only needs to be about one-third to one-half full, depending on how much water is necessary to make the container stable. If the container is filled to the top, a greater proportion of the flower stems are submerged and subject to rotting, thus becoming a breeding ground for bacteria. Since water uptake is mainly through the cut stem base, only that part needs to be submerged to supply adequate water to the blossoms.

Use of commercial flower preservative can double flower life. However, the most economical preservative and one fully satisfactory for home use is a solution of one-third tonic water and two-thirds plain tap water. Tonic water acts as a floral preservative because:

- It contains a sugar to supply the cut flower with carbohydrate for respiration.

- It is acidic, which reduces the rate of bacterial growth in the water.

- It contains a preservative intended to retard microbial growth in the beverage, but which also retards microbial growth in the water in which it is mixed.

Tools and Supplies

For cutting flower stems, a sharp kitchen knife is satisfactory. However, if you do floral arrangements using wild or shrub materials you will find investing in floral shears worthwhile (Figure 19-16). The blades are very strong and capable of cutting woody stems easily. Floral shears are available through floral supply houses listed in the phone book.

Figure 19-16 Flower shears of the most popular type sold for commercial use. (*Photo courtesy of Larry Beck Company and Clause Cutlery Co.*)

For holding flowers in position, there are several alternatives. Simple "natural"-type mixed flower arrangements do not require an aid to arrangement, as the stems hold each other in position with a little careful placement.

One solution for holding flowers in position is a pinholder or dome type *frog* (Figure 19-17). Before the arrangement is started, the frog is pushed down on a pad of *floral clay,* a sticky, waterproof material that temporarily secures a frog to the base of the container and keeps it from tipping under the weight of the flowers it is holding. The stems are then pushed onto the spikes or between the metal wires to secure them in position.

Floral foam (Figure 19-18) allows an arranger greater creativity. The foam soaks up and holds water which can then be absorbed by the flowers. It is sold in blocks at craft and hobby stores for one-time use and can be cut to the desired size and held in position in the container with waterproof florist tape designed to secure under wet conditions. The advantage of the floral foam is that it can extend above the mouth of the container and flowers can be inserted into it at many angles, even pointing downward. This allows the creation of weeping-style arrangements. The stemmed glass containers shown in Figure 19-14 would be ideal for use with floral foam.

Principles of Flower Arrangement

Many different styles of arrangement are practiced throughout the world.

Traditional American Arrangements

Commercial florist arrangements made by professionals are commonly one of five or six types identified by *line,* meaning the primary visual orientation of the key flowers in the arrangement.

Circular design (Figure 19-19) is essentially a hemispherical shape designed to be viewed from all sides. Depending on its height and diameter, it would typically be used as a dinner table arrangement or small table arrangement.

Radiating design is similar to circular design. The essential difference is that while a circular design has a rounded, even surface, a radiating design has a stronger, more dramatic look because spike-type flowers are used in its construction. A radiating design can be viewed from either one or all sides, depending on personal preference and the amount of flowers available.

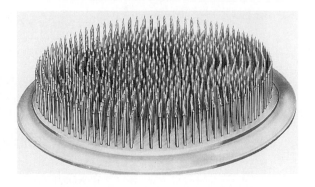

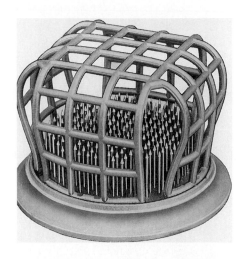

Figure 19-17 Two types of floral frogs: pin and dome. (*Photos courtesy of Beagle Industries*)

Figure 19-18 Floral foam can be cut to size to fit a particular container. (*Photo courtesy of Smithers-Oasis U.S.A.*)

Figure 19-19 Circular design of roses, lilies, and carnations. (Photo courtesy of Syndicate Sales; design by Denise Lee, AIFD, PFCI)

Vertical design (Figure 19-20) sometimes has few flowers, but striking or expensive ones. The emphasis in a vertical arrangement is similar to Ikebana (next section).

Horizontal design (Figure 19-21) style also adapts well for use as a centerpiece arrangement and is designed to be viewed from all sides and from the top as well. It is normally arranged in a shallow container. Creation of a horizontal design generally requires use of floral foam to position flowers horizontally and still supply them with water.

Triangular design (Figure 19-22) is, as it sounds, based on the shape of a triangle with a flat base and a pointed top. The primary lines are visually defined by the flowers that point to the tips of the triangle.

Hogarth or "S" curve is an uncommon and more difficult design to make because it requires creating a curved line to define the "S" curve. Wild plant materials such as vines can be used for this purpose as can flowers that have developed a natural curve in the stem during growth.

Asian-Style Floral Arrangements

Among the most popular of floral arrangement styles is Japanese (Figure 19-23). True Japanese Ikebana style is a carefully constructed arrangement of natural materials symbolic of elements in life: heaven, earth, and man, at descending levels within the arrangement, and sometimes other symbolic representations as well.

Figure 19-20 *Vertical design. (Photo courtesy of Syndicate Sales; design by Denise Lee, AIFD, PFCI)*

Figure 19-22 Triangular floral design. (*Photo courtesy of Syndicate Sales; design by Denise Lee, AIFD, PFCI*)

Figure 19-21 Horizontal design in a wire basket. (*Photo courtesy of Syndicate Sales; design by Denise Lee, AIFD, PFCI*)

Figure 19-23 Asian-style design. (*Photo courtesy of Syndicate Sales; design by Denise Lee, AIFD, PFCI*)

Figure 19-24 A modified vegetative design showing elements of Hogarth curve. (*Photo courtesy of Syndicate Sales; design by Denise Lee, AIFD, PFCI*)

Mastery of the art of Japanese flower arranging takes many years of study and practice. However, one can create a Japanese-style arrangement not strictly conforming to the rules of Ikebana arrangement, but still visually pleasing, by following a few simple principles.

First, all types of living materials can and should be used, not just flowers. In fact, a Japanese-style arrangement may not use any purchased flowers at all; it may be composed of twisted stems, seed pods, individual leaves, grasses, reeds, and fruits attached to branches. When choosing materials, the arranger should select and incorporate unusual and wild plant materials. In winter, leafless branches, evergreen boughs, and branches with clinging berries can be used.

Second, Japanese arrangements are almost always asymmetrical, meaning they are not even in shape all around as American-style circular and radiating designs typically are. In addition, the arrangement is also often designed to be viewed from one direction only instead of from all sides.

Japanese arrangements are normally made in neutral-colored glazed containers in brown, black, beige, or the like. A woven straw basket can also be used. Most interestingly, a shallow plate, provided it is adequately deep to hold some water and a frog, can be used quite effectively. The frog is generally essential in such a container to hold the floral materials in position.

European Arrangement Styles

In recent years, several floral arrangement styles originating in Europe have become popular, including the vegetative and landscape-style designs (Figure 19-24).

Vegetative Design. This design style often uses a shallow horizontal container. The container visually symbolizes the earth and serves as a *foundation,* as it is called. Moss, bark, and similar plant materials are used to create the effect of the earth and cover the foundation of the arrangement which is normally floral foam, but can also be one or more pinholder frogs.

The design concept is that the arrangement should symbolize a small slice of nature. Flowers are grouped by species in separate areas of the arrangement as they would grow naturally, instead of being spaced throughout the arrangement as is common in American designs. The materials are displayed at the angles at which they might grow naturally, instead of being positioned evenly and formally throughout the design as, for example, in circular design. The floral materials appear, in effect, to be growing out of the ground, and although the effect is not totally natural, the concept is conveyed.

Figure 19-25   Mille fleur design. (*Photo courtesy of American Florists Marketing Council*)

Mille Fleur Design

Translated as "1,000 flowers," this style of design (Figure 19-25) has been very popular for at least 10 years. It is a "mixed bouquet" in which at least four, and preferably many more, different types of flowers are combined in one arrangement in a seemingly random fashion to give a profusion of many colors. The design does not have the "line" of traditional American arrangement styles. The shape is loose and free, and the arrangement is designed to look as close as possible to a bouquet of garden flowers.

Mille fleur arrangements can be done without frogs or floral foam, although a more refined and formal effect can be achieved with a foam foundation which allows more precise positioning of the individual flowers.

Contemporary Designs

The amateur arranger can experiment with waterfall, *pavé,* sheltered, phoenix, and other relatively unusual contemporary designs. In waterfall design, the primary line and focal point of the arrangement is, as the name implies, downward sweeping. In *pavé* (meaning "paved") design, flowers are arranged so that the blooms

actually touch. This style was popular in Europe in the previous century and is experiencing a limited revival. In phoenix design, the classic story of the phoenix bird rising from the ashes is expressed in flowers, with strong emphasis on a vertically positioned flower or flowers in a visual nest of other flowers and greens.

Techniques for Making Arrangements

The technique of constructing an arrangement can be broken down into several steps, keeping in mind the design principles italicized and explained below.

Deciding on the Arrangement Style. Ideally, arrangement type is decided on before selecting flowers. But when flowers are a gift, this may not be possible, and the arranger will have to work with the flowers that are available. In either case, strive for visual harmony with the container and among the flowers. The shapes and colors should complement each other.

Arrangement style will also be limited by the types of vases available and by the purpose of the arrangement. If the arrangement is to be, for example, a table centerpiece, the choice of design is limited by height.

Preparing the Container. This includes cleaning and, if necessary, positioning a frog or foam block for stabilizing the flowers.

Developing the Line (Figure 19-26). The principal or "line" flowers, those that are to establish the outermost points of the design, should be put into the design. Strive for the proper proportion between the container and the height of the flowers. For most arrangements, the tallest flowers generally should be 1½ to 2½ times the height of the container.

Achieving "Visual Balance." Visual balance within an arrangement is achieved by placing the smaller flowers higher up in the arrangement and the larger ones lower to project visual stability and avoid a top-heavy look. Balance from side to side is achieved in symmetrical arrangement styles by finding an imaginary center line which extends through the visual center of the arrangement. The proper number of flowers are then positioned on each side of this imaginary line to give equal visual weight to each side.

Establishing "Unity." Unity within the design is established in traditional American arrangements by not bunching flowers of the same type together but instead by positioning them individually at varying levels throughout the entire arrangement. Using odd numbers (3, 5, 7, etc.) of each species of flower avoids a rigid, lined-up look.

(a)

(b)

(c)

Figure 19-26   The steps in creating a right triangle design: (a) inserting line flowers, (b) filling in with foliage and spike flowers, and (c) completing with focal flowers and fillers. Note that the visual line of balance is slightly to the left of center and not directly through the middle of the vase. (From *Floral Design and Arrangement* (3rd ed.) by Gary L. McDaniel. Englewood Cliffs, NJ: Prentice-Hall, 1996, pp. 93–94.)

Figure 19-27 Dried floral materials used in several arrangements.

"Filling In." This is the final part of constructing the arrangement. When the line flowers are positioned, fill in with greens, leftover buds, and the like. The goal is to hide all "mechanics" (floral foam, tape, etc.) to make it appear that the flowers are positioned without aids.

Dried Flowers

Dried flowers are arranged using the principles and design types of fresh flowers, although the arranger is re-stricted somewhat by the inflexible nature of the dried stems in regard to arrangements having curved lines (see Figure 19-27). References at the end of this chapter list several books on working with dried flowers. Table 9-6 lists a number of flowers suitable for drying.

Selected References for Additional Information

American Horticultural Society. *Houseplants* Vol. 1 of the AHS *Illustrated Encyclopedia of Gardening.* Mt. Vernon, VA: American Horticultural Society, 1980.

Austin, Richard L. *Designing the Interior Landscape.* New York: Van Nostrand Reinhold Co., 1985.

Elbert, George, A., and V. Elbert. *Foliage Plants for Decorating Indoors: Designs, Maintenance for Homes, Offices and Interior Gardens.* Portland, OR: Timber Press, 1989.

Fitch, Charles M. *Fresh Flowers, Identifying, Selecting and Arranging.* New York: Cross River Press Ltd./Abbeville Press, 1992.

Gaines, R. L. *Interior Plantscaping: Building Design for Interior Foliage Plants.* New York: McGraw-Hill, 1977.

Hunter, Norah. *The Art of Floral Design.* Albany, New York: Delmar Publishers, 1994.

Kramer, J. *Indoor Trees.* New York: Universe, 1984.

Manaker, G. W. *Interior Plantscapes.* 2nd ed. Englewood Cliffs, NJ: Prentice-Hall, 1987.

Packer, June. *The Complete Guide to Flower Arranging.* New York: Doring Kindersley, 1992.

Snyder, Stuart D. *Building Interiors, Plants and Automation.* Englewood Cliffs, NJ: Prentice-Hall, 1990.

20

Greenhouses and Related Climate-Controlling Structures

A home greenhouse is regarded as the ultimate luxury by many gardeners, conjuring up visions of exotic or bright-light-requiring plants which cannot be successfully grown indoors. But then the realities of expensive construction and heating costs are remembered. With fuel costs rising each year, a greenhouse can be a costly hobby or even a frivolous waste of energy. However, it is possible to have a greenhouse or related structure that is both practical and affordable if the selection is made wisely. The key is to balance the advantages and disadvantages of each of the following structures with regard to range of usefulness, initial construction, labor, materials cost, and heating and cooling expenses. The gardener can then select the structure appropriate for his climate, budget, and hobby interest.

Outdoor Plant-Growing Structures

Cold frames, hotbeds, sun-heated pit greenhouses, traditional greenhouses, and shade houses are included in this category. With the exception of shade houses, each structure provides some heat, whether by the sun, decomposing manure, oil, gas, electricity, or a combination.

Cold Frames

A cold frame (Figure 20-1) could be called the "poor man's greenhouse" because it is the least expensive plant-growing structure. Once a common feature of almost every gardener's yard, it is less used now, having been succeeded by more sophisticated structures.

Sunlight heats a cold frame by a phenomenon called the *greenhouse effect,* which occurs in any translucent or transparent structure. Sun passes through the covering, falls on plant leaves or other surfaces, and is changed into infrared (heat) energy. This heat will then be radiated or conducted back into the air and is held in by the covering.

Uses of Cold Frames

Cold frames have the most limited usability but are practical for some plants. They can be used to harden off greenhouse-grown transplants, giving cold and wind protection during the transition period and reducing shock after transplanting. They can also be used for raising transplants of cool-season vegetables such as cabbage, cauliflower, and lettuce.

In areas with cool, short, or foggy summers, melons, cucumbers, and sweet potatoes can be grown to maturity in the frame and will produce exceptional yields because of the extra heat. In fall the growing season can be extended for short-term crops of lettuce, radishes, or beets, and they can be harvested into early winter.

In cold climates, seedlings of cool-season flowers like the pansy can be grown to transplant size before being moved into a greenhouse. Hardwood cuttings can also be rooted and overwintered. A cold frame is excellent for precooling (vernalizing) bulbs to be forced into bloom in winter and for protecting semi-cold-hardy

Figure 20-1   A peaked-roof cold frame with a plastic covering which has been rolled up. (*USDA photo*)

plants, which would otherwise freeze out over the winter.

Cold Frame Construction

A cold frame is relatively simple to construct because it does not require electricity or a foundation. Some resemble miniature greenhouses (Figure 20-1, whereas others have only one area of glass or plastic facing south or southeast (Figure 20-2).

   The site selected for the cold frame should be shielded from winds, unblocked by shade from trees or buildings, and well drained. The size decided on will depend on the types of plants that will be raised. For a few transplants, cuttings, or potted bulbs, a 3 × 6-foot (1 meter × 2 meters) frame is adequate and can be readily covered with a 3 × 6-foot (1 meter by 2 meters) window sash. If film plastic covering is used, size can be flexible because the plastic can be cut to fit. The maximum span of a peaked plastic frame should not, however, be over 7 feet (2.3 meters) and its length 12 feet (3.7 meters) (without additional support) so that

water will not collect and cause it to sag. The height of the frame at the tallest point can be 18 inches (0.5 meters) to several feet (1 meter) depending on the size of the plants being raised. But its lowest wall should never

Figure 20-2   A bank of south-facing, glass-glazed cold frames open for ventilation. (*Photo from* Plant Propagation: Principles and Practices, *5th ed. by Hartmann and Kester, 1991. Reprinted by permission of Prentice Hall*)

be less than 12 inches (30 centimeters), as a minimum height for transplants.

The base of a cold frame can be constructed of wood using 1 × 12-inch (3 × 30 centimeter) boards, block, or concrete. Two by 4-inch (5 × 10 centimeter) rafters are sufficiently strong to support either glass or plastic covering.

Because air leakage causes heat loss, the frame should be tightly constructed and banked with soil for insulation. The covering should fit snugly, with strips of foam padding used as a seal around the glazing.

Beyond the stipulations stated there are no building absolutes, and the gardener can generally build a frame by looking at Figures 20-1 and 20-2. However, books on greenhouse construction or Cooperative Extension Service bulletins will have materials lists and building plans, and provide specific directions.

The term "glazing" refers to the clear or translucent materials used to cover cold frames, greenhouses, and similar structures. It can include a variety of coverings such as glass, fiberglass, or plastic films and also refers to the application of the covering.

Glazing used for cold frames can be of any of the common commercial materials, but window sashes and plastic are most frequently used. Sashes are easier to install (being simply laid over the frame) and permanent, but they are expensive if purchased new. However, salvage yards frequently carry used sashes with broken panes of glass, which are easily replaced. Plastics are inexpensive but are more work to install. Depending on the thickness, plastic or vinyl film may last one to ten years before requiring replacement.

Cold Frame Operation

Temperature is controlled in a cold frame by raising and lowering the glazing to ventilate with cooler outside air. Sash glazings are propped open and plastic coverings propped or rolled up.

Frequency of ventilation will depend on weather. In late winter the days may be relatively cold, and the cold frame should remain closed all day. In spring, sunlight and warmer air temperature will heat the inside of the cold frame rapidly, and daytime ventilation will be necessary. Although many gardeners "guesstimate" when the glazing should be raised, a thermometer in a shaded area of the frame is a better indicator. Anytime the temperature rises above 70°F (21°C) (for cool-season flowers and vegetables), the frame should be ventilated. Eighty-five degrees Fahrenheit (30°C) is optimum for warm-season plants.

In general, the hours of ventilation on sunny spring days are from midmorning (around 10 o'clock) until late afternoon (about 3:30 to 4 o'clock). Closing the frame several hours before sundown allows heat to build up inside to warm the plants through the night. Delaying venting until 10 A.M. makes the bed warm quickly after the cool night.

The question of how cold the outside night temperature can drop without injuring frost-tender plants in the frame is not easily answered. In a relatively airtight frame, accumulated heat will normally allow plants to survive down to 25°F (−4°C). When temperatures are predicted to fall lower, additional insulation such as burlap, canvas, or carpet should be placed over the glazing. If frost-tolerant plants are being raised, the night temperature can drop as low as 20°F (−7°C) before insulation is required. Banking soil around the edges of the frame will also give extra insulation.

Hotbeds

A hotbed is structurally and operationally the same as a cold frame. However, a hotbed has a chemical or electrical heat source in addition to heat from the sun. Because of the additional heat, the frame will stay warmer at night, and there is less chance of frost damage. Hotbeds are thus superior to cold frames for growing warm-season plants.

Manure-Heated Hotbeds

The original source of heat for hotbeds was decomposing manure. The decomposition process produced heat which kept the hotbed warm at night. The old method of hotbed heating is not commonly used today because of the inconvenience of hauling manure and its objectionable odor, but it is an economical heating method.

Horse manure is usually used for hotbeds, and it must be fresh. For localities where the temperature is not expected to drop below 12°F (−11°C), enough manure should be hauled to make a layer 12 to 15 inches (30–40 centimeters) thick. For colder areas, add 1 inch additional manure for each degree Fahrenheit lower anticipated temperature (5 centimeters per degree centigrade).

The manure should be laid in a flat pile outside the hotbed for two or three days until it begins to heat. It should then be turned, sprinkled when dry, and left several more days. By this time it should be heating uniformly and can be forked into a shallow pit dug in the hotbed.

A thin layer should be spread first, with any large lumps shaken out with the pitchfork. The layer should be tamped and water added if needed to make it slightly moist. The layering and tamping process should be repeated until the correct depth is reached, and the top layer covered with 5 to 6 inches (13–15 centimeters) of soil. Additional manure should be banked around the sides of the frame for insulation.

When the manure bed is young, it will generate a large amount of heat and the soil layer may be 90 to 100°F (32–38°C). Planting should be delayed until the temperature falls to 85°F (30°C) or lower. A thermometer buried in the soil at seeding depth will determine this.

Cornstalk-Heated Hotbeds

Although manure is the usual source of chemical heat for hotbeds, cornstalks can also be used. The stalks are chopped, moistened, and packed the same way as the manure. The decomposition will provide heat for one to two months. Combinations of manure and cornstalks can also be used.

Electrically Heated Hotbeds

Electrically heated hotbeds use soil-heating cables buried in the bed. A thermostat regulates the amount of heat emitted from the cable (Figure 20-3). Because this type of heating requires electrical power, the hotbed should be located near an electrical outlet. After the current is routed to the bed, the cable is placed on the soil or a 2-inch (5-centimeter) bed of sand. Sand is the preferred material because it will drain water away from the bed and conserve heat. The cables should be spaced according to manufacturers' directions without kinks and without touching the frame. About 3 inches (8 centimeters) of soil should cover the cable, and thermostat should be set at the recommended temperature (see Table 20-1).

Table 20-1 *Thermostat Temperatures for Electric Hotbeds*

Plant	Thermostat setting	
	During germination	*For growing*
Cole crops	70°F (21°C)	60–65°F (15–18°C)
Onion	70°F (21°C)	60–65°F (15–18°C)
Lettuce	60–65°F (15–18°C)	60°F (15°C)
Eggplant	70°F (21°C)	65°F (18°C)
Pepper	70–75°F (21–24°C)	65°F (18°C)
Tomato	70–75°F (21–24°C)	62–65°F (16.7–18°C)
Annuals	70–75°F (21–24°C)	65°F (18°C)
Cucumber	75–80°F (24–27°C)	70°F (21°C)
Summer squash	75°F (24°C)	65°F (18°C)

Hotbed Operation

Electrically heated hotbeds are easier to operate than manure-fueled beds or cold frames. Because they are thermostatically controlled, there is less chance of over- or underheating. Manure-heated beds, on the other hand, produce a large amount of heat initially but decline rapidly as decomposition is completed. Covering electrically heated hotbeds at night is not essential. However, the heating costs can be reduced by about 30 percent by this practice.

Pit Greenhouses

A pit greenhouse (Figures 20-4 and 20-5) includes features of both a cold frame and a regular greenhouse. You can walk upright in it like in a greenhouse, but it must be covered each night with insulation like a cold frame. As the name implies, it is partially underground, with the lower part constructed in a pit 3 to 4 feet (1–1.2 meters) deep. The glazed portion extends from ground level up to about 4 feet (1.2 meters).

The unique feature of pit greenhouses is their use of the heating and cooling properties of the earth. The depth to which the ground freezes (frost line) averages 24 inches (60 centimeters) over all of North America. In winter, unfrozen soil below that level will conduct warmth onto the floor and walls of a pit greenhouse and prevent them from freezing. In summer the temperature of the ground is lower than that of the air, and the earth helps keep the greenhouse cool. A pit greenhouse is the ideal design for cold-winter or hot-summer areas, requiring little or no supplementary heat or cooling and having a moderate construction cost.

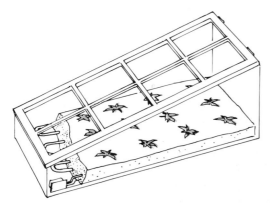

Figure 20-3 An electrically heated hotbed.

Figure 20-4 A lean-to pit greenhouse. Note the ventilation door on the end and the rolled-up insulation. (*Photo courtesy of George Taylor*)

Pit greenhouses can be heated exclusively by the sun in most climates, and the temperature will stay above freezing at night and in the range of 40 to 50°F (4–10°C) or warmer during the day. Plants that can be grown at this range are limited to those classified as "cool-season" species. Most tropical houseplants are excluded, but many flowers such as snapdragons, bulbs, primroses, pansies, and azaleas can be raised along with many common vegetables. Table 20-2 gives a partial listing of these cool-season plants.

Figure 20-5 The interior of a pit greenhouse. (*Photo courtesy of George Taylor*)

With minimal additional heating, the temperature of a pit greenhouse can be raised to a moderate (55–60°F; 13–16°C) or warm (65–70°F; 18–21°C) daytime range. At these temperatures the variety of plants which can be grown includes most houseplants and vegetables.

Pit Greenhouse Styles

A pit greenhouse can be considered a full-glass pit or a half-glass pit, depending on whether all or only half the roof is glazed.

A full-glass pit is brightest inside since the roof is entirely glass. However, the glass roof also loses heat rapidly, and supplemental heat might be required to maintain a temperature above freezing. Therefore a full-glass pit is recommended only for climates where the average winter night temperature is not below 15 to 20°F (−9 to −7°C).

A half-glass pit has only the south facing portion of the roof glazed and could be used in any climate. Direct sunlight will enter from the south, but the north slope of the roof is solid and insulated. Heat loss is reduced substantially with minimal reduction in light. A half-glass pit will remain above freezing without supplemental heat in most climates and could be

Table 20-2 *Greenhouse Temperature Requirements of Selected Ornamental Plants*

Botanical name	Common name	Temperature requirement	Code[a]
Aphelandra squarrosa	Zebra plant	Warm	Fl, Fo
Justicia brandegeana	Shrimp plant	Intermediate	Fl
Begonia semperflorens hybrids	Wax begonia	Cool-warm	Fl
Mammillaria, Opuntia, Rebutia, and other genera	Cacti	Cool	Fl, Fo
Calceolaria herbeohybrida	Pocketbook plant	Cool	Fl
Celosia cristata	Cockscomb	Cool	Fl
Senecio × hybridus	Cineraria	Cool	Fl
Cyclamen persicum	Cyclamen	Cool	Fl
Fuchsia × hybrida	Fuchsia	Cool	Fl
Pelargonium × hortorum	Zonal geranium	Cool	Fl, Fo
Sinningia speciosa	Gloxinia	Warm	Fl
Impatiens spp.	Impatiens	Cool-warm	Fl
Chrysanthemum frutescens	Marguerite daisy	Cool	Fl
Euphorbia pulcherrima	Poinsettia	Warm	Fl
Primula spp.	Primrose	Cool	Fl
Saintpaulia ionantha	African violet	Warm	Fl
Salvia splendens	Salvia	Cool	Fl
Schizanthus pinnatus	Butterfly flower	Cool	Fl
Solanum pseudocapsicum	Christmas cherry	Warm	Fl
Matthiola incana	Stock	Cool	Fl
Schlumbergera truncata	Christmas cactus	Cool	Fl, Fo
Fatsia japonica	Japanese fatsia	Cool-warm	Fo
Begonia × rex-cultorum	Rex begonia	Warm	Fo
Chlorophytum comosum	Spider plant	Cool-warm	Fo
Cissus species	Kangaroo vine, grape ivy, and others	Warm	Fo
Coleus × hybridus	Coleus	Warm	Fo
Codiaeum variegatum	Croton	Warm	Fo
Dieffenbachia maculata cultivars	Dumbcane	Warm	Fo
Ficus spp.	Fig, rubber plant	Warm	Fo
Hedera helix varieties	English ivy	Cool-warm	Fo
Maranta leuconeura varieties	Prayer plant	Warm	Fo
Peperomia spp.	Peperomia	Warm	Fo
Pilea spp.	Pilea	Warm	Fo
Sansevaria trifasciata	Snake plant	Warm	Fo
Saxifraga stolonifera	Strawberry begonia	Cool-warm	Fo
Selaginella spp.	Irish moss	Warm	Fo
Tradescantia, Zebrina	Wandering Jew	Warm	Fo
Iris spp.	Bulbous or Dutch iris	Cool	Fl
Freesia × hybrida	Freesia	Cool	Fl
Agapanthus africanus	Lily of the Nile	Cool	Fl, Fo
Crocus spp.	Crocus	Cool	Fl
Hyacinthus orientalis	Hyacinth	Cool	Fl
Tulipa spp.	Tulip	Cool	Fl
Narcissus spp.	Daffodil	Cool	Fl
Lathyrus odoratus	Sweet pea	Cool	Fl
Viola × wittrockiana	Pansy	Cool	Fl
Viola odorata	Violet	Cool	Fl
Dianthus caryophyllus	Carnation	Cool	Fl
Abutilon megapotanicum	Trailing abutilon	Cool	Fl
Calendula officinalis	Pot marigold	Cool	Fl

(continued)

Table 20-2 *Greenhouse Temperature Requirements of Selected Ornamental Plants* (continued)

Botanical name	Common name	Temperature requirement	Code[a]
Camellia japonica and *Camellia sasanqua*	Camellia	Cool	Fl, Fo
Citrus spp.	Citrus	Cool	Fl, Fo
Daphne spp.	Daphne	Cool	Fl
Osteospermum ecklonis	Osteospermum	Cool	Fl
Gazania ringens	Gazania	Cool	Fl
Gerbera jamesonii	Transvaal daisy	Cool	Fl
Lantana camara	Lantana	Cool	Fl
Antirrhinum majus	Snapdragon	Cool	Fl
Plumbago auriculata	Cape plumbago	Cool	Fl
Petunia × *hybrida*	Petunia	Cool-warm	Fl
Rosmarinus officinalis	Rosemary	Cool	Fl, Fo
Anemone coronaria	Anemone	Cool	Fl
Clivia miniata	Clivia	Cool	Fl, Fo
Ranunculus asiaticus	Ranunculus	Cool	Fl
Cypripedium spp.	Lady slipper orchid	Cool	Fl
Cymbidium spp.	Orchid	Cool	Fl
Adiantum spp.	Maidenhair fern	Cool-warm	Fo
Asplenium spp.	Mother fern, birdnest fern	Cool-warm	Fo
Cyrtomium falcatum	Holly fern	Cool-warm	Fo
Nephrolepis spp.	Sword fern	Warm	Fo
Phyllitis scolopendrium	Hart's-tongue fern	Cool	Fo
Polistichum spp.	Shield fern	Cool	Fo
Muscari spp.	Grape hyacinth	Cool	Fl
Lobularia maritima	Sweet alyssum	Cool	Fl
Tropaeolum majus	Nasturtium	Cool	Fl
Oxalis spp.	Bulbous oxalis	Cool	Fl

[a]Fl—plant grown primarily for its flowers; Fo—plant grown primarily for its foliage.

maintained without heat in zones 6 and 7 and with minimal supplemental heat in zone 5.

More energy efficient still is a lean-to, half-glass pit (Figure 20-4) constructed against a south-facing house wall. Its efficiency is derived from the warmth received from the adjacent heated building and from the wind protection the wall provides.

Pit Greenhouse Construction

Site selection is as important to pit greenhouses as to cold frames. In full-glass houses the roof must face north and south. In half-glass houses the glass portion must face south. No variation in direction can be allowed or the house will not heat properly. Sites with obstructions that would block light are not suitable.

Plumbing is an optional but valuable addition to a pit greenhouse. A single pipe buried below the frost line will enter the greenhouse at a convenient level of about 1 to 2 feet (30–60 centimeters).

Electricity is essential because it is infrequently but indispensibly required for fan ventilation and emergency heating.

The benches built for a pit greenhouse should take advantage of the sunlight entering the house (Figure 20-6), and different bench arrangements can be used depending on the plants being grown. For medium-sized potted flowers, the north bench can be stepped. For overwintering larger potted plants, a bench can be optional, and the plants can be set on the drainage gravel. If a gardener wanted to raise cut flowers, a ground planter 12 inches (30 centimeters) high could be planted with tall-growing carnations or snapdragons.

Benches can be made from wood or wood with pipe supports. Hardware cloth benches are also recommended. Poured concrete, concrete block, or wood is used for ground beds.

The insulation used for nighttime covering of the greenhouse should be thick and efficient. Fiberglass insulation with a paper or foil backing that can be rolled up with pulleys during the day (Figure 20-4)

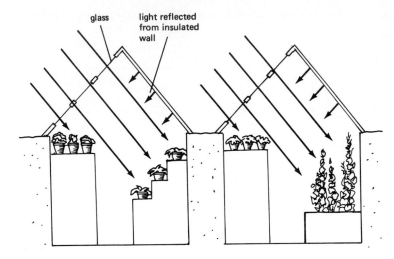

Figure 20-6 Two styles of benches for a pit greenhouse. The ground bed on the right enables the gardener to grow tall vegetables and flowers.

can be used. Styrofoam or flexible foam is another excellent insulator.

Pit Greenhouse Operation

Daily maintenance of a pit greenhouse involves covering and uncovering the glazing and ventilating by opening the doors on either end of the house. As with a cold frame, the covering should be taken off by early morning when the sun is shining and replaced in late afternoon. On cloudy days with below-freezing temperatures the coverings should not be removed, and the house may remain covered for several days during long stretches of cloudy weather.

Watering may be required daily, although plants lose water slowly in the cool humid atmosphere. Careful watering with a narrow spout is recommended to minimize spillage. Cool temperatures and high humidity make the fungus disease mildew (Chapter 13) a constant threat, but one that can be lessened by cautious water use.

One advantage of a pit greenhouse is that it can be left unattended for several days, if necessary, by leaving the insulation over the glass. The temperature will normally stay above freezing.

Summer care of the house includes covering the glazing with a "shading material" to decrease the light entry and heat buildup. Lath, bamboo blinds, and Saran (a plastic, netlike covering) can be used for this purpose. Because of its southern exposure, a pit greenhouse can overheat easily, so all vents should be left open except on very cool nights. Air circulation with a fan may also be necessary.

Conventional Greenhouses

A conventional greenhouse offers its owner the advantage of almost complete environmental control. Although partially heated by the sun, gas, electricity, or oil is the main heat source, and the heat is thermostatically controlled to keep the house cool, moderate, or warm. Depending on the budget, venting, watering, cooling, shading, and fertilizing can also be automated.

Of the structures discussed, conventional greenhouses are the most expensive to operate because of heat loss through the large glazed area. However, conventional greenhouses are the most common for commercial and home use because of the ease of construction and superior light penetration.

Many options in shape, glazing, and framing materials, and heating, watering, and cooling systems are available when choosing a greenhouse. The gardener should think of his greenhouse as a component system and select each component according to climate, budget, convenience, plant, and aesthetic requirements.

Most greenhouses are ordered from catalogs, and with competition encouraging lower prices, good quality cannot always be assumed. Some inexpensive greenhouses are essentially toys, having only sheets of fiberglass screwed to a flimsy wood frame. Such greenhouses are not suitable for use in harsh climates. Cold air would enter through the poorly designed junctions of the glazing and the first snowfall would collapse the roof. Look for guaranteed quality and durability when selecting a greenhouse. Information given by the manufacturer should include:

Frame construction

    Framing material (wood or metal)

    If wood, type and dimensions of framing members

Glazing

    Type of glazing material (glass, fiberglass, PVC film, and so on)

    Guarantees of glazing against breakage and discoloration, and expected life

    Options of buying replacement glazing

    Airtightness of glazing

Ventilation

    Type and location

Greenhouse Shapes

Span Roof Freestanding. This type is the most common and is used frequently by commercial growers. The design is angular, with a pitched roof and straight or sloping sides. Three variations on the span-roof greenhouse are the standard, Dutch light, and curvilinear (Mansard).

Standard greenhouses (Figure 20-7) have straight sides and a two-sided roof that slopes down to the sides of the house. It may be glazed to the ground or have sidewalls up to 30 inches (80 centimeters) high under the benches. Sidewalls decrease the rate of heat loss and give greater stability but shade the area under the benches.

A Dutch-light greenhouse (Figure 20-8) is similar to a standard span roof except that the sides are glazed to the ground and are constructed at a slight angle rather than being vertical. The angle of the sides causes winter sun to strike the glass at closer to 90°, the optimum angle for complete light penetration. Due to the absence of sidewalls and because of the angle of the glass, Dutch-light houses are slightly brighter than standard houses.

The mansard or curvilinear greenhouse (Figure 20-9) takes this principle a step further. Each section of glass attaches to the next to form a tunnellike structure.

Figure 20-7   A standard greenhouse with brick sidewalls. (*USDA photo*)

Figure 20-8   A Dutch-light greenhouse. Note that the sides are entirely glass and are not quite vertical even at the base. (*USDA photo*)

Because the sun's angle changes with the seasons, each set of panes will eventually be at the optimum angle for light penetration. Mansard greenhouses are not readily available and their light transmission is not significantly superior to that of a full-glass standard or Dutch-light houses.

Figure 20-9   A mansard-type greenhouse. Note the different angles of all the rows of glass panes.

Lean-to. Lean-to greenhouses are split, span-roof greenhouses of one of the types discussed. Their appeal lies in convenience and aesthetics. Since they can often be built to cover a sliding glass door (Figure 20-10), they can be warmed with the household heat. A lean-to can also extend the living area to form a garden room or hobby area.

Although most lean-to greenhouses are attached to the south side of the house for maximum light, they can be situated on another side for convenience. East- or west-facing greenhouses receive moderate light and grow almost any plant, and even north-facing greenhouses can be used for shade-loving ferns and many houseplants.

Quonset. These greenhouses are usually framed of bent pipe and glazed with one or more layers of plastic film (Figure 20-11). Although less attractive than some styles, they are inexpensive and can be recommended for home use.

Dome. Although they appear faddish, geodesic domes (Figure 20-12) constructed of interlocking triangles are actually very well designed. The angles of the glazing permit excellent light penetration, and hanging plants

Figure 20-10   A lean-to greenhouse. (*Photo courtesy of Burnham Greenhouse Co. and the USDA*)

can be hung from the structure to make maximum use of the light.

Greenhouse Framing Materials

The frame of a greenhouse, the part which supports the glazing, can be of almost any material. A strong, weather-resistant material is essential for even a temporary structure. The relative strengths of greenhouse framing materials are discussed below.

Wood was once the main greenhouse framing material. It is a good insulator, moderately durable, but requires some maintenance such as painting. Redwood, western red cedar, and pressure-treated softwoods are all satisfactory. Creosoted wood should never be used in a greenhouse; the creosote will vaporize into the greenhouse and injure the plants.

Aluminum alloy is lightweight and moderate in price. It is durable and requires little maintenance but loses heat more rapidly than wood.

Steel bars or pipe can be used to frame a greenhouse. The material is strong and relatively inexpensive but unless galvanized, it must be painted regularly to prevent rust. Steel also transmits heat rapidly, making it a poor insulator.

Foundations and Sidewalls

To withstand wind without glass breakage or slippage, a greenhouse should have a foundation. A foundation will also keep the house level. Many prefabricated greenhouses state that no foundation is needed. However, unless the greenhouse is to be used only as a temporary spring structure, a foundation is recommended. As a minimum, cinder block can be laid on level soil. But a concrete slab is preferable, poured with bolts to anchor the walls and the water and electrical pipes in place.

If the design includes sidewalls, a concrete base to support them should go at least 1 foot (30 centimeters) below the frost line. The walls can then be built of the chosen material.

Glazing Materials and Methods

For many years, wooden sashes with panes of glass puttied into place were the main glazing material. Glazing

Figure 20-11   Quonset-style plastic greenhouses. These are exceptionally tall to house citrus trees. (*USDA photo*)

used today includes a wide range of plastics and fiberglass as well as glass.

Glass. Still preferred for looks, durability, and ease of cleaning, glass is and will remain a popular glazing ma-

Figure 20-12   A geodesic dome greenhouse for home use. (*Photo courtesy of J and D Solar Products*)

terial. Its light transmission ability is excellent, and although breakage can occur, it will not be a problem in a well-constructed house. The expense is moderate to high but is compensated for by its long life.

Modern methods of glazing with glass utilize several systems. In "dry glazing," the glass is inserted into a channel in wood or between metal bars. Glass applied by dry glazing is easy to install, but air leakage can be a problem.

A second glazing system places the glass on a strip of putty and holds it in place with a clip. The junction is then covered with a cap to weatherproof the seal.

Film Plastic (Polyethylene). These thin (2- to 6-millimeter thickness) plastics can be stapled to a wood frame or run in a continuous piece over a quonset house. They are inexpensive but last only one to two years in most climates. When signs of tearing or discoloration appear, they should be replaced.

Greenhouses covered with plastic film are very airtight. Consequently, the humidity can climb very high, causing condensation runoff from the plastic and creating conditions favoring disease development. Proper ventilation is therefore essential in a plastic house.

Vinyl Plastic (Polyvinyl Chloride or PVC). Vinyl plastics are thicker than film plastics (3–12 millimeters) but still moderately flexible. They are generally stapled onto a wooden frame.

Vinyl plastics are moderate in cost and usually have a guaranteed life of up to 10 years. Both transparent and translucent types are sold.

Fiberglass. There are many brands and grades of corrugated fiberglass used for glazing greenhouses. The durability of some is excellent, and they are guaranteed for 10 to 20 years. Most require resealing of the surface with a plastic resin every one to two years to prevent abrasion of the protective coating. The manufacturer should provide information regarding the durability and maintenance of the glazing material.

Double Glazing. The loss of heat through all glazing materials is relatively rapid, from both conduction through the material and air leakage at junctions. In double glazing, two layers of glazing are applied to the greenhouse, eliminating air exchange and creating a "dead air" space.

The easiest double glazing for home greenhouses involves supplementing the present glazing with a layer of plastic film. The film can be applied either inside or outside the house and will result in up to a 50 percent heat saving,.

Heating Systems

Self-contained forced-air heating units are the best for home greenhouses. They can be oil or gas fired or electric. Electric units do not require venting and are easy to install. They are installed at one end of the greenhouse and aimed so the fan circulates warm air throughout the house.

Gas and oil heaters set up inside the greenhouse require venting. In several models the heater fits into the wall of the greenhouse. The combustion chamber is outside, and no venting is required.

Heating by household heat is used frequently with lean-to greenhouses. However, a fan will still be required to circulate air between the two if the greenhouse is very large.

Cooling Systems

Cooling of greenhouses is accomplished mainly by venting and shading. Many inexpensive greenhouses are constructed to be ventilated only by opening the doors; however, vents on the roof or at the eve are more efficient because the warmer air will rise and flow out through them. Greenhouse models with these vents are preferable.

Shading of a greenhouse in summer will restrict direct sunlight and decrease heating of the interior. Bamboo blinds, cheesecloth, or Saran placed over the glass will function very effectively. A coating of white-wash can be used on glass, plastic, or fiberglass greenhouses and can be removed in fall with water and a brush.

In areas with very hot summers, an evaporative cooler can decrease the temperature of the greenhouse to less than the outside air temperature. Evaporative coolers are widely used in commercial greenhouses, and small units are available from home greenhouse suppliers.

In this cooling system, a fiber pad is installed at one end of the greenhouse with water trickling through it at a slow rate. A fan at the opposite end of the house pulls outside air in through the pad. The water evaporates into that air, absorbing heat in the process and reducing the air temperature.

Interior Greenhouse Equipment

Most greenhouses are equipped with benches that support plants at a convenient working level. For greenhouses without sidewalls, double layers of benches (Figure 20-13) are practical, with shade or dim-light plants grown on the bottom level.

Benches framed with wood and covered with galvanized hardware cloth have much to recommend them. They are rustproof, permit free air and heat circulation, and allow light to pass through to plants growing in the area beneath.

When large potted plants must be accommodated, a section of bench can be left out and the pots placed on the floor of the greenhouse. If vegetables such as tomatoes or trellised cucumbers are being raised, a ground bed (Figure 20-14) will be needed.

Other equipment used for operating a greenhouse is minimal. A thermometer that specifies the minimum and maximum temperature reached is necessary and should be placed out of direct sunlight and at the same level as the plants.

A humidity chamber for rooting cuttings is almost indispensable and can be easily made by draping vinyl or plastic film tent fashion over wire supports (Figure 20-15).

A mist nozzle (Figure 20-16) is useful although not essential. Likewise, a fertilizer proportioner (Figure 20-17) will enable the greenhouse gardener to fertilize

Figure 20-13 A double-layer bench arrangement in a greenhouse. Note that low-light-requiring ferns are being grown on the lower bench. (*Photo by Rick Smith*)

Figure 20-14 A ground bed for raising large foliage plants. (*Photo by Rick Smith*)

Figure 20-15 A rooting chamber made of plastic sheeting draped over wire supports.(*Photo by Rick Smith*)

Figure 20-16 A mist nozzle. (*Photo courtesy of Vocational Education Productions, California Polytechnic State University, San Luis Obispo, CA*)

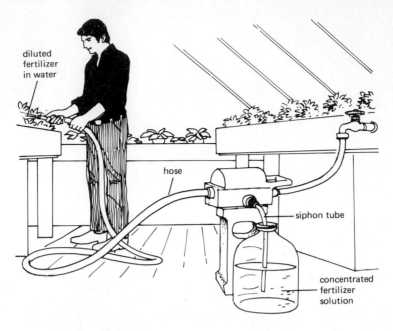

Figure 20-17 A simple fertilizer proportioner system that works by water pressure.

plants while watering by mixing concentrated fertilizer into the water supply.

Automatic systems for shading and venting are relatively expensive; however, an automatic watering system can be easily and inexpensively installed. Commercial greenhouses use this "spaghetti system," consisting of a flexible ½-inch (13-millimeter)-diameter plastic pipe inserted at various points with "microtubes" about the thickness of spaghetti. The water flows at low pressure through the main line and out each of the microtubes to a pot (Figure 20-18). The system is identical to the trickle irrigation discussed in Chapter 11.

Growing Plants in a Greenhouse

Ornamentals

Growing ornamental plants in a greenhouse is much like growing them in the house. They require the same essentials of water, fertilizer, and potting medium and the same maintenance. However, in the greenhouse they receive brighter light and humidity conditions, and consequently they will grow faster and may be healthier. But a greenhouse should not be expected to guarantee success with plants. The person who has successfully grown plants indoors will probably be even more successful in a greenhouse, but one who has had a number of problems with plants indoors will have the same problems in the greenhouse. The greenhouse will not substitute for knowledge or experience.

Greenhouse environments can be divided into two types by temperature, although some plants can be grown in both temperature ranges. A cool greenhouse maintains a night temperature above freezing and generally in the range of 40–55°F (4–13°C). Daytime highs can reach up to 60°F (16°C). In a tropical or warm greenhouse the minimum night temperature is 55°F (13°C) with daytime around 65–70°F (18–21°C). The drop to a lower nighttime temperature is a standard practice to promote healthy growth by slowing respiration during the night hours.

Table 20-2 gives a listing of ornamental plants that are commonly grown in greenhouses and their suggested temperature ranges. Because of the improved growing conditions that a greenhouse provides, many beautiful flowering plants from semi-tropical climates can be grown, most of which will not be familiar to temperate-climate gardeners. As with any unknown plant, it is worthwhile to research the specific growing requirements of a species to assure the best performance.

Vegetables and Fruits

For some people, a greenhouse is seen as the means for growing fresh produce out of season. A number of vegetables and fruits can be successfully grown in a greenhouse, although their growth rate and taste may not equal what they would be outdoors due to the reduced light intensity and shorter days. Vegetables that can be most successfully greenhouse grown are listed

Figure 20-18 A spaghetti watering system for potted plants in a greenhouse. The lead weight at the end of the microtube keeps the tube from falling out of the pot. (*Photo courtesy of Chapin Watermatics, Watertown, NY*)

here, along with a brief discussion of their growing requirements.

Beans. Snap beans should be grown in a warm greenhouse and can be seeded in February or March in ground beds for a May crop. Pole bean cultivars grown on a trellis work well, but bush types can be used if space is limited.

Beets. Beets can be grown for both their tops and their roots. They are a cool-greenhouse crop and can be grown throughout winter.

Carrots. Short cultivars such as 'Oxheart' or 'Short 'n Sweet' can be grown in ground beds from a February sowing and harvested in early spring. Carrots for fall and winter use are more easily obtained by mulching in the remaining carrots in the vegetable garden and pulling them as needed. See Chapter 6 for details.

Cucumbers. Cucumbers should be grown on trellises and can be harvested through most of the winter if managed correctly. Seed sown in November, January, and March should yield a nearly continuous supply of fruits.

Cultivars of cucumbers that do not require pollination to set fruit (called *parthenocarpic*) are best for greenhouse culture. Growing temperature should be at least 60°F (16°C) and preferably 70°F (21°C) for best production.

Herbs. A number of leafy herbs can be grown successfully year-round in either a cool or a warm greenhouse, including mint, parsley, rosemary, chive, thyme, basil, and sage.

Lettuce. Either 'Bibb' or leaf lettuce will thrive under cool conditions. A continuous winter harvest can be expected from November from seed sown about once per month.

Onions. Green onions can be grown easily from seed in a cool greenhouse. Two to three sowings starting in September will provide onions through spring.

Radishes. This quick-maturing vegetable will be ready one month after the seed is sown and can be grown throughout winter in a cool house.

Tomatoes. Tomatoes can be propagated from seed or from cuttings taken from plants in the garden and grown in large pots or a ground bed (Figure 20-19). A warm greenhouse (60°F [16°C] or higher) is best. As the plants grow taller, they should be staked upright. Hand pollination is necessary, and lightly tapping the flower clusters during the bloom period will assure good fruit set. Choosing tomato cultivars bred for greenhouse growing will help ensure success.

Figure 20-19   Growing tomatoes in a greenhouse ground bed. (*Photo courtesy of Vocational Education Productions, California Polytechnic State University, San Luis Obispo, CA*)

Hydroponic Greenhouse Gardening

Hydroponic gardening involves supplying water and minerals to plants by intermittently flooding their roots with a liquid nutrient solution. No soil or potting medium is used. Instead the plant roots are supported in beds of gravel, sand, or similar unreactive material, which is filled and drained several times per day with the water-based solution.

Hydroponic gardening is considerably more complicated than conventional culture. However, the practice achieved considerable popularity recently as a hobby and even as a commercial enterprise. Many books deal with the science of hydroponics and offer growing instructions including recipes for nutrient solutions.

Shade Houses

Shade houses (Figure 20-20) provide a growing area of filtered shade for plants that are not tolerant of full summer sunlight. Fuchsias, impatiens, and many begonias and ferns grow well only in a shade house during the summer months but can tolerate full sun in the greenhouse during the winter months when the sun is dim. A shade house can also be used for hardening transplants and for summering houseplants outdoors. If large enough, it can be used as an outdoor living area (see Chapter 10).

Shade houses are usually constructed with a wooden frame and covered with lath or Saran. The covering may or may not extend to include the sides of the structure. Lath houses shaded by pieces of lath or 1×2-inch (25×50-millimeter) wood strips form an attractive, durable structure. The spacing of the lath will depend on the intensity of the sunlight in the area and the amount of shade required by the plants. North-to-south orientation of roof lath is suggested because it casts alternate sun and shade on plant leaves through the day. However, orientation is not crucial, and many gardeners prefer to orient lath east-west or in decorative crisscross patterns.

Plastic Saran can also be used to cover a shade house. It is less durable, lasting one to two years, but provides an even shade throughout the structure. The depth of shade will be determined by the weave of Saran selected.

Figure 20-20 A lath-covered shade house for a home garden. (*Photo courtesy of George Taylor*)

Indoor Plant Growth Units

Greenhouse Windows

A greenhouse window is one way to expand the indoor growing area for houseplants or to raise garden transplants. Although the heat cannot be controlled as in a regular greenhouse, the light is better and the humidity often higher than in an ordinary window.

When buying or constructing a greenhouse window, a major criterion is ventilation. A side or top vent should be included in the design for excess heat dissipation in summer. Beyond this requirement, the considerations of price and aesthetic appeal become important. A large window is not usually correspondingly higher in price than a small one, but it will provide valuable extra space for additional plants at the same time that it becomes an attractive addition to a room.

Environment-Controlling Growth Units

These indoor units (Figure 20-21) monitor and regulate heat, humidity, and light continuously for optimal plant growth. They are useful in homes with insufficient natural light and for species which are difficult to grow out of a greenhouse. They are also attractive for displaying and maintaining collections of exotic or miniature plants. However, for growing ordinary houseplants their expense is usually prohibitive.

Terrariums

The primary plant-growing benefit of terrariums is their ability to hold in humidity, the lack of which causes many houseplant problems. But the glass that holds in the moisture also excludes some light, making terrariums suitable mainly for medium- and low-light plants.

Almost any container, aquarium, fish bowl, water carboy, or bottle can be used as a terrarium. Those with narrow openings are more difficult to plant, but the main consideration is that the container be made of clear or only slightly tinted glass. Colored glass containers will exclude light, and plants growing in them will eventually die.

In making a terrarium, first provide a layer of drainage material at least 1 inch (3 centimeters) deep in the bottom of the container. This layer will be a reservoir for excess water if the terrarium is accidentally overwatered. Most prepackaged or homemade potting media will be satisfactory and should be moistened, if necessary, to slight dampness.

Assembly of an open-mouth terrarium can be done quickly using your hands, but narrow-necked terrariums require special tools and more time. The drainage layer and potting medium should be poured in through a funnel to avoid dirtying the inside of the glass. Each layer is then smoothed out with a tool fashioned by taping a spoon to a piece of wooden molding, a dowel, or a straightened wire coat hanger.

Plants are inserted by soaking the media from the roots, folding the leaves upward, and sliding them

Figure 20-21   An environment-controlled plant growth chamber for indoor use. (*Photo courtesy of General Aluminum Products*)

through the opening. They are then positioned and planted with the spoon tool and a plain dowel. Unrooted cuttings can also be used and will root easily in the humid atmosphere.

The plants chosen for the terrarium should be medium- or low-light-requiring and slow growing. Table 20-3 lists several dozen recommended species. Cacti and succulents are not suitable for terrarium planting. The stagnant air and restricted light provide poor growing conditions for these and other xerophytic plants.

When assembly of the terrarium is complete, the new plants should be watered in to assure contact of the medium with the roots. Trickle several teaspoons (10–15 milliliters) of water onto each plant by pouring it down the dowel. Rinse any media off the glass with a trickle of water or by wiping with a wet paper towel attached with a rubber band to a dowel.

Terrarium maintenance involves pruning and occasional watering. Fertilizer is not used because it encourages rapid growth, and the plants will outgrow the container. Even without fertilizer, pruning may be needed to control vigorous species and to encourage branching. If the container has a wide mouth, pruning can be done with a scissors and the prunings taken out and discarded. In a narrow-neck terrarium a razor blade taped to a dowel can be used, and a pickup tool will remove the trimmings and any dead leaves or flowers.

How frequently a terrarium will require watering depends on the size of the opening. In a sealed terrarium no water is lost to the outside air, and watering should never be required. A narrow-neck terrarium will have minimal water loss through the opening and will require infrequent watering, on an average of every two to six months. Wide-mouth terrariums, however,

Table 20-3 *Plants for Small and Moderate-Sized Terrariums*

Botanical name	Common name	Remarks
Adiantum spp.	Maidenhair fern	Many species suited to terrarium culture
Araucaria heterophylla	Norfolk island pine	Seedlings make attractive miniature trees
Begonia spp.	Begonia	Miniature species are hard to find but work well, including *Begonia foliosa*
Chamaedorea elegans	Parlor palm	Several seedlings are usually sold in a small pot and can be separated and used for small trees
Dracaena surculosa	Gold dust dracaena	Formerly known as *Dracaena godseffiana*; shrubby plant with deep green leaves irregularly spotted yellow
Episcia dianthiflora	Lace-flower vine	Fuzzy-leaved relative of the African violet bearing white tubular flowers
Ficus pumila	Creeping fig	Small-leaved creeping vine; the varieties 'Minima' and 'Quercifolia' are smaller foliaged but rarer
Fittonia verschaffelti 'Minata'	Miniature nerve plant	Slow growing and low-light-requiring; leaves are about 1″ (2.5 cm) long
Maranta leuconeura varieties	Prayer plant	Low-light-requiring; young plants are ideal but may outgrow the terrarium in 6 months
Pellaea rotundifolia	Button fern	Deep, blue-green pea-sized leaves on a flat-growing plant
Peperomia caperata	Emerald ripple peperomia	One of the smaller and more attractive peperomias; can be propagated by a single leaf pushed into the soil
Peperomia rubella		Small-leaved peperomia with red stems and leaf undersides; grows up to 6″ (15 cm) tall
Podocarpus macrophyllus	Southern yew	Slow growing and useful as a small tree; deep green ribbon-type foliage
Pteris spp.	Table fern	Most species are excellent; some are plain green, others ruffled or lined in silver
Saxifraga stolonifera	Strawberry begonia	Small plants will root readily in the terrarium; produces interesting runners
Schlumbergera spp. cultivars	Christmas cactus	Normally sold as *Zygocactus*; Christmas, Thanksgiving, and Easter cacti; all prefer moist soil and will thrive in a terrarium
Selaginella spp.	Irish moss	Many species are excellent, either creeping or forming small mounds
Sinningia pusilla	Slipper plant	Tiny flowering plant only 3″ across; flowers produced sporadically in pastel shades
Soleirolia soleirolii	Baby tears	Commonly sold as *Helxine soleirolli*. Has tiny ¹⁄₁₆″ (2-mm) leaves and a matlike growth habit; becomes tall and stringy in insufficient light

have less restricted air exchange and may require watering every one to four weeks.

Overwatering causes the death of terrarium plants in many instances, so caution is advisable. In wide-mouth terrariums the medium can be checked with the fingers and should always feel slightly damp. In narrow-neck containers water should be applied only when the top of the medium begins to appear dry and never as long as there is condensation on the glass or water in the pebble reservoir. Watering by teaspoonfuls (5 milliliters) is suggested to lessen the danger of overwatering.

Garden Rooms

Garden rooms combine the features of a greenhouse and living room, creating an area useful for both growing plants and dining or relaxing. Solariums, Florida rooms, atriums, and conservatories are all variations of garden rooms.

Unlike a greenhouse, a garden room is normally not completely glazed. A completely glass room becomes too hot during the day and cools off too rapidly at night to be comfortable for people. Instead it is partially glazed, perhaps with large windows and several skylights. Ventilation and electricity are also present along with furniture and, usually, a waterproof floor. Plumbing and drainage facilities are optional.

Location of the garden room should be governed by convenience rather than orientation. A large garden room designed for family use will be most conveniently situated near kitchen, living, or dining areas. A more intimate room could be placed off a master bedroom or bath. If several locations seem agreeable, orientation can be considered. An east-facing garden room is ideal in many ways. It provides morning sunshine yet remains relatively cool and shady the remainder of the day. A south-facing room might be ideal for cold or low-light climates since it receives direct sunlight all day. However, a north room would be better suited to warm areas of the country. A west-facing room is brightest from midday until sunset and would be a cheery retreat late in the afternoon.

The cost of a garden room is approximately equal to that of a regular room on a cost per square foot (square meter) basis. Since most garden rooms are additions, the services of an architect will probably be required. Consult, if possible, an architect with experience in designing garden rooms, and bring pictures from magazines or books to help explain the project and what you have in mind.

Selected References for Additional Information

*Ball, Jeff. *How to Grow Plants in a Greenhouse*. Irwindale, CA: KVC Entertainment.

Food and Agriculture (FAO) Organization of the United Nations. *Soilless Culture for Horticultural Crop Production*. Plant Production and Protection Paper 101. Rome: FAO, 1990.

Nicholls, Richard. *Beginning Hydroponics*. Philadelphia: Running Press, 1990.

Smith, Shane. *Greenhouse Gardener's Companion*. Golden, CO: Fulcrum Publishing, 1992.

Sunset Book Editors. *Greenhouse Gardening*. Menlo Park, CA: Sunset Publications, 1976.

Taylor, K. S. and E. W. Gregg. *Winter Flowers in Greenhouse and Sun-Heated Pit*. New York: Charles Scribner's Sons, 1976.

Walls, I. G. *The Complete Book of the Greenhouse*. London: Wardlock, 1993.

Williams, T. Jeff, and Larry Hodgson. *Greenhouses*. San Ramon, CA: Ortho Books, 1991.

*indicates a video.

Appendix of Plant Societies

The addresses of the societies change frequently and can be obtained by writing to the American Horticultural Society, 7931 East Boulevard Drive, Alexandria, VA 22308. Telephone (800) 777-7931. The list is courtesy of the Society.

United States

African Violet Society of America, Inc.
American Bamboo Society
American Begonia Society
American Boxwood Society
American Calochortus Society
American Camellia Society
American Conifer Society
American Daffodil Society, Inc.
American Dahlia Society
American Fern Society
American Fuchsia Society
American Ginger Society
American Gloxinia and Gesneriad Society
American Gourd Society
American Hemerocallis Society
American Herb Association
American Hibiscus Society
American Hosta Society
American Iris Society
American Ivy Society
American Orchid Society
American Penstemon Society
American Peony Society

American Poinsettia Society
American Primrose Society
American Rhododendron Society
American Rose Society
American Willow Growers Network
Aril Society International
Australian Plant Society of California
Azalea Society of America, Inc.
Bromeliad Society, Inc.
Bromeliad Study Group of Northern California
Cactus and Succulent Society of America
California Rare Fruit Growers
Chestnut Growers Exchange
Colorado Water Garden Society
Cryptanthus Society
Cycad Society
Cymbidium Society of America
Dwarf Iris Society of America
Epiphyllum Society of America
Eucalyptus Improvement Association
Gardenia Society of America
Gesneriad Hybridizers Association
Gesneriad Society International
Hardy Fern Foundation
Hardy Plant Society
Hardy Plant Society of Oregon
Heliconia Society International
Herb Society of America
Heritage Rose Foundation

Historic Iris Preservation Society
Holly Society of America, Inc.
Home Orchard Society
Hoya Society International, Inc.
Indoor Citrus and Rare Fruit Society
Indoor Gardening Society of America, Inc.
International Aroid Society, Inc.
International Camellia Society
International Carnivorous Plant Society
International Geranium Society
International Golden Fossil Tree Society, Inc.
International Lilac Society, Inc.
International Oak Society
International Oleander Society
International Ornamental Crabapple Society
International Palm Society
International Tropical Fern Society
International Water Lily Society
Los Angeles International Fern Society
Magnolia Society, Inc.
Marigold Society of America, Inc.
National Chrysanthemum Society, Inc.
National Fuchsia Society
North American Fruit Explorers
North American Gladiolus Council
North American Heather Society
North American Lily Society, Inc.
Northern Nut Growers Association
Northwest Fuchsia Society
Northwest Perennial Alliance
Pacific Orchid Society
Peperomia Society International
Perennial Plant Association
Plumeria Society of America
Rare Fruit Council International
Rare Pit & Plant Council
Reblooming Iris Society
Rhododendron Species Foundation
Rose Hybridizers Association
Saintpaulia International

Sedum Society
Sempervivum Fanciers Association
Society for Japanese Irises
Society for Louisiana Irises
Society for Pacific Coast Native Iris
Society for Siberian Irises
Solonaceae Enthusiasts
Southern Fruit Fellowship
Species Iris Group of North America
Spuria Iris Society
Texas Rose Rustlers
Tropical Flowering Tree Society
Tubers
Walnut Council
Woody Plant Society
World Pumpkin Confederation

Canada
African Violet Society of Canada
Alpine Garden Club of British Columbia
British Columbia Begonia and Fuchsia Society
British Columbia Lily Society
Canadian Chrysanthemum and Dahlia Society
Canadian Geranium and Pelargonium Society
Canadian Gladiolus Society
Canadian Iris Society
Canadian Orchid Society
Canadian Peony Society
Canadian Prairie Lily Society
Canadian Rose Society
Dahlia Society of Nova Scotia
Newfoundland Alpine & Rock Garden Club
Ontario Herbalists Association
Ontario Rock Garden Society
Rhododendron Society of Canada
Royal Saintpaulia Club
Saskatchewan Perennial Society
Toronto Cactus & Succulent Club

Glossary

abscisic acid a growth-inhibiting chemical that induces abscission and dormancy and inhibits seed germination, among other effects.

abscission the dropping of leaves, fruits, flowers, or other plant parts.

absorption the uptake of any material by a plant or seed, such as water by the roots or pesticide through the leaves.

accent plant a landscape plant used to call attention to a particular feature of an area, for example, an attractive shrub planted near the front door of a house.

acid-forming a material that reacts to acidify the soil or acidifies the soil as it decomposes; for example, peat moss, agricultural sulfur, and oak leaves.

acid-loving describes plants that grow best in acidic potting media or soil.

active growth a phase of the plant life cycle characterized by rapid stem lengthening and leaf production and sometimes flowering or fruiting.

active ingredient the ingredient in a pesticide which causes its effectiveness. Other ingredients are called inert ingredients.

adhesion the attraction of liquids for solids, for example, the attraction of water for soil.

adventitious bud a bud produced on a part of a plant where a bud is not usually found, such as on a leaf vein or root.

adventitious root a root produced by other than primary root tissues, such as from the stem.

aeration a method of increasing the water and air penetration into compacted soil by the removal of slices or cores of soil throughout the compacted area.

aggregation the clumping together of soil particles into large groups.

air layering the rooting of a branch or top of a plant while it is still attached to the parent.

algae a single-cell plant which often grows in colonies.

allelopathy the injury of one plant by another through excretion of toxic substances by the roots; the black walnut has allelopathic properties.

amendment a material added to soil that renders it more suitable for plant growth; for example, sand, lime, compost, and peat moss.

anatomy the internal structure of an organism.

angiosperm the botanical designation of a plant that produces its seeds in ovaries.

annual a plant that grows to maturity, flowers, produces seed, and dies during one season; often refers to frost-tender flowers.

anther the top, pollen-producing portion of the stamen.

anthocyanin a red pigment found in many plants, including the fruits of cherries and strawberres.

antidesiccant a spray used on transplanted plants to slow transpiration and prevent burning of the foliage.

aphid a small, soft-bodied insect that feeds on young leaves and buds and sucks plant juices, causing curling and distortion of new growth.

arboretum (arboreta) a place where trees are grown—either alone or with other plants—for scientific and educational purposes.

arboriculture the branch of horticulture that deals with the care and maintenance of ornamental trees.

artificial medium a growing medium that does not contain soil.

ash content   the amount of mineral matter present in a soil amendment.

atomizer   an instrument that breaks a liquid into small droplets through the use of forced air, for example, a pesticide applicator.

auxin   a plant hormone involved in dormancy, abscission, rooting, tuber formation, and other activities.

axil   the angle formed where a petiole joins a stem.

axillary bud   a bud located in a leaf axil.

bactericide   a chemical used to control bacterial disease.

bacterium (bacteria) a single-celled, nonchlorophyll-containing plant; some bacterial species assist in the decomposition of organic matter and the fixation of nitrogen, whereas others cause such plant diseases as fireblight.

balanced fertilizer   a fertilizer containing equal proportions of nitrogen, phosphorus, and potassium.

balled and burlapped   describes plants (usually trees) grown in the ground, dug, and sold with the root ball wrapped in burlap.

bare root   refers to deciduous plants grown in the ground, dug, and sold while dormant with no soil around the roots.

bark ridge   the raised ring of bark where a branch emerges from its parent branch.

bear   to "bear" is to produce fruit.

bearing age   the age at which a tree produces its fruit.

bed   a soil area prepared for planting.

bench   a raised platform in a greenhouse for growing plants.

berm   a mound or ridge of soil created to add interest or screening in a landscape.

biennial   a plant that produces vegetative growth during one season, flowers during the second season, and then dies; examples are carrots and parsley.

binomial system   the system of naming plants developed by Carolus Linnaeus whereby plant names are given in two parts, the genus and species.

biological control   the use of another organism to kill a plant pest, for example, the use of *Bacillus thuringiensis* disease to kill caterpillars.

black frost   a frost that kills plant tissue without forming ice on the outside of the plant, so named because of the resulting blackening of injured plants.

blanch   to exclude light from a vegetable to make it more tender, less strong tasting, or more attractive.

blight   a bacterial or fungal disease that kills portions of a plant and may eventually destroy the entire plant.

block garden   a form of vegetable gardening that involves planting vegetables in groups or blocks rather than rows.

borer   any insect that burrows into the stems, roots, leaves, or fruits of a plant.

botanical insecticide   an insecticide derived from a plant; examples are pyrethrum, rotenone, and ryania; *see also* organic insecticide.

bottom watering   a technique for watering potted plants in which water is poured into the drainage saucer and absorbed through the drainage hole by capillary action.

bract   a small, underdeveloped leaf that functions as a flower petal; a flower often arises in the axil of a bract.

brambles   the collective name given to the fruits in the genus *Rubus;* examples are raspberries, blackberries, dewberries, and loganberries.

branch collar   the ridge of bark tissue at the base of a branch.

branch spreader   a piece of wood inserted between the trunk and branch of a fruit tree to widen the crotch and create a stronger branch.

breaking dormancy   causing a plant or seed to change from a dormant to an active growth phase.

broadcast   to distribute (for example, fertilizer) by throwing handfuls to scatter it on the ground.

broadleaf weed   a dicot weed.

bud   a rudimentary shoot or flower; may be either vegetative (containing leaves, stem, and so forth), flowering (producing one or more flowers), or mixed (containing both vegetative and reproductive parts).

budding   implanting a bud of one plant onto another plant so that the resulting shoot bears the characteristics of the bud parent on the roots of the host.

bulb   an underground storage organ composed of a shortened stem covered with modified leaves called *scales.*

burn   browning of a plant due to lack of water, exposure to high temperatures, or exposure to a toxic chemical or excess fertilizer.

cactus   a member of the plant family Cactaceae.

candle   the succulent new growth of a needle evergreen such as spruce or pine.

cane   the long shoots/weeping stems such as those produced by grapes, brambles, or roses.

cane cutting   *see* **stem section cutting.**

canker   a disease particularly affecting woody plants which destroys the cambium and vascular tissue of a generally localized area of a plant.

capillary action   the sideways and upward movement of water through a growing medium owing to the adhesion of water molecules to the medium.

capillary water   the water that moves in soil or another growing medium owing to the forces of adhesion and cohesion.

carbohydrates   the sugar and starch required for the growth, structure, and metabolism of a plant.

carotene   an orange plant pigment that gives color to oranges and persimmons, among other plants.

casting   soil that has passed through an earthworm.

cation-exchange capacity (CEC)   the ability of a growing medium or soil to attract and hold nutrients.

Cell-pack®   a plastic container of transplants in which each transplant has its own root packet.

cellulose   a chemical compound that forms a main part of plant cell walls and provides strength to stems and other parts.

central leader   the main, upright shoot of a tree.

central leader form   a training method for fruit trees that leaves the central leader intact (as opposed to modified central leader and vase forms); also refers to the natural form of many ornamental trees.

chelated nutrient   an essential element for plant growth chemically formulated to be nonreactive in soil or potting media.

chemotheraputant   a fungicide that kills existing fungus and prevents future spore germination.

clay   one of the three primary soil particles; clay particles are platelike, extremely small, and retain nutrients well.

clod   a hard lump of clay soil that is difficult to break.

clone   any plant vegetatively propagated from another is a clone of the parent plant.

coarse aggregate   a material such as sand or perlite, for example, made up of relatively large particles and added to a growing medium to improve its drainage.

cohesion   the attraction of identical molecules for each other, for example, the cohesion of water molecules.

cold hardiness or tolerance   the minimum temperature at which a plant can survive.

cole crop   vegetables of the genus *Brassica,* including broccoli, kale, cauliflower, and mustard.

colloids   an interspersed system of finely divided soil particles.

compaction   excessive packing or settling of a potting medium or soil.

companion planting   the planting of insect-repellent plants next to susceptible ones to repel pests.

compatible   describes two species that will cross-pollinate or that can be grafted together.

compensation point   the point at which photosynthesis produces sufficient carbohydrate for respiration but does not produce enough carbohydrate for growth.

complete fertilizer   a fertilizer containing nitrogen, phosphorus, and potassium; *see also* **balanced fertilizer.**

compost   an organic soil amendment made from decomposed garden refuse.

composting   the piling of garden refuse to allow it to decompose.

compound leaf   a leaf composed of more than one leaflet; *see also* **simple leaf.**

container gardening   the growing of fruits, vegetables, and ornamentals outdoors in containers; also called *pot gardening.*

container-grown   refers to landscape plants grown to selling size in containers such as cans or plastic pots.

cool-season vegetable   a species of vegetable that grows best at daytime temperatures ranging from 50 to 65°F (10–18°C) and is tolerant of frost; includes peas, lettuce, broccoli, and turnips.

cork cambium   the layer of cells that produces bark on woody plants.

corm   a thickened underground storage organ (stem) found on such plants as gladiolus and crocus.

Cornell Mixes   the Cornell University potting media.

corner planting   a group of trees, shrubs, or groundcovers used to visually soften the corner of a house.

cottage garden   landscape style characterized by informal massing of flowers, herbs, and sometimes vegetables.

cotyledon   part of an embryo of a seed made up of one or more simple leaves; serves as a storage organ in some cases, for example, in peas and beans.

cover crop   a grass or similar fast-growing plant sown over a vegetable garden in fall and turned under in spring to supply organic matter to the soil.

critical photoperiod   the daily dark/light ratio necessary for various plant responses to occur, particularly flowering and germination.

crop rotation   planting a different species in an area of the garden each year to prevent buildup of diseases or insects associated with particular crops.

crown   the part of a plant that enters the soil.

crusting   a soil surface condition typified by a smooth, glazed appearance and common on silty soils or after rain.

cucurbit a group of warm-season vegetables having similar cultural requirements; included are melons, cucumbers, squash, and pumpkins.

cultivar a combination of words "cultivated variety," designating a manmade plant variety.

cultural control alterations in plant care (watering, fertilizing, and so forth) for the purpose of controlling disease, insect, or weed problems.

curative a fungicide that kills existing fungus and prevents future spore germination.

cut flower a garden grown specifically for cutting and displaying in a vase.

cuticle the thin, waxy layer on the exterior of stems and leaves that retards water loss.

cutting a vegetative plant part that regenerates roots and forms new plants.

cytokinin a naturally occurring plant hormone involved in stem elongation, bud formation, breaking dormancy, and other activities.

damping off a disease in which a soilborne fungus attacks seedlings at the soil line, constricting the stems and killing the plants.

day-neutral plant a species that flowers irrespective of the dark/light ratio.

deciduous a plant that drops its leaves in winter; the opposite of evergreen.

decline the general decrease in vigor of a previously healthy plant.

decomposition the breakdown or rotting of plants and animals.

defoliation the dropping of most or all leaves from a plant.

desuckering removal of the suckers.

determinate describes tomato varieties that remain short and bushy; *see also* **indeterminate.**

dethatching removal of thatch from a turf area by a machine or by hard raking.

dicot a subdivision of angiosperms characterized by the presence of two cotyledons, net-veined leaves, and the like; most cultivated plants are included in this group; *see also* **monocot.**

dioecious a species in which the male and female organs of reproduction are produced on separate plants; examples are holly, dates, and bittersweet.

direct seeding sowing seed in the area where the plants will grow to maturity.

disease a plant malady caused by a microorganism or an environmental factor; the latter diseases are called *physiological* diseases.

division a method of vegetative propagation involving separation of a plant into two or more pieces, each containing a portion of the roots and crown.

dormancy a phase of a plant life cycle characterized by slowed or stopped growth; leaf drop and death of the aboveground portions may also occur.

dormant oil a lightweight oil sprayed on dormant plants for control of insects such as scale; the oil coats the insects to exclude air, suffocating them and their eggs.

double glazing the application of two layers of glazing to a greenhouse to reduce air leakage and to conserve heat.

double leader two competing central leaders in a tree.

drainage the rate at which water flows through a soil or potting medium; also, the percentage of the applied water the material retains; poor drainage is recognized by slow water flow and high retention, whereas good drainage is recognized by fast water flow and low retention.

drain tile interlocking clay tiles buried in an area with poor drainage to carry excess water away from plants.

drill hole fertilization method a technique for fertilizing deep-rooted plants such as trees by pounding holes into the ground throughout the root zone of the tree, filling them with granular fertilizer, and covering with peat moss or topsoil.

drip irrigation *see* **trickle irrigation.**

drought tolerance the ability of a plant to live without much water in the soil or to absorb the small quantity of capillary water remaining in a dry soil.

dry rot (of cacti) the stem of cacti characterized by discolored and sunken stem areas.

dwarf the miniature varieties of a species; *see also* **semidwarf** and **standard.**

dwarfing rootstock a rootstock that makes the scion grafted onto it grow less vigorously.

eelworm a nematode.

elongation lengthening.

embryo the immature plant in a seed.

encapsulated fertilizer a slow-release fertilizer composed of beads coated with a plastic-like resin that is permeable to water, which enters through the coating, dissolves the fertilizer, and carries it back out through the coating.

endosperm the part of a seed where carbohydrate is converted into energy to enable germination.

enzyme a chemical that regulates an internal chemical reaction in an organism.

espalier the stylized training of a tree or shrub to grow flat against a wall or trellis, sometimes with a precise branch pattern.

essential elements the 16 micro- and macronutrients required for plant growth.

established describes a seedling or other plant whose roots have penetrated well into the soil.

ethylene a gas produced by senescing plants or plant parts and ripening fruits; used commercially in pineapple production and to induce flowering of bromeliads (see Chapter 3).

everbearing describes strawberry and raspberry varieties that bear two crops annually: one in early summer and one in fall.

explant disinfected meristem of a shoot used as the primary propagation material in tissue culture.

exposed aggregate paving surface design created by imbedding gravel in wet concrete and then hosing down the partially set slab to leave the gravel raised from the surface but imbedded in it.

fall vegetable gardening the practice of seeding cool-season vegetables in summer to mature and be harvested in fall.

family a unit of plant classification grouping related genera.

fan and pad cooling a technique for cooling a greenhouse whereby a fiber pad is installed at the end of the greenhouse with water trickling through at a slow rate; a fan at the other end pulls outside air through it into the house; water then evaporates into the air as it passes through the pad, absorbing heat in the process and reducing the air temperature.

fast-release fertilizer a granular, powdered, or liquid fertilizer containing nitrogen only in nitrate form—the form immediately usable by plants.

fertility the ability of a soil to supply good conditions and sufficient nutrients for healthy plant growth. Also the ability of a plant to reproduce sexually.

fertilizer analysis the proportions of essential plant elements contained in a fertilizer which are listed on the container.

fertilizer burn an injury to plant roots owing to an excess concentration of fertilizer; wilting and browning of leaf margins are symptoms resulting from water deficiency.

fertilizer salt a chemical compound that includes one or more of the essential elements for plant growth.

fibrous (netlike) roots a fine, branched root system such as that found in many grasses.

filament the stalk portion of the stamen that supports the anther.

filler material a material such as sawdust added to a fertilizer to increase its volume inexpensively.

flat a shallow plastic, wood, or metal tray used to grow seedlings, carry plants, and the like.

fleshy roots roots thickened with stored carbohydrates or water.

floral foam solid polymer foam cut in blocks and used for anchoring flower stems in arrangements.

floriculture the science and practice of flower growing and arrangement.

flower border a long, narrow flower bed.

foliar feeding a method of supplying nutrients to plants by spraying them with fertilizer, which is absorbed through the leaves.

foot-candle a measure of light intensity; superseded by the metric lux.

force to cause a plant to bloom out of season by altering the environmental conditions, especially temperature.

fork a narrow angle between two branches.

form the overall shape of a plant, such as vase-shaped, pyramidal, or mounded.

formal flower bed a symmetrically designed bed for annual and perennial flowers.

foundation planting shrubs or groundcover planted against the base of a house.

freestanding bed a flower, vegetable, or shrub planting designed to be accessible from all sides.

frond a fern leaf.

frost-hardy describes plants that are able to withstand frost.

frost pocket a low-lying area more susceptible to frost than the surrounding vicinity.

frost-tender describes plants that are killed at temperatures below freezing.

fruit a swollen ovary.

fruit load the amount of fruits that a plant is bearing; a "heavy load" indicates a large quantity of fruits.

fruit set the development of fruits after flowering; few fruits develop in "poor fruit set," as opposed to "good fruit set."

fungicide a chemical used to control a fungus-caused disease.

fungus (fungi) a single-celled or multicelled plant lacking chlorophyll that may be saprophytic (living on dead organic matter) or parasitic (existing on living plants and other organisms); saprophytic fungi are useful in the garden because they decompose compost and release the nutrients in organic matter, but parasitic fungi are the cause of many plant diseases.

fungus gnat a minute, winged insect whose worm-like larvae are found occasionally in unsterilized potting media and damage plant roots through feeding.

gall an abnormal swelling on a plant part; leaf, stem, and root galls are typical examples.

gel polymer hydrating gel granules which can be a potting medium component to hold extra water.

genetic dwarf a fruit tree that is a dwarf owing to its genetic makeup rather than to grafting on a dwarfing rootstock.

genetic engineering the incorporation of genes of one species of plant into an unrelated one by use of a carrier plant.

genus (genera) a group of related species; one or more genera are grouped in a family.

germination the sprouting of a seed, spore, or pollen grain.

germination rate the percentage of seeds that will germinate under favorable conditions.

gibberellin a naturally occurring plant hormone involved in increasing flower, fruit, and leaf size, and elongation; also influences vernalization and other activities.

girdle to restrict the function of the xylem or phloem of a dicot plant; girdling by cutting through a portion of phloem is used to encourage flowering. Girdling can also be caused by rodents eating tree bark in winter.

glazing the clear or translucent material used to cover a greenhouse or related plant-growing structure.

girdling root a root, usually of a tree, that becomes embedded in the trunk of the tree at ground level and will eventually kill part or all of the tree.

grading changing the slope level of the soil in an area.

grafting implanting a branch or bud from one plant onto another.

granular fertilizer a fertilizer in a pelletized form.

greenhouse effect heating the air in a translucent or transparent structure owing to energy from the sun; as light energy passes through the glazing, infrared rays are trapped by the glazing.

green manure crop a grass or similar fast-growing plant sown over a future garden area and turned under before it reaches maturity to provide organic matter in the soil.

greens foliage used in floral arrangements.

groundcover low-growing plants that form mat-like growth over an area.

ground bed a planter constructed on the floor of a greenhouse for growing tall vegetables or flowers.

growing medium (media) a material used for rooting and growing plants which may contain sand, peat moss, soil, or any other number of ingredients.

growth habit the overall shape of a plant, for example, weeping or pyramidal.

grub the soil-inhabiting larva of many beetles.

gymnosperm a botanical group of evergreen plants with generally needlelike foliage and usually bearing cones; includes pines, juniper, and spruce.

harden off to gradually acclimatize a greenhouse- or house-grown plant to outdoor growing conditions.

hardpan a layer of compacted soil that slows or stops movement of water.

hardwood cutting a cutting made from the mature growth of a woody plant, usually taken in late fall or early spring.

harvest interval the number of days after treatment with a pesticide that a gardener must wait before harvesting his crop.

head the leafy portion of a tree.

heading back a pruning technique to control size and shape of shrubs by cutting back shoots to the parent branch to encourage new growth.

heal in to temporarily cover the roots of a plant with soil to prevent their drying until the plant can be properly planted.

heavy soil a soil composed predominantly of clay particles.

herbaceous not producing woody growth.

herbaceous cutting a cutting of a nonwoody plant.

herbicide a chemical used to control weeds; a "weed killer."

hill a group of seeds planted close together; hill planting is common for melons and squash.

hoar frost a frost that leaves ice crystals on soil and plants owing to condensation of humidity at a decreased temperature.

holdfast a suction-cup-like organ on some climbing vines, anchoring them to walls and other surfaces.

home horticulture the growing of ornamental and edible plants in and around the home.

hose-end sprayer a hose attachment for applying pesticide or fertilizer; the sprayer contains a concentrated chemical that is diluted with water through a siphon mechanism as it is applied.

hot cap   a paper or plastic dome set over warm-season vegetables in early spring to protect them against frost and to increase the daytime growing temperature.

humic acid   an acid formed by the decomposition of organic matter.

humidity   the amount of water vapor in the air.

hydroponics   a method of growing fruits, vegetables, and flowers in which the plants are planted in beds of sand, gravel, or a similar material and the roots are supplied with nutrients in a water-based solution.

ikebana   Japanese style of flower arrangement.

implant   *see* **trunk implant.**

incompatible   describes two plants that will not graft together or that form a weak and unsatisfactory graft union; also refers to two species that will not cross-pollinate.

indeterminate   describes tomato varieties that grow into long vines; *see also* **determinate.**

integrated pest management (IMP)   a system of insect and disease control stressing thorough understanding of the life cycle of the pest as well as cultural and biological controls; chemical control is used only when cultural methods fail.

interface   the division between two types of growing media of different quantities or sizes of pore spaces.

internode   the portion of plant stem between two successive nodes.

interplanting   sowing seeds of a quick-maturing vegetable with seeds of a slower one or between transplants; the fast-maturing vegetable is then harvested before it begins to crowd the main crop.

interstock   a section of stem inserted between the stock and scion in a graft union; the interstock may contribute a characteristic to the tree to make an otherwise incompatible stock and scion graftable.

invasive roots   a vigorous root system that can clog sewer lines and wells.

irrigation   watering.

juvenility   the first phase of plant development in which the plant is strictly vegetative.

lankiness   unattractive lengthening of stem internodes, usually owing to insufficient light.

larva (larvae)   the immature stage of an insect; larvae are frequently wormlike, whereas the adult may be winged.

latent heat of the earth   the relatively uniform, above-freezing temperature of the soil; a pit green-

house makes use of this latent heat by having its walls below ground.

lateral   a side branch arising from a vertical shoot.

lath house   a garden structure covered with thin strips of wood and used for growing shade-loving plants.

lava rock   a coarse aggregate made from ground volcanic rock.

lawn substitute   a groundcover plant tolerant of foot traffic and planted in lieu of grass in a lawn area.

layering   a method of propagating shrubs or vines with a trailing growth habit or flexible branches; a shoot is bent to the ground, held in place with a wire loop or stone, and covered with soil; after it generates roots, it is then severed from the parent plant.

leaching   a succession of waterings used to dissolve and flush out excess fertilizer or other chemicals from a soil or potting medium.

leaf apex   the tip of a leaf.

leaf base   the section where the leaf blade joins the petiole.

leaf blade   the flattened portion of a leaf that is responsible for the majority of photosynthesis.

leaf bud cutting   a cutting that includes a short section of stem with a leaf attached; leaf bud cuttings are taken after the stem tip is used.

leaf cutting   a cutting made from a leaf and its attached petiole.

leaflet   one of the expanded, leaflike parts of a compound leaf.

leaf margin   the edge of a leaf blade.

leaf miner   a wormlike insect larva that tunnels between the upper and lower surfaces of a leaf, leaving a scribble pattern.

leaf mold   a soil amendment or potting medium component composed of partially decayed leaves.

leaf vegetable   a vegetable raised for its edible leaves, such as lettuce, chard, or Chinese cabbage.

lean-to greenhouse   a greenhouse constructed with one wall in common with a house or other building.

legume   a member of the Leguminosae (Fabaceae) (bean or pea) family.

LD$_{50}$ an indication of pesticide toxicity established by experimentation on laboratory animals; abbreviation for "lethal dose that will kill 50% of a population" and measured in mg/kg of body weight: the higher the LD$_{50}$, the lower the assumed toxicity of the pesticide.

light duration   the total number of hours of light a plant receives per day.

light intensity   the brightness of light.

light soil   a soil composed predominately of sand.

limbing up   the removal of the lower branches of a tree.

lime   any of several calcium-containing materials used to raise soil pH.

loam   a soil having large proportions of both sand and clay particles.

lobe   a section of a leaf that protrudes from the main part. Figure 2-8(b) is a lobed leaf with three lobes.

long-day plant   a species that flowers only with a short daily dark period.

low-maintenance landscape   a landscape that is planned for minimal upkeep.

lux   a metric measure of light intensity, equal to the illumination on a surface 1 m from a standard candle (1 lumen/square m).

macronutrient   one of the essential elements needed in relatively large amounts for plant growth; these include nitrogen, phosphorus, calcium, magnesium, and sulfur.

macropore space   a large pore space found between large particles or aggregations of particles; *see also* **micropore space.**

maleic hydrazide   a growth-inhibiting chemical occasionally used to slow the growth of lawns and hedges.

maturity   the phase of plant development characterized by flowering and fruit production.

mealybug   a destructive plant pest that is about 1/8–1/4 in. (3–6 mm) long and covered at maturity with white waxy fibers which form a protective covering.

medium (media)   a material used for rooting or supporting the roots of plants. Media are subclassified by function as rooting media, potting media, seeding media, etc.

meristem   an area of rapidly growing and dividing cells.

microclimate   a small area with a climate differing from that of the surrounding area because of such factors as exposure, elevation, and sheltering structures.

micronutrient   one of the essential elements needed in minute quantities for plant growth; these include iron, copper, zinc, boron, molybdenum, chlorine, and cobalt.

microorganism   an organism that can only be seen through a microscope, for example, fungi, bacteria, and mycoplasmas.

micropore space   a small pore space found between minute particles such as clay; *see also* **macropore space.**

mildew   a foliar fungus disease characterized by a white powdery layer on plant leaves.

mille fleur   flower arrangement style consisting of a mixed bouquet of many species of flowers.

mineral soil   soil made up principally of weathered rock particles such as sand or clay; opposite of an organic soil.

mite   an eight-legged, near-microscopic plant pest that damages foliage by sucking cell contents.

miticide   a chemical used to control mites.

modified central leader form   a training method for fruit trees in which the central leader is removed back to the nearest scaffold branch to restrict the height of the tree and facilitate fruit harvesting.

monocarp   a plant that grows vegetatively for more than one year but only flowers once and then dies.

monocot   a subdivision of angiosperms that includes plants with a single cotyledon in the embryo and mostly straplike leaves with parallel venation; *see also* **dicot.**

mulch   any material used to cover the soil for purposes such as moisture conservation or weed suppression.

muscadine grape   *Vitis rotundifolia*, an American grape species native to the South.

mutation   a spontaneous change in the genetic makeup of a plant, for example, variegated shoots on an otherwise green plant.

mycoplasma   a microscopic organism intermediate between a virus and a bacterium that causes several plant diseases.

mycorhizzae   soil fungi that live in conjunction with plant roots to the benefit of both roots and fungi.

needle feeding   a method of fertilizing and watering deep-rooted plants such as trees: consists of a container of fertilizer connected to a pipe with a pointed tip; a garden hose is in turn connected to the pipe; the pointed tip is pushed into the ground around the tree, and when the water is turned on, dissolved fertilizer is forced into the ground around the roots.

nematicide   a chemical used to control nematodes.

nematode   a microscopic or near-microscopic worm-like plant pest that damages roots and occasionally burrows in foliage.

nitrate   a nitrogen-containing compound and the form in which most plants absorb nitrogen.

node   the point of leaf attachment to a stem.

nonselective herbicide   an herbicide that injures or kills plants regardless of species.

nutrient   an essential element for plant growth.

nutrient deficiency   the lack in a plant of a sufficient quantity of one or more of the essential elements.

offset a young plant produced at the base of a parent plant; in the case of cacti and a few other plants such as pineapple, the offset may develop on top of the parent plant.

olericulture the science and practice of vegetable growing.

open-center form *see* **vase form.**

organic defined by the laws of chemistry as a chemical that contains carbon atoms; also, a fertilizer or pesticide derived from a naturally occurring material.

organic matter the decomposing bodies of dead plants and animals.

organic soil soil composed mainly of decayed plant and animal remains. Opposite of mineral soil.

ornamental a plant grown for its beauty.

ornamental horticulture the branch of horticulture that deals with the cultivation of plants for their aesthetic value; this designation includes landscape horticulture, floriculture, indoor plant growing, and sometimes turfgrass growing.

osmosis movement of water across a membrane.

ovary the bulbous base of the pistil which contains egg cells; the ovary enlarges to become the fruit.

overwatering applying an excess volume of water at too frequent intervals for healthy plant growth.

packing *see* **compaction.**

parent stock the plant(s) from which material is obtained for propagation.

parthenocarpy the growth of a fruit without fertilization of the included egg; also, the continued development of a fruit after the abortion of the embryo.

pathogen a fungus, bacterium, virus, or mycoplasma which causes a disease.

patio tree a small ornamental tree planted near a patio or deck for shade and beauty.

pavé flower arrangement style in which flowers are tightly arranged so that the blooms nearly touch.

peat moss any of several bog plants, particularly sphagnum moss, which are harvested, dried, and used as components in growing media or for soil amending.

peat pot a pot made of compressed peat moss and used for growing seedlings, which are then planted in the garden in their container.

pelleted seed a vegetable seed coated with a dissolving material that makes it larger and easier to sow.

perennial a plant that lives for an indefinite number of years; often refers to herbaceous flowering perennials.

perfect flower a flower with both male and female parts.

perlite a course aggregate made from expanded volcanic rock.

pesticide a chemical used to control an undesirable organism; *see also* **bactericide, fungicide, herbicide, miticide, nematicide.**

petiole the stalk that attaches the leaf blade to the stem.

pH a measure of the acidity or alkalinity of a soil, potting medium, or liquid.

phloem the portion of the vascular system in which carbohydrates are moved throughout the plant.

photoperiod the dark/light ratio in a day.

photoperiodism the effect of the daily dark/light ratio on the growth and flowering of plants.

photosynthesis the chemical process by which green plants manufacture carbohydrates in the presence of light.

phototropism the hormone-induced leaning of a plant toward light.

physiological disease a plant malady due to insufficient amounts of essential nutrients or excessive (toxic) amounts of certain elements such as sodium in the soil.

phytochrome the chemical involved in plant response to photoperiod.

pinching removal of the terminal bud of a plant to promote branching.

pistil the female parts of a flower: style, stigma, and ovary.

plant breeding the science of controlled pollination of plants to develop new varieties.

planter a defined area, usually raised and constructed of wood or concrete, for growing plants.

plant food a lay term for fertilizer.

plastic mulch a layer of clear or black plastic used as a mulch in vegetable or ornamental plantings.

plugging a method of lawn establishment involving transplanting small cores of sod into a prepared area.

pollarding a formal training method applied to deciduous trees whereby branches of the tree are pruned back almost to the main trunk every year.

pollination the deposition of pollen on the flower stigma.

pomology the science and practice of fruit growing.

pony pack a plastic package of four to eight flower or vegetable seedlings.

pore spaces   the gaps between the particles of soil or other growing media; the gaps may be filled with either air or water.

postemergence herbicide   an herbicide that kills established weeds.

pot-bound   describes a containerized plant allowed to grow in a pot too long; the over-abundance of roots in the pot jeopardizes the health of the plant.

pot up   to dig up a seedling and transfer it to a pot.

potting compost   British term for a growing medium.

preemergence herbicide   an herbicide that kills weed seedlings as they germinate but has no effect on established plants.

propagation   plant reproduction by sexual or vegetative methods.

protectant   a fungicide that prevents germination of fungal spores (see curative).

pruning   removal of parts of a plant for size control, health, or appearance.

public area   the area of a landscape in front of a house and viewable from the street; includes the front door, driveway, and walkway.

pup   popular name given to the offset of a bromeliad.

radiation frost   a frost caused by the loss of heat from plants by radiation during the night.

radical   the primary root that emerges from a seed.

recommended variety   a variety of fruit or vegetable well adapted to growing in a particular region and recommended by the local Cooperative Extension Service.

refoliate   to grow new leaves after being leafless; for example, a deciduous tree refoliates each spring.

relative humidity   the amount of water vapor present in the air compared to the total amount the air could hold at its present temperature.

renovation pruning   severe, heavy pruning used to rejuvenate a neglected plant; used on fruit trees, grapes, strawberries, ornamentals, and sometimes houseplants.

resistance   nonsusceptibility of a plant to insects or disease.

respiration   the breakdown of carbohydrates to yield energy for growth and cell activities.

rhizome   a usually thickened, horizontal, underground stem.

rock garden   a combination of rocks and plants that creates a natural-looking area for the cultivation of plant species native to rocky or alpine regions.

roguing   eliminating certain plants from a group, usually those considered the least desirable; may also refer to the removal of diseased or infested plants to prevent the spread of the condition.

root-bound   *see* pot-bound.

root cutting   a cutting made from sections of roots alone.

root hair   a tubelike outgrowth of a root epidermal cell through which water and dissolved nutrients pass.

root mealybug   root-parasitizing insects resembling foliar mealybugs.

root rot   the death and decay of plant roots owing to unfavorable growing conditions and resulting infections by microorganisms.

root stock   the plant that functions as the root system in a grafted plant.

root tuber   a thickened root that stores carbohydrates and is used in propagation.

root vegetable   a vegetable grown primarily for its edible root, such as radishes, turnips, and carrots.

root zone   the volume of soil or growing medium containing the roots of a plant.

rooting hormone   a plant hormone used in promoting rooting of cuttings.

rootstock   the plant that bears the roots of the new plant in a budding or grafting operation.

rosette   a plant form in which the plant has a shortened stem with leaves that overlap and radiate outward from a central point; examples are African violets and strawberries.

runner   an aboveground stem produced as a natural means of vegetative reproduction of plants such as ferns and strawberries; one or more plants grow on the stem, generate roots, and become independent of the parent.

rust   a fungus disease characterized at one point in the life cycle by rust-colored pustules on the leaves of the affected plant; rust diseases require two host plant species for a complete life cycle.

Saran®   a plastic, screenlike material used for covering a shade house to reduce light penetration.

scaffold branch   a main branch growing from the trunk of a tree.

scale   a destructive plant pest that is attached to a plant and exudes a shell-type covering.

scarification   the mechanical scraping or damaging of a seed coat to facilitate germination.

scion   the twig (or bud) to be grafted onto the roots of another plant, the stock, in a budding or grafting operation.

scorch   the death of plant tissue from excess heat or insufficient water.

scoria a coarse aggregate made from ground volcanic rock.

screening plant a plant used to block a view or to separate areas in a house or landscape.

scurf the white or silver scales that cover some leaves, for example, bromeliads.

seed coat a covering that helps prevent injury and drying of the seed.

seed leaves the first leaves (cotyledon) produced by a seedling; they generally do not resemble the mature leaves of the species; *see also* **true leaves.**

seed potato a section of a potato tuber containing at least one bud that is planted to start a potato plant.

seed tape a strip of water-soluble plastic tape embedded with vegetable seeds.

selective herbicide an herbicide that harms some species but not others, for example, 2,4-D, which is selective for dicot weeds in turfgrass.

self-incompatibility the inability of pollen to fertilize eggs of the same plant.

semidwarf a plant (usually a fruit tree) intermediate in size between dwarf and standard (full size).

semihardwood cutting a cutting made from the partially matured new growth of a woody plant.

senescence the aging and death of a plant or any of its parts.

sepal the often green, leaflike outermost part of a flower that encloses and protects the unopened flower bud.

service area the area of a landscape used for storage, laundry lines, pet runs, and other utilitarian purposes.

shade house a garden structure covered with Saran® or lath and used for growing shade plants such as ferns.

shading material (whitewash, lath, or bamboo blinds, for example) applied to glazing to decrease light entry and heat buildup in a greenhouse or related structure; also, the act of applying the material.

shock wilting and leaf drop owing to root loss during transplanting; in houseplants, shock may result from a rapid decrease in light intensity.

short-day plant a species that flowers only with a long daily dark period and a short daily light period.

shoulder ring the raised ring of bark where a branch emerges from its parent branch.

sick building a building with poor air quality as a result of inadequate ventilation.

sidedressing a method of fertilizer application that involves sprinkling a narrow band of fertilizer on both sides of a plant row over the root area.

sidewall a concrete block or wood wall extending partway up the side of a greenhouse in lieu of glazing and used to reduce heat loss.

silt one of the three basic soil particles, *see also* **clay.**

simple leaf a leaf with a single blade; *see also* **compound leaf.**

single element fertilizer a fertilizer that contains one of the elements essential for plant growth; urea is a single-element fertilizer.

slip a synonym for cutting.

slow-release fertilizer a fertilizer chemically formulated or encapsulated to release its nutrients over a period of several weeks or months.

small fruits the collective name given to the plants that bear edible fruits of a relatively small size; included are grapes, strawberries, and raspberries.

softwood cutting a cutting made from the new growth of a woody plant.

soil ball the soil and the contained roots of a balled and burlapped plant.

soilborne refers to a disease or insect pest that lives in the soil; many wilt diseases and nematodes are soilborne.

soil drench a liquid poured over the root zone of a plant to kill soil insects or to cure root rot, damping off, or to kill soil insects.

soil layer the topsoil or subsoil portion of soil.

"source" to "sink" movement the translocation of material (hormones, metabolites, and so forth) from their point of origin or manufacture to the sink, generally meristems, roots, and storage tissues.

spaghetti system an automatic watering system involving a flexible plastic pipe inserted at intervals with spaghetti-sized microtubes which lead to plants and apply water at a slow rate.

species a group of individuals with many characteristics in common and usually interbreeding freely.

specific epithet the second word in a botanical name.

specimen plant a showy ornamental plant grown solely for its beauty.

spore a simple reproductive cell that develops into a plant; usually used with reference to fungi and ferns.

spp. abbreviation for species (plural) designating collectively all species within the genus; for example, *Quercus* spp. refers to all oaks.

springtail a minute, harmless insect characterized by an unusual jumping motion.

stamen the male parts of a flower: anther and filament.

standard a tree or shrub with a tall stem that stands alone; also, the normal size of a species; *see also* **dwarf, semidwarf.**

starter solution a liquid fertilizer high in phosphorus content and applied to seedlings to hasten growth.

stem section cutting a cutting made from a short piece of thickened leafless stem containing at least one node.

stem tip cutting a cutting made from the top of a growing stem.

stigma the knoblike top of the pistil that receives pollen.

stock abbreviated form of rootstock or stock plant.

stock plant the plant from which cuttings or meristems are obtained for propagation.

stolon *see* **runner.**

stomate (stomata, stomates) an opening between leaf cells that functions in gas exchange.

stone fruit a fruit that contains a single, large pit; included are plums, peaches, apricots, and nectarines.

stratification the chilling of seeds to aid in breaking their dormancy.

stress a harmful growing condition that inhibits healthy plant growth, for example, drought (water stress), root boundness, and excess heat or cold (temperature stress).

stringiness unattractive lengthening of stem internodes, usually due to insufficient light.

style the portion of the pistil connecting the ovary and stigma.

submersion watering a watering technique for potted plants in which the container is submerged in water for several minutes and then allowed to drain; useful for water-repellent potting media and sphagnum-lined hanging baskets.

subsoil the layer of soil directly beneath the topsoil which is characterized by a lower proportion of organic matter and less fertility than the topsoil.

substrate a growing medium of any type, for example, potting soil or gel for tissue culture.

subsurface irrigation an irrigation technique in which water is applied to an area through seepage from underground pipes.

succession planting planting vegetable seeds several weeks apart to assure plants maturing at different times and therefore continuous harvest.

succulent a plant having thickened leaves or a stem adapted for retaining water; also the tender new growth of plants.

sucker a vigorous, vertical-growing shoot produced at the base of a plant; also, a young plant produced around the base of a mature one.

sucking insect an insect that damages plants by puncturing the leaves and sucking the sap.

summer bulb a bulbous-type plant that is not cold-hardy; includes tuberous begonia, canna, and gladiolus.

sun scald an injury to a plant from excess sun or a sudden increase in light intensity.

surface root a plant (usually a tree) root that grows along the surface of the soil; also, the very shallow roots of certain plants.

systemic pesticide a pesticide absorbed into a plant, killing the organisms that prey on it.

taproot a single root that grows vertically into the soil; for example, carrots and parsnips.

taxonomy the scientific classification of plants and animals according to their presumed natural relationships.

teepee training tying stakes together in pyramid fashion as a trellis for vining vegetables.

terminal bud a bud located at the tip of a stem.

terrarium a transparent glass or plastic container retaining high humidity used for displaying and growing plants indoors.

texture the visual impact of a plant owing to the size of its leaves: "coarse-textured" plants have large leaves, and "fine-textured" plants have small leaves; also refers to the proportions of sand, silt, and clay in a soil.

thatch a layer of stems and roots (both dead and living) that accumulates between the surface of a turf and the soil.

thinning a pruning technique for shrubs which involves removing the oldest stems to promote new growth; also, the selective removal of the excess fruits on a fruit tree to improve the size and quality of the remaining fruits; also, removal (usually by pulling out) of excess seedlings spaced too closely for optimum growth.

thrips minute, winged insects that damage plants by rasping the surface of the leaves and lapping the exuded sap.

tilth a characteristic of soil referring to its aggregation and workability.

tip burn the death and browning of leaf tips, common on indoor plants in low-humidity environments.

tomato cage a wire mesh cylinder used to support a tomato plant without tying.

topdressing a method of fertilizer application that involves the sprinkling of fertilizer over the soil surface.

topiary the stylized training of a shrub by pruning or shearing to grow in a controlled shape.

topping removal of the central leader of a tree to make the head fuller (not a recommended pruning practice).

topsoil the uppermost layer of soil, usually characterized by a higher quantity of organic matter and nutrients than the subsoil.

trace element a micronutrient.

training a system of pruning and tying plants to conform them to a particular size or shape; used on fruits to aid harvesting and on ornamentals for decorative effect.

translocation the movement of carbohydrates, water, minerals, and other materials within the vascular system of a plant.

translucent characteristic of a material that light can pass through but not images.

transpiration the loss of water from a plant, usually through leaves, in vapor form.

transplant a seedling plant grown in a cold frame, in a greenhouse, or on a windowsill for later planting outdoors; also, to dig up and move a plant to another location.

trap crop the planting of an expendable insect-attracting plant near a crop plant to draw the pests away from the desired crop (not proved to be an effective pest control method).

trickle irrigation an irrigation technique for outdoor plants involving flexible water pipes inserted at intervals with small microtubes that run to the base of nearby plants.

true leaves the second set of leaves produced by a seedling; these leaves resemble the normal leaves of the species, whereas seed leaves do not.

trunk implant capsule of pesticide inserted into a tree trunk to release pesticide through the vascular system.

tuber a thickened underground stem in which carbohydrates are stored; for example, the potato; *see also* **root tuber.**

utility area *see* **service area.**

U-C Mixes the University of California potting media.

underplanting planting shorter plants under taller ones, for example, groundcover under a tree.

underwatering applying water in insufficient quantity or at insufficient frequency for healthy plant growth.

variegation the genetic patterning of leaves with white or yellow markings.

variety a subdivision of a species; usually differs from the species in one or more minor ways.

vascular bundle the bundle consisting of xylem and phloem; the conducting system in plants.

vascular cambium the layer of cells that gives rise to xylem and phloem cells.

vascular system the network of cellular pathways that moves water, minerals, and carbohydrates within a plant; includes the xylem and phloem.

vase form a training method for fruit trees in which from two to five scaffold branches grow from a short trunk and the central leader and other scaffold branches are removed to keep the tree short and make harvesting easier; also refers to any tree that grows naturally in this fashion.

vase-shaped a tree or shrub without a central leader and having a V- or Y-shaped form.

vegetative describes the parts of a plant not involved in the sexual reproduction process.

vermiculite a coarse aggregate made from expanded mica and having a high cation-exchange capacity.

vernalization the subjecting of a plant to cool temperatures for the promotion or enhancement of growth or flowering.

vertical mower machine with vertical blades that cut and pull out thatch in a lawn.

virus a microscopic noncellular organism dependent on living hosts for its nutrition and reproduction and the cause of many plant diseases.

volunteer a seedling produced by natural reseeding of a garden plant.

warm-season vegetable a vegetable that grows best at daytime temperatures ranging from 5 to 90°F (18–32°C); includes tomatoes, eggplant, peppers, and corn.

waterfall design floral design style in which the primary line and focal point of the arrangement extend downward.

water holding capacity the ability of a growing medium to retain water.

water logged refers to soil that is too wet for healthy root growth.

water sprout a vigorous, vertical-growing shoot produced on the trunk or branches of a tree.

water stress decline in the health of a plant due to a lack of water.

wet-dry watering cycle a technique for watering houseplants whereby the plants are only watered when the medium becomes dry and just before the plants wilt.

wetting agent a chemical added to pesticide spray, peat moss, or soil to decrease surface tension and increase water adhesion or penetration.

whiteflies white insects resembling tiny moths which cluster under plant leaves and cause damage by extracting sap.

wick watering a watering technique that moves water from a reservoir to the potting medium through a fabric wick to keep the potting medium constantly moist.

winter annual a frost-hardy flower planted in the fall in mild-winter climates for winter bloom; for example, stock and calendula.

winterkill death of normally hardy plants over the winter due to excessively cold temperatures or damage to the roots due to heaving of the soil from repeated freezing and thawing.

wrapping the practice of covering the trunk of a young tree with paper tape or burlap to prevent sun scald.

xylem the portion of a plant vascular system in which water and minerals, taken in by the roots, move throughout the plant.

yield the amount of produce obtained from a fruit, nut, or vegetable variety; "good yield" indicates heavy production.

zone of elongation the section of a root, just behind the apex, in which the cells produced in the meristem enlarge and lengthen the root.

Index

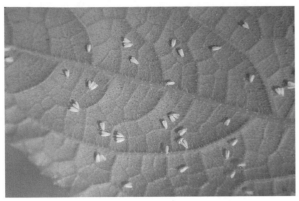

A closeup of whiteflies, which are often a pest on fuchsias.

A mixed flower border with strawflower and lavender.

Summer- and winter-type squashes.

Winter squashes with hard rinds suitable for storage.

A caper plant flower.

Euphorbia resembles a cactus, but is actually in the poinsettia family.